KU-626-518

WITHDRAWN
FROM STOCK

COGNITIVE RADIO ARCHITECTURE

COGNITIVE RADIO ARCHITECTURE

The Engineering Foundations of Radio XML

JOSEPH MITOLA III

A JOHN WILEY & SONS, INC., PUBLICATION

Published by John Wiley & Sons, Inc., Hoboken, New Jersey.
Published simultaneously in Canada.

For general information on our other products and services please contact our Customer Care Department within the U.S. at 877-762-2974, outside the U.S. at 317-572-3993 or fax 317-572-4002.

Wiley also publishes its books in a variety of electronic formats. Some content that appears in print, however, may not be available in electronic format.

Library of Congress Cataloging-in-Publication Data:

Mitola, Joseph.
Cognitive radio architecture : the engineering foundations of radio XML / by Joseph Mitola.
p. cm.
Includes bibliographical references and index.
ISBN-10: 0-471-74244-9
ISBN-13: 978-0-471-74244-9
1. Software radio. 2. XML (Document markup language) I. Title.

TK5103.4875.M58 2006
621.384–dc22

2005051362

Printed in the United States of America

10 9 8 7 6 5 4 3 2 1

CONTENTS

PREFACE

On 14 October 1998, I coined the term "cognitive radio (CR)" to represent the integration of substantial computational intelligence—particularly machine learning, vision, and natural language processing—into software-defined radio (SDR). CR embeds a RF-domain intelligent agent as a radio and information access proxy for the user, making a myriad of detailed radio use decisions on behalf of the user (not necessarily of the network) to use the radio spectrum more effectively. (This is the first of several informal definitions of cognitive radio. The technical definition is given in a computational ontology of the ideal cognitive radio, the iCR.) CR is based on "software radio." (See J. Mitola, *Software Radio Architecture*, Wiley, Hoboken, NJ, 2000).

Between 1998 and 2000, I refined cognitive radio concepts in my dissertation research. At that time, I built a research prototype cognitive wireless personal digital assistant (CWPDA) in Java—CR1—and trained it, gaining insights into cognitive radio technology and architecture. While working on my dissertation, I described the ideal CR (iCR) for spectrum management at the Federal Communications Commission (FCC) on 6 April 1999 (see the companion CD-ROM or web site for the text of this statement) and in a public forum on secondary markets in a layperson's version of a core doctoral program (FCC, *Public Forum on Secondary Markets*, Washington, DC, 21 May 2000). It showed the potential economic value of iCR in secondary radio spectrum markets. I first presented the technical material publicly at the IEEE workshop on Mobile Multimedia Communications (see J. Mitola III, "Cognitive Radio for Flexible Mobile Multimedia Communications," Mobile Multimedia Communications (MoMUC 99), IEEE Press, New York, 1999).

The FCC uses the term cognitive to mean "adaptive" without requiring machine learning. This text coins the phrase "ideal cognitive radio (iCR)" for a CR with autonomous machine learning, vision (not just a camera), and spoken or written language perception. There will be an exciting progression across aware, adaptive, and cognitive radio (AACR). Enjoy!

DISCLAIMER

This text was prepared entirely on the author's personal time and with personal resources. The author is an employee of The MITRE Corporation on loan via the provisions of the Interagency Personnel Act (IPA) to the U.S. Department of Defense (DoD). This document has been "Approved for public release; Distribution unlimited" per DoD case number pp-05-0378 and MITRE case number 06-0696. "The author's affiliation with DoD and The MITRE Corporation is provided for identification purposes only, and is not intended to convey or imply MITRE or DoD concurrence with, or support for, the positions, opinions, or viewpoints expressed by the author."

Joseph Mitola III

ACKNOWLEDGMENTS

In 1999 and 2000, MITRE Corporation supported the author's final year of doctoral research at KTH, The Royal Institute of Technology, Stockholm, Sweden on which this text is based. The author would like to acknowledge the truly supportive environment of MITRE, which is a world-class resource for the creation and application of information technologies for the public interest.

Without Professor Chip Maguire's vision, imagination, incredible technical depth, professional reputation, and unbending support, the cutting edge research in cognitive radio wouldn't have happened at all, at least not by me and not in 1997–2000. KTH and Columbia University couldn't do better. Thanks, Chip. Thanks also to Professor Jens Zander, a KTH advisor, who kept asking all those hard radio engineering questions and offering insights that have stood the test of time.

Finally, my wife, Lynné, is a saint to have been so supportive not only through the doctoral work, but in support of my passion for the public benefit of radio technology over the decades, starting with teaching in the 1970s, graduate work in the 1970s and 1980s, my first book—*Software Radio Architecture*—in 2000, continuing through the cognitive radio research at KTH, and finally the publication of this book about *Aware, Adaptive, and Cognitive Radio*. Lynné not only is the wind beneath my wings, she is my wings. Thanks for your years of sacrifice and support, Hon.

J. M. 3

CHAPTER 1

INTRODUCTION

This book is about making radios so smart that they can autonomously discover how, when, and where to use radio spectrum to obtain information services without having previously been programmed to do so. Cognitive radio integrates machine perception software into wireless systems—radio nodes and networks. Radios today are evolving from awareness (e.g., of location) toward cognition: the self-aware radio autonomously learns helpful new wireless information access and use behaviors, not just sensing the radio frequency (RF) spectrum but also perceiving and interpreting the user in the user's environment via computer vision, speech recognition (speech-to-text), and language understanding (Figure 1-1).

This progression of awareness and adaptation toward cognitive radio (AACR) leverages traditionally nonradio technologies: computer vision, navigation, speech recognition and synthesis, and the semantic web [1]. Machine perception grounds the ideal cognitive radio's "self" and its perception of its user's communications needs, priorities, and intent in the world of space, time, and situation so that the ideal cognitive radio (iCR) more transparently and efficiently accesses *useful information* via whatever wireless means might be made available. The wireless mantra "always best connected" (ABC) is transformed by the iCR focus on quality of information (QoI) to "always better informed" (ABI). This transformation is facilitated by semantic web technologies like the eXtensible Markup Language (XML) and the Ontology Web Language (OWL) [2] adapted to radio applications via a new metalanguage, Radio XML (RXML).

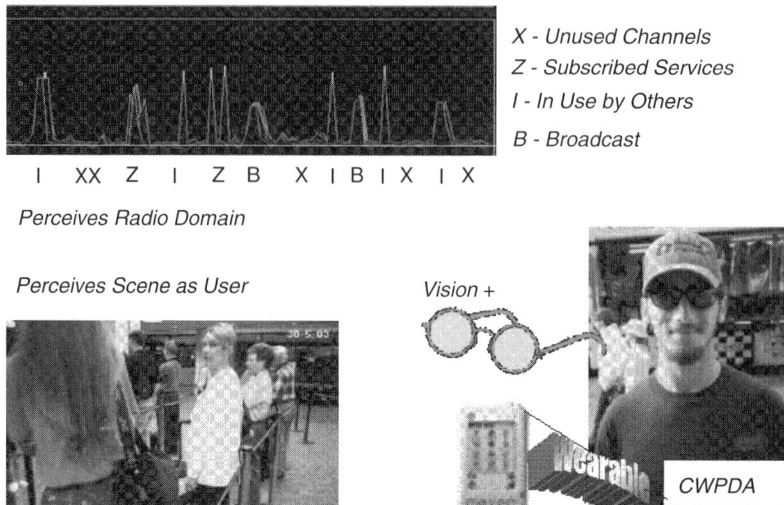

FIGURE 1-1 Notional cognitive wireless personal digital assistant (CWPDA).

The iCR is a far-term vision. The path suggested in subsequent use cases evolves increasingly from *aware* and *adaptive* radios toward *cognitive radio*, the *AACR* revolution. AACR technology can also power increasingly autonomous cognitive wireless networks (CWNs). The cognitive radio architecture (CRA) defines functions, components, and design rules by which to evolve software-defined radio (SDR) toward the iCR vision. The core technology of the CRA evolution is the <Self/>,[1] defined in RXML, perceiving the radio spectrum, enabling vision and speech perception with embedded autonomous machine learning (AML) for RF awareness, cooperative networking, and mass customization of information services for the <Self/>'s own <User/>.

This initial chapter draws important distinctions among similar AACR concepts and sets the perspective for the balance of the book. The foundation chapters then further develop the use cases and technical ideas from radio technology, machine perception, machine learning, and the semantic web, organizing the approach into the CRA and illustrating this architecture with the research prototype CR1. The Java source code of CR1 illustrates the CRA principles in a simulated cognitive wireless personal digital assistant (CWPDA). Subsequent chapters on radio and user-domain skills develop the ideas further. Exercises engage the serious reader.

[1] Terminated XML tags like <Self/> are ontological primitives of Radio XML.

1.1 PERCEPTION

SDRs sense specific radio bands but lack broad RF, audio, and visual perception. Perception technologies enable AACR to autonomously take the user's perspective, to understand referents in speech and vision, recognizing QoI features of both RF and user sensor-perception domains with a goal of zero redundant instructions from the user to the AACR for information access. The iCR accesses information as presciently as the legendary Radar O'Riley of 4077 MASH®.

1.1.1 RF Perception

RF perception goes beyond the detection of expected signals on known frequencies. It includes the extraction of helpful information from broadcast channels, deference to legacy (noncognitive) radios, reduction of noise, and minimization of interference not just by running the right SDR modules, but by autonomously constructing the RF behavior most appropriate to the setting. RF perception enables the iCR to characterize the significant entities and relationships in the RF environment. RF perception goes beyond the traditional radio-domain sensing of signal-to-noise ratio (SNR), bit error rate (BER), code space, and the like. For example, to be most effective in the recently liberalized U.S. TV spectrum bands, an AACR not only senses broadcast channels but also computes the likelihood of hidden legacy TV receivers ("hidden nodes"), for example, based on detection of the TV below the noise level [3], directing energy away from hidden nodes.[2] Such RF perception grounds the iCR's <Self/> with its <User/> in the domain (*space* × *time* × *RF*). The iCR's computational models of RF entities include legacy transmitters, aware–adaptive radios (AARs), iCRs, multipath reflectors, sources of noise or interference, and other relevant entities. The continuously increasing digital hardware capacity per gram enables increased wearable sensing with embedded RF scene perception from algorithms that model RF relationships. Thus, spectrum sharing of TV channels can evolve toward the iCR "radio etiquette," autonomous polite use of available radio resources tailored to the situation.

Although it is possible to embed RF sensing and perception in hardware-defined radio, the value proposition of iCR use cases accrues most dramatically via SDR. For example, the iCR negotiates with alternative bearer networks on behalf of the user, downloading specialized air interfaces, and validating them before enabling them in the <Self/>'s embedded SDR. The iCR behaves as an autonomous RF access management agent.

[2] This comment relates to an important use case supporting FCC policy referred to as the TV-spectrum use case.

1.1.1.1 GSM–DECT Priority

Network operators may not see the value of the CWPDA negotiating on behalf of the user. Sometimes the needs of the user contradict the needs of the service provider. Researchers have shown user-centric RF behaviors to be both easy to implement and valuable. For example, in 1997–1999 Ericsson® provided dual-mode [4] GSM–DECT (Digital European Cordless Telephone) wireless badges to KTH, The Royal Institute of Technology at Kista (pronounced "sheesta"), a suburb of Stockholm. When initialized inside the Elektrum building, the badges used DECT's free air time for network access. As the user lost DECT connectivity elsewhere on campus, the badge switched to GSM as planned. KTH paid Telia, the GSM service provider, for air time. Returning to the building, the badges stayed in GSM mode since GSM propagates well at Elektrum, so the badges rarely switched back to DECT, which cost the project a bundle, at least on paper. Reprogramming the badges to reacquire DECT whenever possible avoided the cost of GSM air time while indoors, reducing cost by a substantial fraction: Telia lost revenue from the displacement to a free RF band of what could have been cell phone traffic.

1.1.1.2 Closer to Home

Past may be prolog. Suppose your 3G cell phone has IEEE 802.11 hot-spot capability and you have your own 802.11 networks at home and at work. Would you like your cell phone to switch to your free 802.11 network when possible, reducing cellular air time? I would. Why have cordless phones at home or a desk set at work when your 802.11-enabled cell phone can act as a cordless handset (for free)? Cellular service providers might not smile on such a phone. The hardware of a 3G hot-spot phone could access your free 802.11 networks, saving cellular air-time costs. The software personality of that cell phone almost certainly would not allow that, however, for a mix of social, economic, and technical reasons. But a future AACR with flexible 802.11 access could use either the for-fee hot spot or your for-free home and work wireless access points, for example, via Voice over IP (VoIP). An iCR with sufficient prior training and AML would not have to be programmed for that specific use case. It would discover the free RF access points through its ability to perceive the RF environment. It would discover the availability of your access points and autonomously synthesize a lowest cost (if that is your criterion) network interface that met your needs. How would the iCR know your needs? Such knowledge may be based on <Scene/> perception, the iCR perceiving itself, <Self/>, and its <User/> in a space–time–RF <Scene/>.

1.1.2 User <Scene/> Perception

Multisensory perception grounds the iCR's <Self/> and its <User/> to the everyday world of physical settings with associated events, for example, defined as <Scenes/> in radio XML. Thus, the iCR manages wireless resources as an information services agent. Such an agent requires real-time perception and correlation of the current <Scene/> to similar <Scene/>s experienced previously, indexed efficiently to infer the <Scene/>-dependent needs of the user.

To detect changes in the user's communications needs, iCRs perceive the <Self/> and <User/> in the RF <Scene/> via vision, sound, email, and speech. The focused leveraging of knowledge representation, spatial–temporal task planning, and AML enables responsiveness without user tedium or expensive network customization staffs. AML technology thus offers mass customization of iCR behavior. The sharing of knowledge among AACRs on behalf of their <Users/> creates ad hoc information services without the mediation of a for-fee service provider. This vision of the self-extending iCR may take decades to fully mature, but the radio knowledge, mutual grounding, and open architecture developed in this text assist more rapid technology evolution in this direction.

1.2 AWARE, ADAPTIVE, OR COGNITIVE?

There is a continuum from SDR to iCR with potentially many discrete steps, a few of which establish the technical foundation for evolution. Aware radios (ARs) incorporate new sensors that enhance wireless QoI. Embedding GPS in a cell phone, for example, enhances location QoI of the cell phone user. If, in addition, the cell phone assists the user with GPS navigation, then the cell phone itself is location aware.

Definition: A radio entity <Self/> is GPS aware if and only if (iff) an algorithm in the <Self/> uses the GPS data for <RF/> or <User/>-QoI tasks.

As shown in Figure 1-2, the degree of location awareness ranges from convenient to cognitive.

1.2.1 Convenient

GPS may be embedded, but the radio's location awareness may be nonexistent: mere integration of GPS into a cell phone with latitude and longitude displayed is not location awareness. In such a configuration, the embedded GPS display has no relationship to the cell phone itself other than sharing the mechanical enclosure. This product is convenient but is functionally equivalent to a distinct GPS receiver in the user's other pocket: convenient, but the radio's <Self/>[3] is not GPS aware.

[3] <Self/> always refers to the radio's own self-referential data structures and algorithms, not to a <User/>.

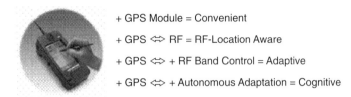

+ GPS Module = Convenient

+ GPS ⇔ RF = RF-Location Aware

+ GPS ⇔ + RF Band Control = Adaptive

+ GPS ⇔ + Autonomous Adaptation = Cognitive

FIGURE 1-2 Wireless PDA plus GPS may be convenient, aware, adaptive, or cognitive.

1.2.2 Aware

For RF-location awareness, the phone must associate some aspect of <RF/> with <Location/>. For example, if the network determines the received signal strength indication (RSSI) at a given location by a query for (RSSI, Location) from the phone, then it is RF (RSSI)-<Location/> aware. The phone associates a <RF/> sensory parameter with <Location/> sensed simultaneously.

The phone is user-location aware if it associates some aspect of the <User/> domain, such as broadcast radio preference, with location. Observations like (WTOP; Washington, DC) learned by the CWPDA support user-location awareness. A user-location aware network may associate user behavior, like placing a call, with user location, for example, to gather statistics on the space–time distribution of demand. Such user-location awareness enables better provisioning and thus better grade of service (GoS) [170]. User-location aware networks are not new.

1.2.3 Adaptive

Adaptivity requires action. Specifically, if the phone itself uses location to optimize RF then the phone is RF-location adaptive. Suppose the phone could automatically change bands from UHF to VHF not when UHF fades, but when <Self/> detects a location and direction of movement where UHF is known to fade based on previous experience. Such a phone is RF-location adaptive, in this case band adaptive. 3G phones typically are mode adaptive, switching from a high data rate, high QoS mode to low data rate, stay-connected modes during periods of weak RSSI.

1.2.4 Cognitive

Suppose the phone had learned RF-location adaptive behavior without having been preprogrammed. For example, the phone could create a database of location-indexed RSSI vectors (Latitude, Longitude, Time, RF, RSSI). Suppose the <Self/> includes a pattern recognition algorithm that detects a sequence of vectors along which UHF fades deeply for several minutes while at the same time VHF has strong RSSI. The pattern recognition algorithm might also determine that it takes 300–750 ms for the cell system to switch bands when UHF fades and that 80% of the time it has lost connectivity in 400 ms. Suppose finally that the phone <Self/> decides that to be always best connected (ABC) it should request handover based on location rather than on RSSI. ABC is a motto of the European Union (EU) wireless research Framework program [5]. The phone might report weak RSSI to the network so it switches bands, not knowing that the phone has strong RSSI but anticipates weak RSSI soon. That phone would be exhibiting cognitive behavior with respect to RF-location because:

1. It observed RF parameters and associated location over time.
2. It associated RF features (e.g., RSSI) with location (i.e., the path over which UHF fades).
3. It detected a relationship among these data associations and its user's need to be connected.
4. It reasoned over time to accurately diagnose that its user was not being connected because of a timing problem with handover.
5. It took effective action to achieve its goal (i.e., it reported low RSSI to obtain timely band handover to keep the user connected).
6. It achieved this specific behavior from general principles, not from having been specifically preprogrammed for this use case.

Professor Petri Mähönen of RWTH Aachen described a "little experiment" in which he integrated a neural network controller into a cell phone and GPS to autonomously learn the association among time of day, vehicle speed, and the location of a long underground tunnel. The phone learned to turn itself off for the 5 minute tunnel transit to save battery life [6]. Network operators already may employ similar learning algorithms to optimize their use of radio resources; what is "best" for the network may not be "best" for the specific user, however.

A cell phone that learns can help the user in ways that do not help the network. Consider the previous example of the KTH GSM–DECT smart badge. Suppose the <User/> told the radio, "It costs 1 € per minute to use GSM, but DECT costs zero, so stay connected, but with cost as low as possible." If from this and only this goal, the radio autonomously learns to use GPS location to switch to DECT when in or near KTH Elektrum, then it is behaving like an iCR. The cost-aware iCR researches tariffs for the user, learning that DECT air time is free while GSM is not. This book develops such entities with perception, planning, decision making, and actions that enable such implicit programming by communicating <User/> priorities via human language.

The iCR of the GSM–DECT example must know that the user's text "GSM" or utterance "gee ess emm" in the instruction of the prior paragraph refers to specific internal signals and software in its own SDR subsystem that might be designated RF1.gsm.6545.v4, not "GSM." A method of organizing such information into categories is called taxonomy. Taxonomy with a comprehensive semantics of the domain is called ontology [7]. If "GSM" invokes a map (<GSM/> <RF1 . . ./>) relating the user's words to the signal path in the chip set, then the radio <Self/> and the <User/> are mutually grounded regarding GSM. Formally [8], ontology is an intensional semantic structure that encodes the implicit rules constraining the structure of a subset of reality. Therefore, ontology defines semantic primitives: data and rules. AACR ontology structures the domains of <Space/>, <Time/>, <RF/>, and <Intelligent-entities/>, especially the <User/> and the iCR <Self/>. To emphasize the ontological role, semantic primitives in this text use XML-style markup, <Semantic-primitive/>. Semantic web enthusiasts are developing tags and ontologies to enhance web access. This emerging semantic web offers foundations, software tools [9], and lessons learned from which the specialized radio ontology kernel Radio XML (RXML) is defined in the companion CD-ROM.

The (Location, Time, RF, RSSI) association sketched above may be realized in a hardware platform with a mix of application-specific integrated circuit (ASIC), field programmable gate array (FPGA), digital signal processor (DSP) or general purpose processor (GPP), and associated firmware or software. The physical realization of AACR requires a mix of hardware–software realizations for behavior that is affordable, efficient, and flexible. The optimal mix changes over time, so this text emphasizes functions and interfaces, not implementation details.

1.3 ADAPTATION

There may be much value to adaptation without cognition. The aware–adaptive radio (AAR) is programmed to adapt itself to some aspect of a <Scene/>.

1.3.1 Adaptation Within Policy

A radio that senses an unused TV channel and adapts its transmission to use that RF channel for a low power ad hoc network is adapting to spectrum availability within a predefined policy constraint. The DARPA neXt Generation (XG) program defined a language for expressing <RF/> constraints to flexibly implement the U.S. Federal Communications Commission (FCC) rules enabling the use of such TV channels for Part 15 networks [64].[4] Many of the myriad other ways of adapting AAR RF behavior autonomously are developed in the sequel.

[4] The use case supporting FCC policy is referred to as the XG or TV-spectrum use case.

1.3.2 Adaptation to the User

Radio adaptation is not limited to RF. A radio with soft biometrics such as face and speech recognition could adapt to an unknown <User/> by protecting the Owner's data.

When my wireless laptop was stolen, there was nothing but a password protecting my personal information from abuse. Suppose somewhere deep in the motherboard were soft biometric models of me at home, at work, commuting, and in recreational settings. The thief might hack the password but might not be able to fool the biometrics. If I were to introduce such a laptop, say, to my daughter to help her with her homework, the iCR laptop would adapt its biometric model of <User/> to include <Barb/>, but it should not let her access my business information without further permission. How can one create such flexible yet trusted devices?

Historically, radio engineers have optimized the graphical user interface (GUI) to classes of users, but not to individual users. Cell phone GUIs are optimized for mass markets and military radios are optimized for military environments. As the complexity of function increases, the GUI complexity continues to increase, particularly in products where the user must set the RF air interface parameters ("modes"). A military iCR, though, may learn the "standard operating procedures" (SOPs) of the military user. Bands and modes for military SOPs may be published in a signal operating instruction (SOI). Instead of requiring the military user to enter parameter sets for an arcane SOP/SOI, the military iCR recognizes the user, time of day, and location, learned the SOP with the user, accesses the SOI, and offers the following dialog between Sgt. Charlie and his iCR Sparky:

Charlie: "Hi, Sparky."

Sparky (recognizing the GI's voice and face): "Hi, Charlie. The schedule says today is a training day. Shall I load the SINCGARS training mode from the SOI?"

Charlie: "OK."

Sparky: "What's today's training password?"

Charlie: "Today we are 'Second Guessing'."

Sparky verifies Second Guessing against the password downloaded via the Army's standard Single Channel Ground and Air Radio System (SINCGARS) secure network.[5] Charlie does not waste time with radio trivia; if encumbered with protective gear he doesn't need to type in the data load, potentially making an unfortunate mistake. Because of the unrealized potential of such speech, vision, and soft biometrics technologies, this book emphasizes such new iCR GUI ideas [10] with perception and AML to adapt to the specific <User/>, Charlie.

[5] This vignette is the SINCGARS–Sparky use case.

1.4 COGNITION

The value proposition of iCR needs further attention. Communications today are increasingly tedious. Commercial cellular users experience greater QoI with a briefcase full of GPS, AM/FM broadcast receivers, triband cellular, VHF push to talk, and cameras. The QoI entails increasingly complex control made transparent by the GUI (e.g., of cellular networks). But the mutual incompatibility of wireless PDAs, home wireless networks, business WLANs, wireless laptops, and so on burdens many users with tedium, limiting market penetration and decrementing QoI. AACR that perceives the user's needs and learns to support them by connecting to *information via any feasible RF* eases the burden of complexity, reduces costs, improves QoI, and enhances market value.

1.4.1 Perceiving User Needs

Is the user jogging or having a heart attack? Multiband cell phones and military radios don't care. But iCR user-perception technologies enable iCR both to sense such user states and to react, supplying contextually relevant personal information services, transforming radio from bit pipe to perceptive RF portal. A wearable iCR that "yells for help" as it detects a heart attack, so a nearby police officer instantly renders first aid, contributes directly to personal health and wellbeing. A user surprised by a massive heart attack cannot dial 911. The iCR that can see and hear—sensing heart rate from the multipath signature of an ultra-wideband (UWB) personal area network (PAN) to infer the impending heart attack asks <User/>, "Are you OK?" and sensing gasping and struggling verifies a health need. The iCR calls for help: "This is an emergency. I am iCR 555-1212. My owner is having a heart attack. He is incapable of communicating. This is not a drill. Please send a medical team immediately."

Wearable cameras are in mass production. Vision subsystems that perceive motion via optical flow are available in chip-sized focal-plane arrays [256]. Thus, CWPDAs that see what the user sees are not far off. An iCR packaged as a CWPDA perceives user communications needs to a degree not practicable with today's radio technology. Some of the technology to make such behavior affordable and reliable is on the frontiers of computer science, so this book offers a radio-oriented introduction to these emerging technologies, suggesting architecture and migration paths for AACR evolution.

1.4.2 Learning Instead of Programming

The iCR might detect other potential sources of bodily harm. To preprogram all such scenes, the way Sparky was programmed to adapt to SOI, is combinatorially explosive. AML of specific user-RF needs, sharing among peer iCRs, and collaboration via CWNs are keys to the mass-customization value proposition. When an iCR first observes a mugging, it extracts the distinguishing semantic features of the scene that precipitated the E911 call by the <User/>, for example, the words of the <Stranger/>. The next iCR that hears "Hey, Buddy, 'c'm'ere; got the time?" from a dark alley might vibrate to warn the elderly <User/> and offer to initiate an E911 call. The architecture and research prototype CR1 illustrate such machine learning in simulated RF, audio, and video sensor-perception domains to enable iCR to learn autonomously instead of being programmed.

1.4.2.1 Learning by Being Told

Suppose in Boston, if bodily harm is imminent, the iCR can "yell for help" on a designated low power radio channel that all police monitor, just as air traffic controllers monitor for "Mayday" distress calls. An iCR from the Midwest could learn such local customs from a Bostonian iCR. Sharing knowledge should be a trustable process to minimize false rumors. The Midwest iCR first learns Boston police E911 RF channels from Scottie, the local iCR, verifying this from a regulatory authority (RA) trusted network.

To share data accurately with peers, iCRs share the semantics of conceptual primitives, like "emergency" as <E911/> and "channel" as <ISM/> in megahertz (MHz) of Figure 1-3. Shared semantics may be implemented (1) by traditional standards that force the developer to hard-code the semantics into the SDR, or (2) by open computational ontologies with standard semantics, for example, as promoted by the semantic web community. Both peer exchange and RA verification mediated by shared semantics are examples of "learning by being told."

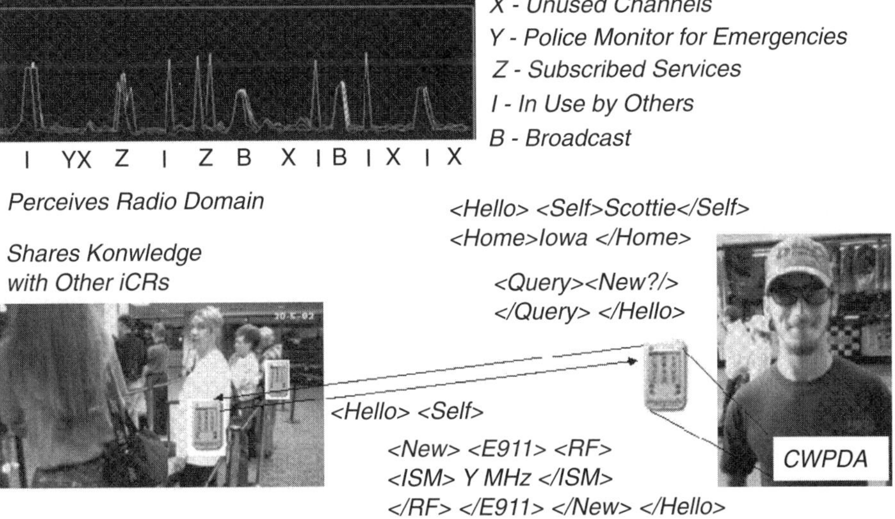

FIGURE 1-3 Shared ontology assures accurate learning.

1.4.2.2 Learning by Observing

Complementing peer knowledge, iCRs also learn local radio-use patterns autonomously. With speech recognition, the iCR could learn radio-use patterns by listening. Suppose a <User/> arrives at an automobile racing event. Racing crews employ pit-crew jargon that differs from radio broadcaster and emergency jargon:

Racing jargon: "We are *a little loose* in that *first turn*."

Broadcast jargon: "*Mikes are hot*; we go to *the booth* after the *commercial*."

Emergency jargon: "We *need rescue* behind the BB grandstand. *Heat stroke*."

Having learned these hugely redundant patterns, the iCR adapts its own <RF/> use patterns accordingly. It plans to "yell for help" on the channel where emergency jargon is most in use without having been told or programmed to do so. It finds the Motor Racing Network's ("MRN") local RF channels offering the <User/> behind the scenes insights.

Both learning by being told and learning by observing the local radio bands reduce user tedium. Speech technology for such AML is brittle. Although 800 directory assistance speech recognition (e.g., TellMe®) is nearly error free, raw error rates remain high in noisy multispeaker environments—often only 50% successful transcription from speech to text, increasing to 70–90% when trained to the user, background, and domain of discourse. The narrower and more redundant the domain, the better. Speech and text natural language follow Zipf's Law [11], exponentially distributing word frequencies as a function of language, domain, and topic.

1. *Language Structure*: "The" is the most common word in written English.
2. *Domain of Discourse*: "Cognitive" and "radio" are the most common words in this text.
3. *Topic Structure*: Each paragraph or section obeys Zipf's Law with surprising consistency.

Thus, in spite of low speech-to-text transcription accuracy, narrow domains exhibit distinctive content words and phrases with such statistical strength that they can be reliably detected in discourse. This text explores whether such brittle technology can detect user communications needs, reducing tedium for the user. Suppose your PDA updated your appointment book when you said, "Yes sir, I will be there next Tuesday at 7 am." The true iCR PDA later autonomously joins an ad hoc 802.11 network to advise the boss that you are stuck in traffic because of a big accident on the Beltway, bypassing cell phone system overload.

1.4.2.3 AML Versus Programming

Computer programming is today's method of synthesizing SDR behavior. A local emergency channel defined in a public XG broadcast can be hard-coded and downloaded. Machine learning isn't needed.

But computer programming is expensive and programming for generic use-cases requires compromises. Network operators can't marshal sufficient programming resources to customize software to narrow situations, so we go for the worst case or average case. For example, the statistics of WLANs in corporate LANs versus rural consumer settings call for different sizes of address space and degrees of protection. iCR autonomously generates protocol variants from experience to optimize for local conditions. Genetic algorithm research shows how to encode wireless features in a digital genome for off-line optimization [75, 76]. With RA supervision, such AML enhances CWNs autonomously.

Software tools reduce the costs of software development and maintenance, but the tools tend not to offer AML as an alternative to programming. Tools tend toward domain independence, speeding programming practice, for example, via refactoring existing code and composable behaviors. In contrast, iCR employs heavily domain-dependent AML, for example, coding wireless features into a genome with radio performance coded in the fitness functions [74]. RF-domain dependence leverages a store of prior knowledge unique to radio for incremental autonomous knowledge refinement and adaptation. As was first encountered in Lenat's AM-Eurisko investigations [320] and Davis' Tieresias [318], and widely proved by expert systems of the 1980s and 1990s [12] and remaining true today [13], autonomous knowledge evolution works well somehow algorithmically "close to" a priori knowledge, but does not extrapolate well. Thus, AML is accurately characterized as brittle.

1.4.2.4 Overcoming Brittleness

Contemporary AML offers incremental methods not fully exploited in SDR, such as case-based reasoning (CBR) with reinforcement. CBR was the first ML technology to formalize experience per se. CBR learning consists of remembering, retrieving, and adapting ("revising") the most relevant historical "case" from a case base of experience, using an adapted case ("reuse") and integrating feedback into the anthology of experience [14]. Although CBR is a form of generalization, CBR can learn from a single example while other ML techniques like artificial neural networks (ANNs) must generalize off-line from large numbers of examples during a training phase. CBR generalizes on-line when confronted with a new situation, learning exceptions from one (validated) instance. In the early to mid-1990s, CBR was commercialized as a type of instance-based reasoning (IBR) [15, 16]. IBR relevant to AACR includes support vector machines (SVMs) [17] for transductive inference, inference from one observed specific case to another without mediating generalizations [18]. Relevance-based learning (RBL) [13, p. 689] formalizes knowledge bootstrapping. Redundant structured domains like emergency radio offer regular patterns and repetition needed for ANNs to learn patterns over time. Hierarchical reactive planning and control systems in robots also learn from the environment [53]. This broad range of AML techniques adapted to SDR enables AACR evolution. The cognitive radio architecture (CRA) of this text facilitates experience aggregation to mitigate the brittleness of AML, enhancing QoI through autonomous use of RF domain knowledge for autonomously perceived user needs.

This book shows how the autonomous customization of AACR may shift from labor-intensive programming to RF- and user-domain-specific AML. The serious reader who does the exercises and experiments with CR1 could contribute to AACR evolution, reducing the cost of tailored services and successfully embedding emerging vision, speech, perception, the semantic web, and AML technologies.

1.4.3 The Semantic Web

The technical foundations of computationally intelligent software are being feverishly developed for semantic information retrieval from the ultimate large data store, the World Wide Web via ontological content tags, not merely text, pictures, and sound [1]. Computational ontologies are a version of the classic parlor game "Twenty Questions." I'm thinking of something and you must guess what it is by asking me not more than 20 questions. The first question is free: "Is it a person, place, or thing?"

<Universe/>:

1. <Person/>
2. <Place/>
3. <Thing/>

Is a cell site a place or a thing? From the network operator's perspective, a cell site is a place near a cell tower. From the equipment manufacturer's perspective, a cell site may be a thing, the tower and associated equipment. A radio-aware user, complaining "Darn, I always get disconnected in this cell site," refers to <Place> <Cell-site/> </Place>.

The recognition of user dissatisfaction depends on shared semantics. The user and the iCR must share the same meaning of <Cell-site/> as <Place/> not <Things/> in the context "disconnected." Shared semantics opens the envelope, defining new relationships among users, regulators, service providers, and network operators. Thus in some sense, this is an "idea generation" book, probing the art of the possible by sketching AACR evolution and identifying key questions, challenges, and the enabling technologies.

Thus, iCR is a semantics-capable software agent embedded in a SDR. The agent learns from users, iCRs, CWNs, and the RF environment. The convergence of radio with the computational intelligence of the semantic web further blurs the distinctions among radio, laptop computer, wireless PDA, household appliance, and automobile, yielding computationally intelligent information environments with AACR throughout.

Since the semantic web is developing rapidly, it is unclear whether the traditional wireless community (think "cell phones") or the traditional computer science community (think 802.11 "wireless LAN") will lead iCR markets. Will the wireless community move from bit pipes to semantic cell phones? If so, then wireless giants like Ericsson, Nokia, Samsung, Lucent, and Motorola may lead the market for billions of new iCR class semantically aware cell phones.

On the other hand, the mobile semantic web may render cell phones to mere commodity hardware like 802.11 nodes from BestBuy$^{®}$ or Kmart$^{®}$, enabling semantic information networks in which Intel, Microsoft$^{®}$, IBM$^{®}$, Dell$^{®}$, Comcast$^{®}$ (home information services provider), or Disney$^{®}$ (content provider) become the market leaders.

Either way, the technical foundations of wireless on the one hand and computational intelligence on the other are developing quickly, driven by complementary market forces.

1.5 COGNITIVE RADIO AND PUBLIC POLICY

Ideal cognitive radios are aware, adaptive radios that learn from experience. AML enables wireless devices to discover and use radio spectrum by "being polite" to each other, employing self-defined radio etiquettes rather than predefined albeit flexible air interfaces and protocols. But will regulators permit such technology to enter the marketplace and if so, when?

1.5.1 FCC Rule Making

The iCR with AML was first proposed in 1998 [19] and presented to the U.S. FCC as "cognitive radio" contemporaneously. The FCC identified the potential of AACRs to enhance secondary spectrum markets. Specifically, the FCC enables TV-aware radios to establish Part 15 (low power) ad hoc wireless networks. The FCC's deliberations included Notice of Inquiry (NOI) [20] and Notice of Proposed Rule Making (NPRM) [21, 22] without requiring the CRs to learn. This is good for the evolution of AACR, authorizing aware–adaptive radios, but it could lead to confusion between iCR and FCC CR, with market hype over FCC CR yielding only the AAR. Thus, in this text the term iCR is reserved for radios that autonomously learn from the environment (user and RF in a specific context or <Scene/>), adapting behavior perhaps beyond current FCC rules.

1.5.2 Global Interest

RAs in the United States, in Europe, and in the Asia-Pacific regions share interest in AACR. The U.S. FCC support of cognitive radio complements other regulatory administrations, such as the U.K. and Japanese RAs and Germany's RegTP, addressing CR [23]. In addition, the European Commission (EC) funded the End to End Reconfigurable (E2R) program with a cognition task that includes the autonomous acquisition of user profiles [24]. Subsequently, RWTH Aachen sponsored the Dagstuhl, Germany workshop [25]. The EC considered CR as a theme of its sixth and seventh research frameworks [26]. Finally, the Software Defined Radio (SDR) Forum formed a special interest group on cognitive radio applications in 2004 [27], meeting in the United States, Europe, and Asia.

1.6 ARE WE THERE YET?

The iCR is a visionary concept. How long will it take to "get there"? A wealth of relevant technologies is rapidly emerging to move the AACR community quickly into the products and services envisioned by the FCC CR and inevitably closer to iCR.

The full realization of the iCR vision requires decades. As illustrated in Figure 1-4, the iCR is a far-term concept, a point on the horizon by which to navigate. The research prototype cognitive radio, CR1, companion to this text, illustrates architecture principles for navigating toward iCR.

The FCC rule for more flexible use of TV band spectrum encourages near-term AACR technology: proactive sensing of the RF spectrum, enhanced detection of legacy users, adaptive creation of ad hoc networks, and polite backoff from legacy users when detected. Such basic AARs were emerging

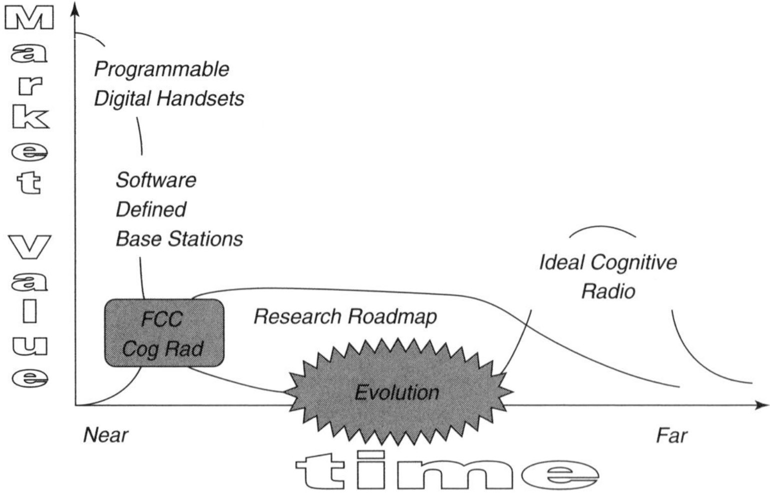

FIGURE 1-4 The vision of the ideal cognitive radio takes time to realize.

in 2003, for example, the Intel® TV band AAR for the PC motherboard [28], leveraging the 2003 Rule and Order (R&O) that made unused television (TV) spectrum available for low power RF LAN applications via a simple predefined spectrum-use protocol [64]. DARPA's neXt Generation (XG) program developed a language for expressing such policies [29]. Other more general protocols based on peek-through to legacy users have also been proposed [145]. But radio communications will not transition instantaneously from AAR to CR. An embryonic AACR may have minimal sensory perception, minimal learning of user preferences, and no autonomous ability to modify itself. RAs hold manufacturers responsible for the behaviors of radios. The simpler the architecture, the easier it is to assure compliant behavior, to obtain certification by RAs, and to get concurrence for open architectures. An autonomous iCR might unintentionally reprogram itself to violate regulatory constraints, with high risk to the manufacturer. Meanwhile, as researchers explore ways for perception and AML to enable new services, the evolution toward AACR will become clearer. Although it is difficult to quantify time to the iCR, further research in that general direction seems valuable. The pace at which markets develop depends in part on the degree to which researchers collaborate to accelerate iCR. One tool toward this end used successfully in the ITU, OMG, TIA, and SDR Forum is the open architecture standard.

1.6.1 Open Architecture Frames Collaboration

Evolution from AAR toward iCR may be accelerated by industry agreement on an open cognitive radio architecture (CRA), a minimal set of AACR functions, components, and interfaces. Standard functions relate to both use cases on the one hand and product components on the other. This text sketches the evolution of functions for RF and (1) user perception via speech, vision, and other sensors; (2) computational semantics; (3) space–time planning; and (4) AML in an open architecture framework.

How will the computational ontologists work with RF designers? When will the speech and signal processing community contribute to better language perception to autonomously determine the wireless information needs of the user in a noisy subway station? Will the speech recognition of the CWPDA fare better than in the speech-capable laptop, where the technology is underused at best? Cell phones of 2006 sport digital video cameras but not digital image perception. To integrate audio, video, and RF perception in managable steps toward the iCR requires an architecture that delineates the common ground of these disparate disciplines. The functional architecture, inference hierarchy, and cognition cycle of this text define that common ground.

Specifically the CRA defines functions, components, and design rules by which families of different designs may rapidly be evolved, employing best-of-breed strategies. This text characterizes the technologies to be integrated for AACR, defining interfaces among hardware–software components from disparate disciplines. Allocation of functions to components and the definition of technical interfaces among these components are major tasks of radio systems engineering. Since computational ontologies are critical for AACR evolution, we're not in Kansas anymore, Toto. So this text draws together disparate technologies to promote radio engineering to rapidly integrate semantic web technical radio knowledge, autonomous agent, and robotic control technologies to evolve AARs toward iCRs. The open CRA is not a final solution but a contribution to academic, government, and industry dialog for iCR sooner rather than later.

1.6.2 Research Prototypes Deepen Understanding

Radio research depends on learning by doing. Thus, CR1, the research prototype iCR, is a working (if not perfect) Java program implementing ubiquitous CBR, learning from every experience, adapting to the RF environment and user situation. CR1's illustrative personalities offer information services perceived through learning, hiding details of radio bands and modes from the user in a simulated environment. The companion CD-ROM includes the Java source code, compiled classes, previously learned/trained personalities, and an integrated runtime system for hands-on experimentation.

1.7 KEY QUESTIONS

Thus, the text addresses the following central questions:

- What is iCR and how does it differ from software radio, software-defined radio (SDR), and aware–adaptive radio (AAR)?
- What new services are enabled by iCR?
- How will emerging AACR services differentiate products and benefit users?
- What is the CRA? How will it evolve through initiatives such as the SDR Forum's CR special interest group [145]?
- What sensory perception and radio knowledge must be embedded into SDR for AAR and iCR? How does computational ontology represent this knowledge, and how is it related to the semantic web?
- What new sensors are needed for FCC CR, AACR, and iCR?
- What skills must a radio system's organization add to its workforce for AACR—natural language processing (NLP), machine learning (ML), ontologists?
- How is regulatory rule making shaping AACR markets?
- What about U.S., European, and Asian R&D?

- How will today's discrete cell phones, PDAs, and laptops merge into the iCR wardrobe?

1.8 ORGANIZATION OF THE TEXT

To address these questions, this text is organized into three parts: foundations, radio competence, and user-domain competence. It includes conclusions, a glossary, references, and a companion CD-ROM with CR1 source code, documentation, and supplementary materials.

1.8.1 Foundations

The foundations part begins with a technical overview of AACR. Since economically viable progress depends on user acceptance, the section develops both radio-driven and user-driven scenarios, motivating an ontological view of data structures that the cognitive entity must define, ground to the real world via sensory perception, and employ effectively. Chapter 3 develops a specific use case in sufficient detail to introduce the main technical ideas. Although most of the use cases could be implemented by hard-coding the use case in C, C++, Java, or C#, the major differentiator between AAR and iCR is the AML technology introduced in Chapter 4, with radio examples. Chapter 5 develops the *OOPDAL* loop, the software flow from stimuli to responses through a perception hierarchy with algorithms to *Observe*, *Orient*, *Plan*, *Decide*, and *Act* while all the time *Learning* about the <Self/>, the <RF-environment/>, and the <User/>. CR1 implementing this architecture is developed in the companion CD-ROM with sufficient detail for experimentation and behavior modification.

1.8.2 Radio Competence

The radio competence part develops radio-domain use cases in Chapter 6. Chapter 7 explicates the radio knowledge into structured knowledge chunks with related methods of using the knowledge, bite-sized for evolutionary implementation. Chapter 8 addresses the implementation of radio-domain competence, formalizing radio knowledge in RXML. It develops reasoning skills—logic, rule-based reasoning, pattern analysis—with autoextensibility through the creation and use of knowledge objects (KOs) evolved via radio-domain heuristics (RDHs). If iCR were a *fait accompli*, you could buy iCRs, not just read books about them, so this research-oriented treatment develops key ideas for radio-domain skills in RXML, KOs, and RDHs so that AACR may bootstrap skill as experience accumulates. This is a snapshot of a work in progress, warts and all.

1.8.3 User-Domain Competence

The user competence part begins with use cases in Chapter 9. The transparent acquisition of knowledge from users depends on sensory perception, enabling iCR to see and hear what the user sees and hears via vision and language technologies discussed in Chapter 10. The emphasis is on the perception of the user in an archetypical setting called a <Scene/>—home, work, leisure, and so on. Chapter 11 develops methods for implementing user-domain competence, grounding symbols, reasoning with user KOs, and evolving via user-domain heuristics (UDHs). Chapter 12 builds bridges to the semantic web community, promoting the autonomous acquisition of knowledge from the semantic web.

1.8.4 Conclusion

The final chapter offers suggestions for the further evolution of industrial strength AACR, with pointers to advanced topics, related architectures, technologies, and components. The main contribution of the companion CD-ROM is to save the reader time in becoming familiar with hardware and software components from relevant disciplines from CR1 to robots to the semantic web. As an interdisciplinary pursuit, the treatment of each discipline has to be light, bordering on superficial to an expert, mitigated by the citations to the Web and the literature.

1.9 EXERCISES

The exercises with each chapter review the key points and explore topics further. After reading this chapter, the interested reader should be able to complete the following exercises.

1.1. Differentiate awareness, adaptation, and cognition as it applies to radio.

1.2. Discuss the difference between network-value-driven behavior and user-value-driven behavior of AACR, explaining examples such as an autonomous CWPDA appliance.

1.3. Is it possible to "define" cognitive radio? If so, give a precise definition, a mathematical definition if that is possible. If not, explain why not. If one could, but it would not be a good idea to try to enforce one, explain that view.

1.4. Informally, what is an ontological primitive? Why should a radio engineer care?

1.5. Find OWL on the Web. Play a game of 20 questions, tracing the evolution of the questions through OWL ontologies. Try something abstract like Superman and something concrete and medical like polio or DNA.

1.6. How is iCR like "customer-premises equipment" (CPE)? When the proverbial black handset was owned by the telephone company and leased to the consumer, there were few choices, prices were high, but technology investments similarly were high, as attested by Bell Labs invention of the transistor. Not unrelated to the breakup of "Ma Bell," the consumer could buy handsets CPE, connect computers to the telephone network using modems, and the like. How is the control of the behavior of cognitive devices similar to and different from CPE?

1.7. Discuss potential cell phone market disruption from iCR PDAs.

1.8. State a narrow definition of iCR from the viewpoint of a cell phone manufacturer and define a roadmap toward iCR for that community based on that definition. The roadmap should specify a sequence of new capabilities over time, with time lines for technology insertion. Do not refer to www.wwrf.org.

1.9. Compare your answer to Exercise 1.8 to the perspective of WWRF.

1.10. State a narrow definition of iCR from the viewpoint of a major supplier of laptop computers such as Dell®. Define a roadmap from a laptop product line on the manufacturer's web site to an iCR laptop for global public safety markets.

1.11. Compare roadmaps and common ground of cellular and ISP markets.

PART I

FOUNDATIONS

CHAPTER 2

TECHNICAL OVERVIEW

This chapter defines iCR in terms of functional capabilities, characterizing the related contributions and limitations of the enabling technologies. Think of this chapter as a needs summary and functional overview of AACR.

2.1 THE iCR HAS SEVEN CAPABILITIES

An ideal cognitive radio (iCR) may be defined as a wireless system with the following capabilities [145], each of which is necessary in evolving AACR toward iCR:

1. *Sensing*: RF, audio, video, temperature, acceleration, location, and others.
2. *Perception*: Determining what is in the "scene" conveyed by the sensor domains.
3. *Orienting*: Assessing the situation—determining if it is familiar. Reacting immediately if necessary. Orienting requires real-time associative memory.
4. *Planning*: Identifying the alternative actions to take on a deliberative time line.
5. *Making Decisions*: Deciding among the candidate actions, choosing the best action.

6. *Taking Action*: Exerting effects in the environment, including RF, human–machine, and machine–machine communications.
7. *Learning Autonomously*: From experience gained from capabilities 1–6.

Capabilities 1 and 2—sensing and perception—may be termed "observing." Together, these seven capabilities comprise a cognitive system. Cognitive systems observe, orient, plan, decide, and act, all the time learning about themselves and their environments in order to be more effective over time. In 2004, DARPA's view was that in order to be termed "cognitive" a system must learn to adapt its behavior through experience [30]. This text addresses these capabilities through a process of use-case development and refinement, with the enabling technologies developed in greater detail later.

2.1.1 Wearability

The focus of iCR is on wearable wireless devices (e.g., CWPDAs) that perceive the world from the user's viewpoint. Chapter 1 postulated the cell phone's video port not on a handheld device but in a pair of glasses so that the CWPDA sees what the user sees, keeping the last few minutes of video from which to save the perfect snapshot or to adjust the 3G equalizer optimally for entering the building that vision perceives approaching.

2.1.2 Distribution of Intelligence

CWPDAs may collaborate in ad hoc CWNs or they may be supported by a CWN operator or service provider. Computational intelligence in networks entails memory, perception, and adaptation pioneered by Theo Kanter of KTH [10]. Petri Mahonen led the 2004 workshop crystallizing the CWN [25]. Network intelligence offsets the need for computational intelligence in wearable devices. This architecture trade-off of network versus device is central to AACR evolution. Initially, AACR behaviors will be supported heavily by network resources. Over time the wearable devices themselves will exhibit increasing distributed intelligence. The iCR vision postulates maximal computational intelligence in the wearable nodes, an intentionally extreme case. Migration paths from today's single-provider networks may evolve iCRs that intelligently obtain access from the most advantageous network provider(s), although market pressures counter that migration. Academic focus on iCR characterizes the technology challenges of that limiting case. The isolation of such limiting cases supports the scientific method, isolating aspects of a problem, generating hypotheses, and testing.

2.2 SENSING AND PERCEPTION: WHAT AND WHOM TO PERCEIVE

The iCR perceives three distinct information spaces: the iCR itself (the <Self>), the <RF/> environment, and the <User/> in its environment.

2.2.1 Legacy Control Does Not Require Machine Perception

The AACR evolution begins with today's mix of legacy hardware-defined radios (like push-to-talk sports and military radios) and emerging SDR devices like cell phones. Control of such traditional radios includes user-oriented (input and output interfaces) and radio-oriented aspects (RF reception and transmission). The radio-oriented aspects are controlled via radio-engineering parameters: layers of the International Standards Organization (ISO) Open Systems Interconnect (OSI) protocol stack from physical (PHY), including media access control (MAC) if any, to applications. A single-band single-mode radio like VHF push-to-talk has null middle layers of the ISO stack with simple band and mode control parameters like on/off, audio volume control, squelch, and channel selection. User controls employ human–machine interface (HMI) principles, traditionally limited to microphone/speaker, dials, buttons, and displays.

Commercial multiband multimode radios (MBMMRs) like most vintage 2005 cell phones perceive RF but hide band and mode from the user except for status such as "Extended" or an icon for digital or analog mode. Military MBMMRs typically use a softcopy HMI display, which may be part of the radio or may be remote (e.g., in the cockpit for airborne radios). User control may include band, mode (e.g., air interface), and related parameters (data rate, voice versus data, etc.). In commercial MBMMRs like triband cell phones, the network sets the parameters for the user, while in military MBMMRs users set many of the parameters.

Contemporary radios may blindly attempt the best communications possible such as 50 W radiated power even if not needed (e.g., between vehicles 20 meters apart). The knowledge of which bands and modes, powers, frequencies, call signs, telephone numbers, and so on yield what kinds of connectivity is not in the radio but in the mind of the user. If an attempt to communicate fails, the process of establishing alternate communications can be labor intensive. The user may give up on the military radio or cell phone, turning to land line, or just wait until later. Such primitive RF control and simple HMI sufficed for inflexible hardware-defined legacy radios. Moore's Law has both increased radio complexity and created technologies for coping with complexity.

2.2.2 Perceptive Control

The increased capability and complexity of radio technology warrant both enhanced HMI and autonomous situation-dependent control. Figure 2-1

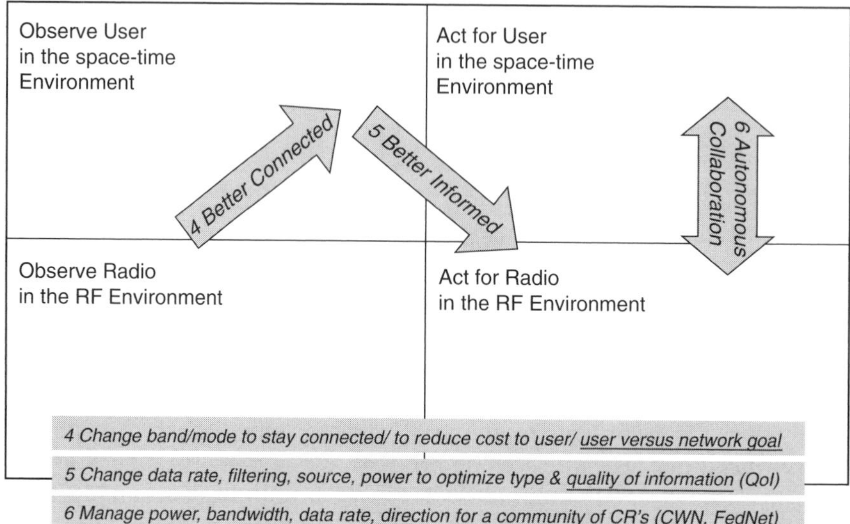

FIGURE 2-1 iCRs enable user-centric, RF-aware, context-sensitive information access.

shows iCRs that observe users in their natural environments to enable situation-dependent information access, employing radio control toward user-optimized ends (not necessarily network-optimized ends), steps that are well beyond the network-optimized WWRF vision of "I" [140].

Unlike prior radio technology, iCRs are computationally aware of the <Self/> in the radio environment of multiple networks, autonomously collaborating to enhance the wireless experience of all users in CWNs. The iCR <Self/> must discover its user, networks, and goals.

2.2.3 From Provider-Driven to Perceived-User Responsive

Items 4, 5, and 6 in Figure 2-1 illustrate the evolution of wireless devices from nonperceptive to <Self/>-, <RF/>-, and <User/>-perceptive. A radio following FCC spectrum-use rules can identify new opportunities for ad hoc networking in <RF> <TV-bands/> </RF>[1] implemented as yet another preprogrammed MAC layer. This is an aggressive regulatory step, enabling user-oriented <TV-band> behavior but not needing the radio to perceive <Self/> or <User/> per se. Similarly, in the GSM–DECT badges at KTH the initial personalities typically stayed in GSM mode inside buildings because GSM penetrates buildings, generating revenue for Telia. With SDR technology, KTH programmed those badges to regularly check the DECT RSSI, switching to DECT whenever

[1] <RF> <TV-bands/> </RF> encapsulates <TV-band/> within <RF/> in any subordinate relationship not just class–subclass or "is-a" but others such as exemplar, partial property inheritance, and space–time ontology.

available, a pro-KTH behavior. This hard-coded behavior implemented the value system of flow 4 of Figure 2-1, needing no <Self/> or <User/> data structures, only the KTH-centric AAR behavior.

2.2.4 Self Perception

Additional wireless badge flexibility could be achieved via a <Self/>, a software object in each badge. Self-awareness data structures would include objects with relationships to the <Self/>, like <Telia/>, <KTH/>, and the <User/>, each noted in the format of a closed XML tag <. . ./> indicating that the name of the data structure defines the ontological primitive (Expression 2-1).

Expression 2-1 XML Data Structure for a Badge's <Self/>

```
<Self>
<Name >Badge-001 </Name >
<Owner> KTH </Owner>
<User> Chip </User>
<Optimization> <Lowest-cost/> </ Optimization>
</Self>
```

This badge's <Self/> behavior is defined not by the preprogrammed pro-Telia nor by the reprogrammed pro-KTH behavior, but by the more general <Optimization/> objective best supporting the <User/> <Optimization/> value system. Normally, the <User/> might assert <Lowest-cost/>, resulting in the pro-KTH behavior. Alternatively, the <User/> might <Optimize/> for <Minimum-handover/> during an experiment that handover would disrupt, and for which the cost of GSM is warranted. Such situation-dependent objectives are realized by *situation-dependent choice* among the fixed pro-Telia and pro-KTH behaviors. Thus, the association of behaviors with relationships among <Self/> and <User/> in a <Scene/> enables greater flexibility than either preprogrammed AAR. The AACR with <Self/> and <User/> data structures can respond to discovered objectives like <Lowest-cost/>.

2.2.4.1 Degrees of Self-Awareness: Aware, Adaptive, Self-Conscious

An AACR containing a <Self/> data structure could be self-aware. Built-in test (BIT) may aggregate self-descriptive data in a way that is tantamount to a <Self/>. The SDR Forum's software communications architecture (SCA), for example, describes the self in terms of resources—components assembled by a factory entity [58, 70, 79]. In 2005, the Forum did not call SCA a <Self/> but the self-describing XML data structures were tantamount to <Self/>. An AACR that modifies its <Self/> data structure depending on its own perception of world states and events is self-adaptive. BIT that overcomes hardware faults implements this level of self-adaptation. The SCA facilitates adaptation to the hardware and software resources available.

If an AACR can modify the <Self/> through learning about explicitly represented behaviors of the <Self/>, then it is self-conscious. Self-consciousness entails explicit introspection. Questions like "Where am I?" are self-referential but not introspective. Questions like "Why did I do that?" are introspective, requiring self-conscious reasoning about the <Self/> over <Memory/> of <Behavior/> of the <Self/>.

Expression 2-2 Retrospective Behavior Trace

```
<Behavior> <Memory>
  <Sensor> <Perception> <Action/> </Perception> </Sensor>. . .
  </Memory> </Behavior>
```

The iCR may reason over a behavior memory trace (Expression 2-2) to infer why it took a specific <Action/>. "Why" here has very precise meaning. Algorithmic introspection entails examining branch points, diagnosing a failure or success, and identifying alternatives. If the <Self/> changes its internal structure to behave differently in similar circumstances in the future, then self-consciousness achieves introspective adaptation. Introspection consumes substantial resources, potentially requiring resolution theorem proving or Turing computability: it may consume unbounded computing resources. Thus, practical iCRs need the ability to suspend introspection, safely performed in a "sleep cycle" of no user activity, with a watchdog timer to wake up the iCR (e.g., for the next day's activities).

2.2.4.2 The Autonomous Self

Radio control thus may migrate toward the autonomous iCR <Self/> using distributed RF control that optimizes situation-dependent <User/> goals. To stretch computational awareness to such an autonomous iCR, the following should be known to the <Self/>:

Expression 2-3 Data Structures for the Autonomous Self

```
<Universe>
  <Self> . . . data structures . . .
    <Autonomous-control> . . . methods. . .</Autonomous-control>
      </Self>
  <User> . . . data structures . . . </User>
  <Environment> <Users/> <RF/> <Self/> <Others/> </Environment>
</Universe>
```

This data structure asserts that the <Self/> has <Autonomous-control/> in a <Universe/> shared with </Users>. Initially, <Environment/> of AAR may be limited to the RF bands and modes of a service provider. The <Self/> shares this <Environment/> with <Users/>, <RF/>, and unspecified <Others/>. Pro-provider <Autonomous-control/> could enable the AAR to be always best connected (ABC, a European Framework objective) with respect to the provider's radio resources. The <Self/> that is autonomously controlled by iCR for the <User/> achieves new ideas like autonomous spectrum rental and open optimization of band/mode alternatives, optimizing ABC across multiple service providers and free access bands like ISM, unused TV, possibly the U.S. Citizens Band for ad hoc voice networking, and so on. Pro-user <Autonomous-control/> could choose a private rental of IEEE 802.11 or DECT air time instead of a conventional service provider.

2.2.4.3 Self-Awareness Tensions

Monolithic network services are an economic engine of today's wireless technology. Nonetheless, RFLAN manufacturers may expand their business models to the private rental of personal 802 radio access points expanding the Wireless Web. With greater AACR autonomy, service providers may expand services to heterogeneous networks. The autonomous <Self/> enables such new perspectives but creates market tension. What cellular service provider programs your cell phone to use your free 802.11 WLAN with VoIP when you are home? Historically that would be anathema. Today, a forward leaning service provider might offer <Home/>WLAN VoIP to gain market share. Some users might like the cost savings while some might not, which is why the phone should learn from the user. Thus, item 4 in Figure 2-1 is not just about mode change; it represents a shift of paradigm from the relatively blind use of a limited set of preprogrammed behaviors to the autonomous synthesis of more flexible, user-adaptive RF capabilities.

The <Self/>, <User/>, <Environment/>, and <Autonomous-control/> with AML may yield aberrant behavior with legal finger-pointing among service provider and third party software developers. Since it is easy to envision the many problems with a shift toward self-awareness, one must clearly define benefits and constraints for a managed evolution along AACR design points toward the iCR. One such sequence emphasizes QoI over mere QoS.

2.2.5 QoS or QoI

Specifically, Figure 2-1 broadens the set of behaviors of the iCR <Self/> from QoS to QoI.

2.2.5.1 QoS

Quality of service (QoS) includes data rate (R_b), bit error rate (BER), delay (dT), delay jitter (σT), and burstiness of the connection. Burstiness may be estimated as the peak data rate $R_{b\max}$ divided by the long-term average data rate including overhead and outages, R_{avg}:

$$B = R_{b\max}/R_{\text{avg}} \tag{2-1}$$

Wireless typically falls short of wireline performance in one or more of these QoS parameters, sometimes by one or more orders of magnitude. In addition, grade of service (GoS), the probability of connection on the first attempt, approximates 1.0 very well on wireline networks but falls off sharply on wireless networks when mobile and during periods of high demand. On the other hand, the probability of connecting to a wireline when you are driving your car is zero, while the probability for wireless is much greater than zero, so when viewed from a mobile consumer's perspective, value of connection is much higher than GoS or QoS alone reflects, suggesting a need to model quality of information.

2.2.5.2 Quality of Information (QoI)

The <User/> data structure must accurately model the user's specific and context-sensitive information value system, QoI. Like QoS and GoS, QoI can be defined. Unlike those metrics, QoI must be defined in user-centric terms. QoI is the degree to which available information meets the specific user's specific needs at a specific time, place, and situation. A mathematical framework for QoI is

$$\begin{aligned}\text{QoI} = {} & \text{Availability} * \text{Quantity} * \text{Relevance} * \text{Timelines} * \text{Validity} \\ & * \text{Accuracy} * \text{Detail} * \text{Need}\end{aligned} \tag{2-2}$$

Item 5 of Figure 2-1, "Change data rate, filtering, source, power to optimize type & quality of information (QoI)," requires a QoI metric that reflects <User/> goals. QoI is defined with respect to a specific <User/> over a finite space–time epoch called a <Scene/>, a Subset of Space x Time for a Specific User:

$$\text{<Scene/>} \equiv \{\text{<User/>} \cap \{\text{<Space/>}\} \cap \{\text{<Time/>}\}\} \tag{2-3}$$

Equation 2-3 makes an ontological statement about referents in the real world, not symbols of a logic system. Specifically, the ontological primitive <User/> refers to the person using the system. Thus, the intersections refer to a subset of physical space and ordinary time along with the person being served. In a programming language, plus in "A + B" operates on memory locations, not on the literals A and B. So too, <User/> refers to the "memory location" out there in the real world, everything inside the skin of that person, including his/her thoughts, perceptions, prejudices, and specifically information needs. Ontological statements in this book express the open set existential potentiality of the real world, not the closed set existentials and universals of symbolic logic. (See Open Sets in the companion CD-ROM for further treatment.)

If there is no information connectivity in the <Scene/>, then Availability = 0 = QoI. If the <Self/> connects the <User/> to all (Quantity = 1.0) the required

(Relevance = 1.0) information (either via wireless connectivity or via a memory of sufficiently recent information) then Availability = Quantity[2] = 1.0. Timeliness may be defined on a time line of information use. For information needed immediately, the Timeliness is inversely proportional to time delay dT. To avoid division by zero, there is no such thing as instantaneous information. Always $dT > 0$ since information transfer takes time even if the information itself is in memory. If shortest time delay is ε, the maximum contribution of timeliness to QoI is ε/dT. Timeliness is normalized by ε so the maximum timeliness is 1.0. If validity is +1 if judged to be true and –1.0 if not true, then QoI may be positive or negative reflecting user judgment, not absolute truth.

Relevance is the degree to which the information corresponds to the need, measured by data mining metrics precision and recall. Recall is the fraction of relevant documents R retrieved from a corpus by a query to the total number T of relevant documents of corpus size $N \geq T$. Precision is the fraction of relevant documents R to the number retrieved X. Recall (R/T) of 1.0 indicates that all relevant documents have been retrieved, while precision (R/X) of 1.0 indicates that no irrelevant documents have been retrieved and $X = R$. Adapting these metrics to QoI, relevance is the product of precision and recall:

$$\text{Relevance} = \left(\frac{R}{T}\right)\left(\frac{R}{X}\right) = \frac{R^2}{TX} \tag{2-4}$$

Accuracy refers to the quantifiable or numerical aspects of the information. Accuracy is the expected or average error, while precision is the smallest error representable in the response. In QoI availability requires the precision of the information to support the accuracy. If the accuracy required by the user is met, the value of the accuracy metric is 1.0; otherwise accuracy monotonically (if practicable) decreases with miss distance or error. Finally, if sufficient detail needed to amplify or justify the information is present, then Detail = 1.0, gradually dropping to zero if required elaborating detail is not available. Each of these parameters may be specified in an information need, such as Expression 2-4.

Expression 2-4 Illustrative Information Need

<Information-need>
<**Query**> "Name of the largest state in the USA" </**Query**>
<**Quantity**> name </**Quantity**>
<**Timeliness**> 2 seconds </**Timeliness**>
<**Validity**> Must be true </**Validity**>
<**Accuracy**> Correct spelling </**Accuracy**>

[2] Quantity (<Scene/>) is understood; QoI may be defined statistically over collections of likely <Scenes/>.

<**Detail**> Null (Do not need any other information) </**Detail**>
</Information-need>

The response "Texas" was valid until "Alaska" became a state; in addition, Texas remains the largest state in the contiguous lower 48 states. Context might render Alaska false, or amplifying detail could name both Alaska and Texas with space–time validity subsets, for example,

<Response>{{<Texas ∩ {{<Time/> <3 Jan 1959} ∪ {<Space/> = <Lower-48-states/>}}} ∪ {Alaska}}</Response>

If the information were provided quickly and were spelled right, then QoI = 1.0. If the query were met a minute later because the CWPDA couldn't reach the cell phone network or WLAN for that length of time, then the QoI is 2/60 = 1/30, substantially less than 1.0.

Intuitively, the degradation from 1.0 also depends on the urgency of the need. If the user were playing Trivial Pursuit with a few friends, then the penalty for time delay might not be great. If the user were playing "Who Wants to Be a Millionaire?" as a contestant on television and asked the CWPDA for help as a phone-in, then even a minute of excess delay could cost a bundle. Need could be expressed as a cost of failure in the need expression and reflected as the inverse of the relative cost normalized to 1.0 in aggregating QoI so that perfect QoI is 1.0.

The iCR uses the QoI equation (2-2) to guide autonomous planning over memories and air interface(s) to maximize context-specific QoI.

2.2.5.3 Military VTC QoI

Consider a military example. After a period of learning about combat teleconferences, an iCR would know that it is better to drop from full motion video to a slow frame rate with superb voice quality than to use limited data rate for video. The iCR computational model of <Combat-teleconference/> would include <Disadvantaged-user>, the <User/> with the low data rate and high BER. Fallback to high quality voice in lieu of video is true except in a <Scene/> when a new map is introduced. The discussion of the map makes no sense unless the remote <User/> sees the map. Detecting the new map requires content awareness. The iCR processes the video not just for MPEG coding, but to detect QoI related events like the presentation of a new <Map/>. <Map/> detection is within the capabilities of supervised ML and pattern recognition.

How does the iCR know to look for a <Map/>? The iCR <Self/> data structure could link the behavior {<Self/>, <Transmit/>, <Map/>} to <Disadvantaged-user/> explicating <Map age = new/>, with a priority <Goal/> to move new <Map/> through the <RF/> with maximum resolution and speed. Later, <Map age = known/> when the <Disadvantaged-user/> has the map. This state change puts the information transfer priority back to high quality voice. A <Map/>-aware iCR might also suppress the hand and arm pointing to the map to a mere mouse pointer for <Disadvantaged-user> to further conserve bandwidth and to make the map easier to see. If data rate and BER improve, <Self/> can enable full motion video.

Like the military VTC, the architect trying to get approval for a design change from a high-end home owner on the other side of the world would learn the same thing about <Blueprints/> as the military iCR learned about <Maps/>. Alternative vignettes are included in the CD-ROM.

Thus, item 5 of Figure 2-1 harnesses computational intelligence to adapt the wireless system to the users' needs for QoI, a paradigm shift from today's relatively embryonic state in which users are constantly struggling to overcome the shortfalls of wireless QoS.

2.2.6 From Hierarchical Control to Cooperation

Perception of the <Self/>, the <RF-environment/>, and the <User/> grounded in space–time enables a transition from pure hierarchical control of the proverbial cellular telephone network to distributed collaboration. Item 6 of Figure 2-1 asserts that if iCRs are smart enough to do items 4 and 5, then armed with a common language (such as a Radio Knowledge Representation Language (RKRL) [145] or Radio XML, RXML), they may autonomously form not just ad hoc networks but user-oriented information sharing federations (FedNets [31]). To realize such collaborative networks, each iCRs needs an internal semantic model of <Others/>.

Expression 2-5 Cooperation Requires Knowledge of Other Intelligent Entities

```
<Entities> <Self/> <User/>
  <Others> <Other-CRs/> <Legacy-radios/> <Other-people/>
  </Others> </Entities>
```

Ad hoc networks share a RF band via specific modes and an ad hoc protocol such as Mesh Network™ [32]. This starting point may be augmented with QoI-driven knowledge about others in the <RF/> <Scene/>.

Expression 2-6 Ad Hoc Networking Using Knowledge of Cooperating Nodes

```
<Other-entities>
  <Other-CR>
    <CR#1> <Band/> <Mode/>
      <Identity> CR5551212 </Identity>
      <Location> GPS coordinates </Location>
      <RSSI> -88 dBm </RSSI>
```

<Will-relay-now> Yes </Will-relay-now>
<CR#2> . . . </Other-CR> . . .</Other-entities>

Specifically, the iCRs exchange RF parameters to enable cooperation. Data structures representing these significant others include the RF parameters with which each can operate; if a legacy user is detected, the iCRs switch bands or modes to defer to the less flexible legacy users. In addition, iCR networks exchange patterns of space–time loci of legacy transmitters (e.g., police use APCO 25 VHF near the bar at 2 am, closing time). A strong architecture validates data to avoid self-replication of errorful rumors. The discovery and reporting of legacy radios among iCRs reduces interference and speeds convergence characterizing <Legacy-users/> as comprehensively as <Other-CRs/> in Expression 2-6.

2.2.7 User Awareness

Perception technologies enable more than better user interfaces. A wearable video port integrated into eyeglasses detects a new person in a <Scene/>. Recognizing the cultural prototype for introductions, the iCR transparently exchanges softcopy business cards. For example:

<Known-male-voice>: "Herrn Mueller, Fraulein Schmidt."
<Unknown-female-voice>: "Guten Tag, Herrn Mueller."
<Owner's-voice>: "Es freut Mich sehr, Fraulein Schmidt."

Even if you do not speak German, you may be aware that a man is called Herr and an unmarried woman is called Fraulein. The repetition of names with the <Owner/> acknowledging the new person provides an algorithmically recognizable spatial–temporal pattern. Because the owner's voice is known with high confidence and because of the structure of introductions, even relatively noisy visual and acoustic scenes may yield sufficient reinforcement to recognize <Introductions/>. The iCR initiates a search of IR, BlueTooth, 802.11, and ad hoc TV bands to find the RF business card for <Unknown-female> Fraulein Schmidt </Unknown-female>[3]. It "pings" these bands to establish <Self/> presence and interest in communicating. The AACR may then establish an ad hoc network with an older AAR of Fraulein Schmidt, exchange business cards, and project her business card into the Owner's eyepiece.

"I see you are from Universität Karlsruhe. Do you know Professor Jondral?" the Owner says, switching to English since the business card shows both languages.

[3] This vignette is referred to as the Introductions use case.

Without the ability to recognize introductions, the AACR either wastes battery life, constantly emitting pings in all the short-range "business card" bands, or waits to be pinged, using less energy but still requiring receivers to be powered up. Recognizing introductions, the iCR knows when to search for "Fraulein Schmidt" wirelessly and can keep a digital snapshot of her for future reference to enhance owner recall of names, a positive skill in business settings.

The level of user context sensitivity described in this abbreviated use case integrates radios more fully with the rest of the user's information spaces so as to become information prosthetics [33]. Without machine perception, Figure 2-1 items 4–6 may not be practical. User adaptation has been promised for over a decade, but the instantiation of user profiles remains a daunting task [34]. Even profiles for location-aware services are not readily adapted to situation-dependent customer interests when users must take time to supply data for their own profiles. The iCR alternative incrementally infers and validates user preferences through perception-driven AML. The research prototype CR1 demonstrates how a CWPDA empowered with simulated sight, sound, and AML learns needs and detects situations. Thus, AML is one technical foundation for low cost mass customization of wireless system behaviors. In addition, the entities perceived in RF and user environments must match both the iCR's own prior knowledge and the knowledge representations used by radio networks via symbol grounding.

2.2.8 Grounding

The process of firmly linking syntactic ("formal") expressions such as the <Self/>, <Owner/>, and <User/> data structures to real-world entities is called grounding [35]. Grounding is fundamental to perception. In logic, grounding is the process of establishing a valid interpretation of the statements of a (formal logic) language with respect to a "model" of those statements in some domain, typically the real world [36]. As Judea Pearl points out, it is not sufficient to simply associate the formal expressions with a model in the world: deep understanding requires an ability to perceive cause and effect—"how things work when taken apart" [37]. This book "takes radio apart," explicating the knowledge needed for CWPDAs grounded in radio domains. It also "takes the user apart," analyzing knowledge at levels of abstraction to realize helpful behaviors. There are several ways to ground the radio's sensory perceptions.

2.2.8.1 *Implicit Grounding*

Legacy radio control includes physical buttons and computer commands, the syntactic expression of the user's intent. Grounding the formal term <Authorized-user/> to people in the real world is informative. Legacy radio systems allow control by an <Authorized-user/>. If the radio has no user authentication system, then any user is an <Authorized-user/>. This type of default grounding allows whoever possesses the radio to control it, no questions asked:

<Authorized-user/> ≡ <Current-user/> (2-5)

Commercial GSM cell phones with built-in SIM cards (not removable ones) employ this kind of implicit grounding of control authority, as do most sports, public service, and amateur radios. <Authorized-user> is the simplest of symbols to ground. <Daughter/>, <Vacation/>, and <Car/> are much more difficult to ground.

2.2.8.2 Explicit Grounding

Explicit grounding of <Authorized-user/> may be based on authentication. Military radios historically have used physical cryptographic keys to limit access to the radio. More familiar to nonmilitary users is the grounding of <Authorized-user/> via a password. In this case:

<Token-correct/> ⇒ <Authorized-user/> (2-6)

If the crypto-key <Token/> is correct, then the radio behaves as if the <Current-user/> is the <Authorized-user/> even if the person who inserted the key stole the key. The GSM cell phone with removable SIM card has this degree of trust. Explicit grounding is more secure than implicit grounding, but still cannot ground <Daughter/>. Perception technology offers an alternative.

2.2.8.3 Perceptual Grounding

Machine vision and speaker identification enable sensory perception grounding. Audio and video soft biometrics ground symbols through machine perception. Suppose a salesperson introduces an iCR, Charlie, to its new owner. "Charlie, this is Joe. Joe, this is Charlie." Through introductions, Charlie extracts a computational model of Joe's face [38] and voice [39]. Although these soft biometric models will not differentiate Joe from billions of other people, such models typically differentiate Joe from his <Daughter>, co-workers, and a thief.

Expression 2-7 Grounding by Learning Soft Biometrics

```
<Authorized-user>  <Name> Joe </Name>
                   <Face> new-face (video-stream) </Face>
                   <Voice> new-speaker (audio-stream) </Voice>
</Authorized-user>
```

<Authorized-user/> with <Name> Joe </Name> is grounded by a computational model of facial features with an associated speaker model. In the

<Introduction/>, Charlie grounds the new-face in its video stream and *contemporaneous* new voice to <Joe/>. The face–voice–name association grounds <Joe/> for continuing AML of the new owner's needs and preferences. If this model is so embedded in the hardware that Charlie can't lose this definition of Joe without being destroyed, then a thief cannot steal <Joe/>'s identity by merely hacking a password. Joe may ground his <Daughter/> by sensory perception in an <Introduction/> protocol. This may be a first step in a process of mutual grounding.

2.2.9 Mutual Grounding

Mutual grounding is the exact match of symbols, semantics, and referents in communications between two entities. The two entities need not employ the same internal symbols for the same things, but each must know what to communicate to obtain desired behavior from the other.

2.2.9.1 Formal Domains

In an XG policy broadcast from a regulatory authority, the expression FCC: <VHFChannel-13/> means that the FCC refers to a particular RF carrier frequency, bandwidth, and signal format (e.g., NTSC or HDTV) as <VHF-Channel-13/>. CR1: <VHF-1.13/> could mean exactly the same carrier, bandwidth, and signal format. A common object request broker architecture (CORBA) Interface Definition Language (IDL) stub could equate the two for the mutual grounding of FCC and CR1 TV channels:

Equals (FCC:<VHFChannel-13/>, CR1:<VHF-1.13/> (2-7)

FCC's VHFChannel-13 is defined by an external (exogenous) source, the policy broadcast, while CR1's VHF-1.13 is CR1's internal (endogenous) representation.

The difference between <VHF-1.13/> and <VHFChannel-13/> is a difference in the assignment of formal symbols to things in the real world, a difference in explicit ontology. Each of these symbols refers to TV Channel 13, a space–time–RF entity of the real world (e.g., WBAL, Baltimore, MD in 2005). In formal communications like XG, ontologies may be reconciled a priori, for example, in the ITU, OMG, and SDR Forum, and in the product's design documentation and source code. A CORBA IDL stub [40] defers the binding of one formal symbol to another often until after both products already exist. Implicit ontologies, on the other hand, need mutual symbol grounding that is much more difficult to verify (e.g., in natural language).

2.2.9.2 Natural Language

Mutual grounding is extremely difficult in natural language. News broadcasts and textbooks employ grammatical natural language with relatively unambiguous antecedents simplifying mutual grounding. Informal language often is not very grammatical, replete with vague or ambiguous references that complicate grounding. <Users/> employ informal language with chronic technology challenges (even given an error-free transcript) including errorful reference, anaphora, ellipsis, ambiguity, vagueness, and constructs that require stylized inferences [41]. Mutual grounding between the <User/> and the iCR <Self/> does not ground every conceivable symbol, but grounds the smallest set that the iCR needs for a given use case.

Sentences that entail a radio-use context may require general world knowledge to properly interpret.

Sentence = "I want to see the Statue of Liberty flying to New York" (2-8)

This sentence has ungrammatical placement of the final clause, implying the statue should fly. The grammatically correct form places "Flying to New York" first, indicating "I" will fly, not the statue. Statue of Liberty and New York are both places and things related to QoI. Typically it is important for iCR to recognize location terms in communications. If the <User/> is interested in the Statue of Liberty or New York, the iCR must retrieve QoI-related information from peers, networks, and the Web, increasing quantity, timeliness, and detail.

Suppose the <User/> wants a traffic and weather radio station "on the 8's." They both should agree on the meaning of "traffic" and "weather" so that traffic and weather are mutually grounded. For example:

User: "Traffic" ⇔ CR1:<Traffic> Cars and trucks on highways </Traffic> (2-9)

User: "Weather" ⇔ CR1:<Weather> Rain, snow, clouds and sun </Weather> (2-10)

The symbol ⇔ indicates mutual grounding of <Domain/>:<term(s)/> ⇔ <Domain/>:<term(s)/>. Thus, the user needing a weather report is mutually grounded with CR1 since both refer to the same sorts of things in the real world. In another <Scene/> to the same user, "traffic" may refer to the rate at which calls are placed on a telephone system, not mutually grounded with CR1: <Traffic>. By differentiating <Commuting/> scenes from <Research/> scenes, CR1 can remain mutually grounded to be both helpful and as transparent as possible, not tedious but asking appropriate grounding questions when necessary.

Grounding remains a research challenge in computer science and robotics. The Semantic Web addresses this by formal ontology resolution. Robots employ fiducials, constrain tasks, and limit vocabulary for mutual grounding. Grounding is fundamental to iCR capabilities to observe (sense, perceive), orient, plan, decide, act, and learn. If internal symbols are improperly grounded the iCR may learn inappropriate behaviors. Mutual grounding of <Self/> in the environment, especially through learning and behavior resolution over time, remains a cutting edge research area. The approach of this text is to limit the mutually grounded symbols to those few with high redundancy that achieve the use case to mitigate the limits of the technology.

2.3 IDEAL COGNITIVE RADIO (iCR) PLATFORM EVOLUTION

An evolutionary AACR strategy realizes the seven core iCR capabilities with incrementally evolving functionality for a sequence of use cases. The hardware–software platform evolves sensors, perception subsystems, and information architecture to ground those critical entities <Self/>, <RF/>, and <User/> with sufficient accuracy to reliably enhance wireless QoI per use case. Platform evolution may be driven by scene perception, RF, or user interface.

2.3.1 Platforms Driven by Scene Perception

A sensor-rich hardware–software scene perception platform perceives the world to ground the <Self/>, <User/>, and other objects in <Scenes/> in two primary domains of discourse: (1) the wireless communications environment and (2) the user's social environment. With this strategy, a communications context consists of a <Scene/> with perception–action linkages for communications services tailored to entities perceived in the <Scene/> given the opportunities and constraints of the <RF/> <Scene/>. More formally, a scene includes the wireless and user aspects:

<Scene/> :: = <RF-environment/> ∪ <User-situation/> (2-11a)

<Scene> <RF-environment/> <User-situation/> </Scene> (2-11b)

<Scene/> is an open-set construct (Equation 2-11a) expressed in RXML in Equation 2.11b. <Scene/> unites two quasi-orthogonal sets, one primarily physical <RF/> and the other the abstract reality of <User/> perception. The <User-situation/> defines perceived <Scene/> boundaries. Thus, <Home/>, <Commuting/>, and <Work/> could be the major scenes of a <Work-day/>, defined not in terms of home WLAN or cell phone coverage but by <User/> perception and associated terminology and visual fiducials of such <Scenes/>.

In a given <Scene/> the constituent subsets usually are not completely orthogonal. If the iCR is planning future connectivity, the current <RF-environment/> is not the <Location/> of the future <Scene/>, so the intersection is empty ("null"). A <Primal-sketch/> locates a <Scene/> in space–time, placing entities in the scene and enabling the iCR to assist the user in planning. Thus <Scenes/> include the space–time coordinates of every experience, plan, action, and response from the environment. A corresponding CWPDA platform (sensors and perception subsystems) continuously locates the <Self/> and <User/> in both physical and conceptual space–time. For example, it is insufficient to assert 10:54 am on 11 November 2004; the iCR must sense location to bind space–time: (10:54 am 11 Nov 04, Rome, Georgia, USA). Although 10:54 seems to be a point in time, there is an accuracy of the measurement implicit in "10:54" that implies a subset of time as well: it is not 10:55 or 10:53, but 10:54, implying a two minute subset of time in which some event occurred.

From these initial considerations, perception-driven platform iCRs should sense <RF-environment/> per Expression 2-8.

Expression 2-8 The iCR Platform Includes Sensors and Actuators

```
<iCR-platform>
  <User-situation> <Scene>
  <RF-environment>
    <RF-capabilities> <RF-sensors/> <Waveforms/>
      </RF-capabilities>
    <RF-knowledge/> </RF-environment>
  <User-sensors> <Vision/> <Speech/> <Others-sensors/>
    </User-sensors> <User-actuators/>
  <Space-time-grounding>
    <Space-grounding> <Vision> Location-reference </Vision>
      </Space-grounding>
    <Time-grounding> <Speech> Time-reference </Speech>
      </Time-grounding>
    </Space-time-grounding> . . . </Scene>
  </User-situation>
</iCR-platform>
```

In this expression, <Waveforms/> are <RF/>-actuators, while displays and speech synthesis are the <User-actuators/>. Waveform is a convenient term for a software-defined air interface {<Waveform/> ⇔ <Air-interface/>}. A mode may be characterized by channel symbol (MSK versus 16 QAM), bandwidth, and so on. The perception-driven platform relates these mode differences in the <RF-environment/> to the <Scene/> perceived—<Home/>, <Office/>, and so on. Platform evolution evolves capabilities per scene possibly beginning with <Home/> where time, home RF, and perceived family members drive QoI. The next platform increment could add <Work/> with the corporate WLAN and perceived places and co-workers driving QoI. <Travel/>, <Sports/>, and other scenes each bring new RF and scene perception needs to the platform.

2.3.2 RF-Driven Platforms

Traditional AACR platforms evolve RF alone via mode chip sets with a simple associated ontology:

Expression 2-9 Self with Explicit Knowledge of Waveforms

```
<Self>
  <RF-capabilities>
  <Waveforms>
    <GSM>
      <RF-low> 860000 <Units> kHz </Units> </RF-low>
      <RF-high> . . .
      <Voice-connection>
        <Data-rate> 13kbps </Data-rate>
        <Isochronous> Yes </Isochronous>
        <Initiate-voice>
          Start-GSM-Voice(<TX-Channel#>, <RX-Channel#>)
        </Initiate-voice> . . . {terminate-voice, etc.} </Voice-connection>
      <RSSI> Get-RSSI() </RSSI>
      <GPRS>
        <Data-rate> 9kbps </Data-rate>
        <Isochronous> <False/> </Isochronous> </GPRS>
        {initiate and terminate data, etc.} </GSM>
    <EDGE/> . . . <802.11g/> . . . </Waveforms> </RF-capabilities> . . .
      </Self>
```

In Expression 2-9, Get-RSSI() senses GSM received signal strength. A sensory function is implemented in a <Waveform/> known to the <Self/> through RXML and further defined, for example, using the SDR Forum's SCA or OMGSRA. Shared RSSI values may be semantically aligned by RXML (e.g., RSSI units are in dBm). RXML also enables the discovery of RSSI units mismatch between dBm and dBW, facilitating ontology alignment [42]. The RXML-enabled traditional AACR platform may evolve first to reason over RF parameters, then 2D maps and ultimately 3D spatial RF <Scenes/>. Higher fidelity RF space–time models require larger data sets and more computational resources to yield more accurate perception of RF, so the fidelity of RF reasoning can have a first-order effect on <iCR-platform/> memory and processing capacity.

2.3.2.1 Reasoning Over RF Parameters

RF perception includes reasoning over RF parameters. Simplified models of first-order objects in the RF scene may suffice for entry-level AACR use cases. The platform implications of RF parameter sensing and use include both RF architecture and computational resources. First, RF chip sets must bring their intrinsic ability to sense the RF environment into the realm of sensory perception through an API (e.g., via RXML declaration). Some chip sets indicate whether a network is present, but provide few engineering details like RSSI or equalizer tap settings to an API. Subsequent use cases show how this level of detail empowers iCRs to enhance QoI. Traditionally, RSSI, BER, and number of taps are the concern of the radio engineer and network, but not of the radio itself, but iCR puts more "engineer" inside the radio.

Next, additional memory speed and processing capacity are needed to store and use the RSSI, BER, equalizer taps, and so on for QoI use cases. Such data might be acquired by AACRs in support of CWNs reasoning about dead spots and relay roles of AACR nodes. The iCR itself also employs this knowledge to fine-tune its own real-time performance, perhaps by distilling the knowledge during sleep periods for rapid use of RF sensory perceptions in real time.

2.3.2.2 Reasoning About Spatial RF Entities

An AACR may reason about its own experience in a 3D spatial <RF-environment/>. RSSI tagged with the place and time of the observation and the estimated location of the distant RF entity enable temporal pattern detection algorithms to find regularities in space–time use of RF, such as high RSSI because of good propagation at certain times of the day (e.g., night propagation of VHF from a distant broadcast station). Such other RF entities are described in the spatial RF scene. As entities are added, additional memory and processing resources are consumed. The TV-band use case may benefit from spatial reasoning about the location of known TV transmitters. The Self's ability to transmit to potentially hidden TV receivers could be modeled in a generic flat earth, $2\frac{1}{2}$D COST 256 statistical model or full 3D cityscape <Scene/>. CAD-CAM rendering of the cityscape explicitly represents large multipath reflectors like buildings at the expense of orders of magnitude more memory and processing resources. Grounding of RF entities in a CAD-CAM cityscape scene may mitigate interference to legacy TV receivers. With additional Moore's Law cycles, the cost–benefit balance will tend to tip in the direction of higher fidelity on-board autonomous 3D spatial reasoning about RF entities.

2.3.2.3 Learning Errors in 3D RF

Learning errors may be difficult to diagnose and to correct in 3D RF settings. Suppose the learning concerns two TV stations A and B in a fringe area of a big city. The stations may be marginally receivable by legacy TV receivers at street level but above the fifth floor people on the North receive TV-A well and on the South receive TV-B well. The AACR with only a 2D flat earth model that has never been on one of the upper floors may learn erroneously that it is OK to create an ad hoc network on the TV channels occupied by TV stations A and B. The AACR with the 3D model is better able to interpret a measurement of zero RSSI taken at street level with a propagation model that shows 10 dB reception, amending the conclusion that it is OK to form an ad hoc network at street level but not on the upper floors of this building. Model errors engender interference. Suppose users report such problems to the RA via the Web. The polite AACRs could check the web site and back off if legacy users experience interference from TV-A ad hoc networking above the fifth floor of that particular building. If a 2D AACR updates its propagation model to reflect TV-A interference, it overstates the legacy reception, precluding a workable ad hoc network.

Initially, trade-offs will favor computational intelligence in the CWNs where the cost advantages of centralized databases and retrospective pattern matching may be amortized over all users. Over time, the computational resources in the iCR nodes will continue to increase to enable initially limited reasoning over simple space–time settings. Initially in networks, later in iCR vehicles, and ultimately in the CWPDA iCR, there will be sufficient memory and processing resources to address full 3D RF with many RF entities within the <iCR-platform/>.

2.3.3 Use Case-Driven Platform

Now consider the platform implications of a specific use case. This case concerns Sue and her four-year-old child, Bernie.[4] The child wears a cell phone to help Sue monitor Bernie. Sue is doing laundry on the first floor of a six-story apartment building. Bernie wants to go outside into the interior courtyard to play with the other kids. Sue would like her cell phone to keep track of Bernie who is only 10 feet away, but on the other side of a wall.

2.3.3.1 Conventional HMI Limits Wireless Timeliness QoI

Sue thinks, "The salesperson said these phones can form a network like my BabyCam, but how do I turn on that feature?" She pushes Help. "BabyCam Mode: Select Mode/ Video/ Short Range. Enter the baby's cell phone number." OK, let's give it a try. The baby's cell phone rings. Nothing. Sue looks at the baby's cell phone. "Please enter your password and PIN to approve an inbound charge to this normally outbound cell phone." "Gee, they didn't tell me it was going to cost me to set it up," Sue thinks, but she keeps on trying. "Download in progress." The network is downloading the TV ad hoc networking protocol to the baby's cell phone. After a few minutes, Sue's video screen shows what Bernie is doing. Bernie walks out of the gate and onto the street. Sue looks up from doing the laundry and screams at the sight of an oncoming car viewed from Bernie's wearable cell phone. Bernie is chasing a ball down the street.

2.3.3.2 Advanced HMI Enhances Time-Sensitive Content

Consider the same vignette with speech, video, and soft biometrics to move these two phones toward iCR. When the devices were sold, the salesperson introduced the two phones, Bert and Ernie, to Sue. The two AACR vision systems isolate Sue in the scene and take soft biometrics of her. Bert says, "Why are you buying two of us?" Sue says, "One of you is for my son, Bernie." "How old is he?" "He's four." "We can do BabyCam if you like." Sue tells the salesperson to enable BabyCam and location reporting.

[4] The subsequent vignettes are referred to as the Bernie–Sue BabyCam use case.

At home, Sue introduces the two phones to Bernie. They take soft biometrics of Bernie. Sue and Bernie go to the first floor to do the laundry. "Don't go out of the courtyard." Bert knows what a courtyard is from scene modeling, uploaded at the point of sale with vision optimized to recognize specifically doors, gates, windows—anything that could pose danger to a four year old. "Ernie, can you and Bert BabyCam for me?" Bert and Ernie check the TV bands. TV-A has negative RSSI here. Bert and Ernie probe the band and agree to use high power from Sue to Bernie so her signal will penetrate the wall, but the asymmetric multipath from Bernie to Sue through the window enables Ernie to put his MIPS into a high performance equalizer, keeping Bert's radiated power low, minimizing interference to TV-A users on the upper floors. Bernie's scene shows in Sue's cell phone display which she perches up with the laundry detergent. Someday it will be wearable, but not in this vignette.

Ernie beeps. Sue looks up to see the face of one of Bernie's friends. "What was that?" Sue asks. Ernie says, "Bert is near a door, gate, or window so he sent an alert." Sue says, "Bernie, you stay away from the gate, OK, Honey?" Bernie says, "OK" but heads for the gate chasing the ball. Bert beeps again, more shrilly as the video flow indicates Bernie is headed out the gate. "Bernie, you get back here right now. Let that ball go," Sue yells as she rushes out of the laundry room to grab Bernie and close and lock the gate. Sue doesn't think twice of the MIPS or battery power needed for that life-saving scenario as she scolds Bernie not to do that again. "You could have been killed if you had gone out into that street." Then a big hug.

Which of these two scenarios is closer to the "killer app"? What does it take in the way of sensors and perception to make this dream a reality? The video didn't break up because the BabyCam was made reliable by iCR collaboration. The legacy TV users didn't complain because of the asymmetrical link created through detailed understanding of the relationship between radiated power, multipath signature, and equalization, trading radiated power for MIPS. They used video scene modeling to analyze the RF environment as well as to keep track of Bernie. The <iCR-platform/> vision, speech, and scene perception made the use case possible.

2.3.4 iCR Objective Platform

Scene perception, RF, and user interface evolution each lead to the iCR objective platform with the capabilities to sense <RF-environment/> and <User-situation/> as outlined above. The iCR CWPDA platform has the radio waveforms, high fidelity RF sensing, and rich suite of audio, video, location, and other perception subsystems, listed in Figure 2-2, the original concept for iCR [145]. Over time, iCR will evolve to a wireless fashion statement, an intelligent wearable information appliance that dramatically enhances its user's capabilities, a RF information prosthetic. The CWPDA personal area network (PAN) interconnects the CWPDA ensemble of glasses and a

RF Bands and Modes
GSM (IS-136, etc)
GPRS (UWC-136 ...)
3G (W-CDMA ...)
RF LAN
AM Broadcast
FM Broadcast
NOAA Weather
Police, Fire, etc.

Environment Sensors
Location:
GPS (Glonass, ...)
Accelerometer
Magnetometer (North)
Positioning:
Environment Broadcast
(Doors, Coke Machines, ...)
Timing:
Precision Clock
GPS Clock Updates
Other:
Ambient Light
Digital Image, Video Clip
Temperature

Local Sensors
Speech Recognizer
Speaker ID
Keyboard, Buttons

Effectors
Speech Synthesizer
Text Display
RF Band/ Mode Control

FIGURE 2-2 An illustrative cognitive radio platform.

belt-pack integrating fourth generation (4G) and beyond wireless [221] with the wireless Web.

With iCR built on a PAN and designed as a fashion statement, readout could be in your sunglasses, the core module worn in the small of your back, powered via micro-power generators in your shoes, mouse-like control via movement of your wrist, and typing by just placing your fingers on some convenient surface and then typing away while sensors in wristbands read out your keystrokes. Cranial implants aren't out of the question, but I don't want one myself. My kid might, though. Forward-looking science fiction in such historically stogy journals as the *IEEE Spectrum* envision mind-expanding use cases [43].

2.4 THE serMODEL OF MACHINE LEARNING FOR iCR

Platforms described above can observe, orient, plan, decide, and act, but not necessarily learn. This section formulates iCR learning as stimulus–experience–response models (serModels). serModels relate <User/> interaction to information from <Waveform/> use, measuring, learning, and enhancing QoI.

2.4.1 The serModel

A serModel is a map from a set of stimulus channels via experience to response channels:

$$\text{serModel: } \{\text{Stimuli}\} * \{\text{Experience}\} \Rightarrow \{\text{Response}\} \qquad (2\text{-}12)$$

This serModel generalizes the artificial neural network (ANN), rule-based inference, logic programming, and reinforcement learning [53]. In the limit, iCR is not programmed at all, but like an infant, learns all behavior through serModels. Such an iCR may learn from stimuli and experience either to react quickly or to formulate sequenced response steps—plans.

The serModel extends SDR development from programming to teaching based on radio engineering methods for SDR including:

1. *Linear transfer functions* $(H(j\omega))$ [45].
2. *State machine/Message Sequence chart (MSC):* ITU Specification and Description Language (SDL) Z.100 [44].
3. *Kalman filter* discrete matrix operations $(\mathbf{x} = A\mathbf{x} + B\mathbf{y} + \cdots)$ over vector spaces [46].
4. *Structured design* [47], structured programming [48], object-oriented design [49], and programming [50].
5. *Unified Modeling Language (UML)* [51] software design tools [52].
6. *serModel:* learning to pair stimuli and experience with appropriate responses.

Methods 1–5 do not require machine learning; even the adaptation of the Kalman filter optimizes the preprogrammed trajectory but cannot learn to initialize itself or to autonomously change modalities. The serModel, on the other hand, establishes iCR as a learning machine. Learning formulations are well known to biologists studying intelligent behavior of animals [54], but typically not to radio engineers developing RF hardware or software. <Scenes/> define stimuli, while experience invokes <Scene/> memory with positive and negative reinforcement in a serModel. Populations of iCRs with different serModels constitute iCR population genetics [55].

The serModel autonomously acquires knowledge by evolutionary adaptation in lieu of programming and design releases. The serModel is more than a new name for the venerable input–output map or transfer function. Input–output maps are programmed by people, while the serModels are learned, either by the iCR or through the evolution of a population of iCRs. Such large collections of complex adaptive systems can solve NP-hard problems at a degree of randomness called the edge of chaos [56].

2.4.2 serModels of Radio

Commercial 3G/4G cell phones have dozens of modes in RF bands [57] approaching the complexity of military radios [58]. For dynamic spectrum use cases, AACRs also access nontraditional bands such as TV. Formulating radio serModels over a large set of <Waveforms/> yields a macrolayer radio control structure relating perceived <Scene/> to <Waveform/> via AML. Orchestrating <Waveform/> behaviors via such high level serModels automates dynamic spectrum management through positive and negative reinforcement of iCR-synthesized behaviors, not requiring all cases to be preprogrammed.

2.4.2.1 Waveform serModels

Consider stimuli that indicate the need for a given waveform. The serModel forms pairs of attributes of waveforms with attributes of the user's information needs. Augmenting the RXML self-description of <GSM/> of Expression 2-9 yields the following:

Expression 2-10 Waveforms Have RF and User-Centric Stimuli

<Waveform><GSM><serModels>
<RF>[5]<Stimulus>GSM-control-channel-active </Stimulus>
<Response>Send (Registration-request) Pay (Airtime) </Response>
</RF>
<User><Stimulus><AND>Desires (User, Speak (Remote-entity))
<NOT>VoIP</NOT></AND></Stimulus>
<Response>Initiate-GSM-voice-call</Response></User>
</serModels>

Expression 2-10 associates the chip set GSM control channel to the registration request explicating air-time charges. The association of connection cost from the <User/> value domain enables this cost-aware iCR to defer the connection until <User/>-sensory-perception indicates a need strong enough to pay for GSM. The serModel reflects the <User/> value that GSM is appropriate when the user wants to speak by voice to a remote entity and VoIP is not possible. The iCR may reason over this data structure to plan a response to initiate a voice call over GSM. The serModels reflect the goal-oriented nature of iCR support to users as well as the causality in radio networks in a way that is readily extensible via AML to optimize the <User/> QoI metric.

The user's expressions for "Speak" are grounded to <Speak/> for the RXML above. In some bands, like marine VHF, speech may be the only available mode. GSM's richer set of capabilities, including Short Message Service (SMS), EDGE, and cell-phone email, may be related to QoI via RXML as above. As new services arise with increasing frequency, iCR serModels learn <User/> preferences for snapshots, video clips, video-phone, chat rooms, and the like as a function of user <Scene/>. The serModels expressed in RXML readily extend to new services relating GoS, QoS, and QoI to <User/> preferences. The serModel thus encapsulates the waveform as a means to an end rather than an end in itself.

[5] <RF/> invoked within <srModel/> appears self-referential but since <RF/> refers to the outside world, not to the <RF/> structure of the <Self/>, there is no logical inconsistency when referents are accurately grounded.

2.4.2.2 serModels in CBR

As introduced previously, pattern matching with machine learning organized as case-based reasoning (CBR) could implement causal relationships. CBR remembers the stimuli, recognizes similarity between <Scenes/> past and present, binds contextually similar referents, and offers to apply the adapted prior serModel-encapsulated experience to the present <Scene/>. CBR-based serModels are the basis for AML implemented in Java in CR1 in the conpanion CD.

In one vignette, someone says, "Let's call Grandma on our new CWPDA" immediately before the first GSM call is placed. The CWPDA observes this and "Hi, Grandma" as the salutation of the call. Its serModel associates the phone number that the new user dialed into a tentative phone number listing under the name <Grandma?>. The salutation reinforces the association of that <Telephone-number/> and <Grandma/>, leading to a question from the CWPDA to make the association permanent. The stimulus speech segment "Let's call Grandma" and the response <Dialing/> the <Telephone-number/> are encapsulated by CBR as a serModel by the successful reinforcement by the <User/>.

Subsequently, the CWPDA observes in the speech sensor-perception channel the phrase "Let's call Uncle Charlie." CBR binds the phrase "Let's call" in the Uncle Charlie and Grandma scenes as a pending <Phone-call/> event. CBR binds <Grandma> in the prior serModel to <Uncle Charlie?>. As a response, it looks up Uncle Charlie's number in the phone book, placing it in the default <Telephone-number/> to <Dial/>, with the text "Uncle Charlie" in the field indicating the name to call. The new CWPDA user would save time if the anticipated call were accurate. If not, the new user dials the correct number, and the CBR is negatively reinforced if the user dialed a different number, turning the attempt into a further learning experience. The interaction of such positive and negative reinforcements is a chronic AML research area, so it will take time for serModels to be fully understood and effectively employed in practical devices. For the detection of speech precursors to phone calls to be practical, the <iCR-platform/> needs high quality acoustics and accurate speech recognition, still at the cutting edge of speech research.

2.4.3 Planning

The serModels suggested above explicate causal links among stimuli and responses. Beginning at least with means–ends analysis [59] and the classic problem of the monkey and the bananas [89], computer scientists have addressed such planning problems in a myriad of ways [290]. The serModel may be thought of as an interactive macro-operator discovery method in which the <User/> performs a sequence of steps that achieve a goal and the <Self/> encapsulates that sequence in a serModel as a means–end relationship. This isn't to say iCRs should not embed a full planning system like the open procedural reasoning system (OPRS), for example [60]. As the complexity of a <Scene/> increases, the time required for planning also increases, often exponentially. The serModels that encapsulate relationships may accelerate the transfer of planning knowledge from the <User/> to the <Self/>, acquiring domain-dependent heuristics augmenting OPRS for real-time planning. Although not the only candidate AML method, serModels illustrate the principles of integrating AML into SDR to enhance QoI.

2.5 ARCHITECTURE

Cognitive radio architecture (CRA) integrates <iCR-platform/>, hardware suite, software components, perception, planning, action, and learning capabilities. AACR architecture augments RF capabilities with location awareness, visual scene perception, speech recognition, and language synthesis needed for autonomous grounding for use cases like Sue's ad hoc baby monitor that kept Bernie out of the street. Toward this vision, the CRA is based on a comprehensive set of RXML formalizations of the seven core iCR capabilities (sense, perceive, orient, plan, decide, act, and learn). An open architecture CRA models the <Self/> as a computationally intelligent software-defined radio (<SDR/>), for example, via the SDR Forum's SCA [27] or the Object Management Group (OMG) [61] software radio architecture (SRA). As cognitive entities, AACRs sense and perceive visually and via natural language. Open CRA thus includes APIs, data structures, and processing flows by which to perceive, plan, and execute responsible actions, "knowing what they are doing." SDR architectures may be extended to the CRA incrementally by describing the <Self/> as <SDR/> with perception, planning, and learning by the addition of perception hierarchies.

2.5.1 Perception Hierarchies

The iCR knows that it is a SDR through the CRA data structure <Self/>. The <Self/> needs additional data structures for iCR to perceive a scene. These data structures must be sufficient for iCR to ground itself in the scene in space–time to learn about radio services and user preferences as a function of place (home, work, leisure) and time (weekend, weekday, noon, midnight, etc.) Candidate architectures range from blackboards [62] to graphs [63] to embedded databases.

The functional organization of such data structures is hierarchical as illustrated in Figure 2-3. Other perception domains like temperature, acceleration, GPS location, and balance may be included in the CRA hierarchy, parallel to the vision perception hierarchy.

This six-layered CRA inference hierarchy has two major groupings: sensory domain perception and situation perception. The three levels below the Phrases/ Expressions structure sensory domain perception. In this simplified view, there are no links across perception domains, but in a comprehensive

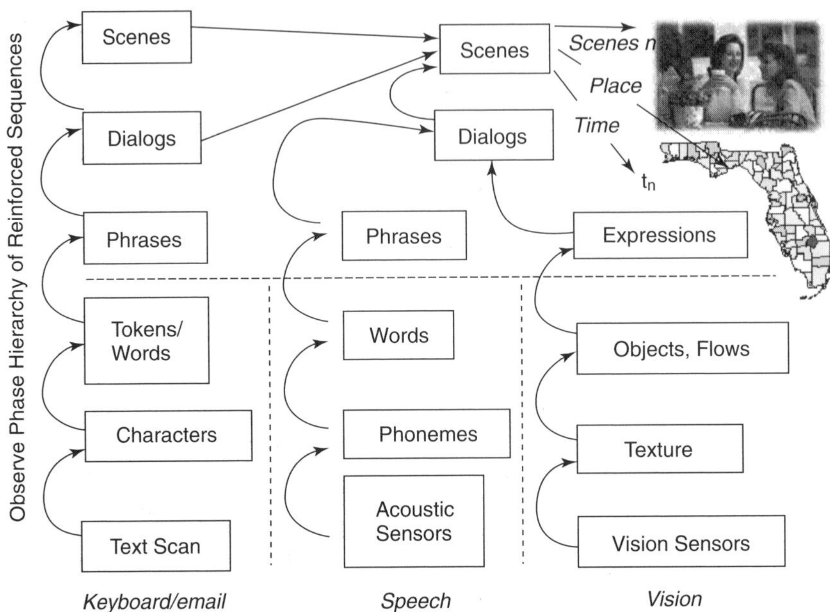

FIGURE 2-3 CRA inference hierarchy of the user view of a scene.

CRA, the sensory domains like vision and acceleration are linked at the lower levels to optimize overall performance. Each perception domain also may be optimized independently. The CR1 research prototype embeds reinforced CBR at every stage of this inference hierarchy, remembering everything and continually determining whether stimuli are known (identically known), familiar (parts have been observed previously), or completely novel. Novelty detection and integrated recognition of knowns occur below the CRA phrase level.

Because of the multidomain multidisciplinary nature of AACR, such an inference hierarchy is essential to the CRA. Phrases are the fundamental units of interaction and hence offer a natural API level for open CRA. A phrase interpreted as a command initiates action or interaction (e.g., for clarification). Some phrases reference space, time, and the user. serModels aggregate stimuli and responses for reinforcement learning. In the CRA inference hierarchy, a dialog is a sequence of phrases. When space, time, or situation changes substantially, a new <Scene/> is asserted. Recall that <Scene/> consists of <RF-environment/> and <User-situation/>. The <Scene/> of Figure 2-3 may be expressed in RXML.

Expression 2-11 Scenes Are Defined by Dialogs in Sensory Domains

```
<Scene> <User-situation>
  <Place> <Time>
```

<Dialog1>
<Phrase1(Keyboard)> <Token1> <Character1/> . . . </TokenN>
</Phrase1>
<Phrase1(Speech)> <Token1/> . . . </Phrase1>
<Phrase1(Vision)> <Object1/> <Area1(TextureX)> . . . </Phrase1>
<Dialog2> . . . <DialogN> </Time> </Place> </User-situation>
</Scene>

In the CRA current <Scenes/> may be registered to prior <Scenes/> and current dialogs to previous dialogs, inferring the current roles of entities from the previously learned associations.

Corresponding to Figure 2-3 is a parallel inference hierarchy for the <RF/> domain. Each band and mode known to the radio comprises an inference hierarchy of its own, ranging from physical layer primitives to applications and information resources available via wireless connections. Extension to wired interfaces is straightforward. In addition, the <Self/> must be perceived, grounded in the User's environment.

The perception hierarchy of Figure 2-4 illustrates these relationships in the CRA. The iCR's Self perception includes the ability to asses its own resources. Resources may be physical or logical, including SDR resources defining waveforms that are known (e.g., by RXML reference), available on the platform,

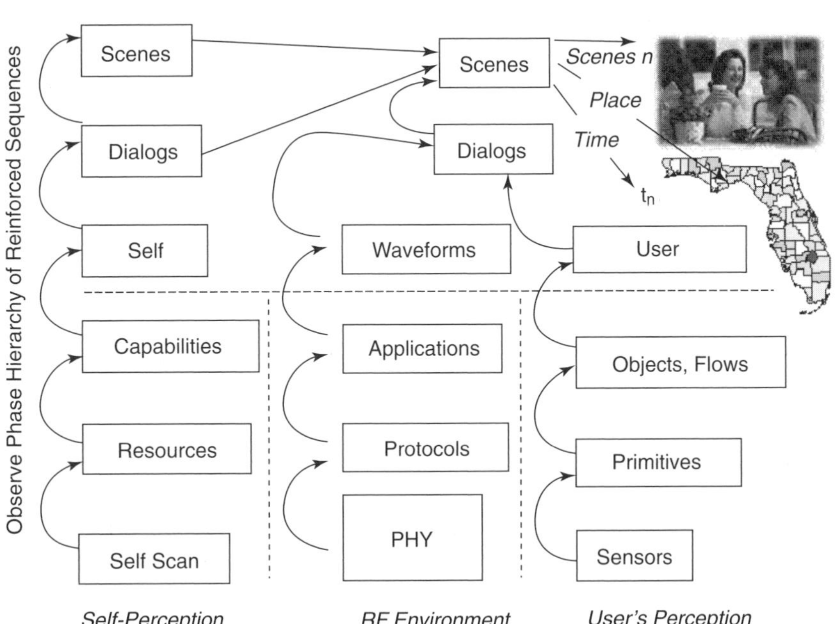

FIGURE 2-4 CRA represents the iCR's Self, User, and RF mutually grounded in a scene.

or active in the RF environment as perceived through the <RF/> inference hierarchy. <Self/>, <RF/>, and <User/> are grounded in space–time at the Scene level of the CRA hierarchy.

2.5.2 Cognitive Processes

In this CRA, the iCR reasons over the <RF-environment/>, the <User/>, and the <Self/> with roles summarized as follows:

Core Perception–Action Process in the CRA

1. Sensing
   (a) <RF/> physical and protocol stacks
   (b) The local visual scene
   (c) The local acoustic scene
   (d) Other sensing modes such as temperature and acceleration
2. Perceiving
   (a) The <Self/> in Space–Time–RF
   (b) The <User/> in the acoustic–visual Space–Time–Scene
   (c) Other Objects contributing to the Scene
3. Orienting
   (a) Detecting known and new patterns of stimuli
   (b) Reacting immediately (if necessary)
4. Planning
   (a) Generating alternative response plans with associated actions
   (b) Assigning QoI metrics to plans
5. Decision Making
   (a) Choosing among alternative actions based on QoI value
   (b) Forcing choices within time and other resources
6. Acting in both
   (a) RF and
   (b) Physical domains (e.g., speech synthesis, composing email)
7. Learning
   (a) Learning by being told
   (b) Assimilating positive and negative reinforcement

This is the sequence of perception–action in the CRA, ontologically referred to as the <Cognition-cycle/>, although a strict time ordering of these functions is not needed and may be counterproductive.

The CRA, then, rests on the following pillars:

1. The <Self/> in the <Universe/> described in formal semantics (e.g., RXML).

2. The <iCR-platform/>, with SDR, sensors, perception, memory, and computational resources needed to ground the <Scene/> to experience.
3. AML illustrated by serModel acquisition and use.
4. The seven core perception–action capabilities of iCR organized into a cognitive process.
5. A comprehensive inference hierarchy of abstractions that integrate perceptions, enabling well informed responsive information actions that enhance QoI for the <User/>.

This text develops one candidate CRA that includes these pillars in which the iCR remembers everything and constantly acquires incremental knowledge by resolving newly acquired experience against its experience base. CR1 implements the representation hierarchy, the seven core perception–action capabilities, a sequential cognition cycle, and continuous learning via ubiquitous CBR. CR1 has many limitations, but it illustrates the architecture pillars. Industry may define more than one CRA. If such a CRA accommodates evolution to iCR learning from experience, that CRA has the full potential of the iCR vision. Such a CRA advances the goal of mass customization but presents challenges of self-referential loops, QoI semantics, and adherence to social norms including privacy.

2.5.2.1 Self-Awareness Without the Gödel Flaw

Since radio communications are isochronous, CRA RF and network-oriented behavior must be isochronous, with actions that occur on temporal boundaries precisely defined by radio protocols, for example, 10 millisecond code division multiple access (CDMA) frames. Thus, the cognition cycle must be deterministic with respect to the realization and control of radio resources. Similarly, its perception of and interactions with the <User/> must be isochronous on temporal boundaries defined by the user's expectations, typically seconds to minutes, with detailed expectations learned per-user. Thus, the cognition cycle for the user also must have deterministic bounds. Such isochronous behavior may be obtained from bounded-recursive functions—computer programs with iteration, but not with recursion or with Until or While loops [142]. The prohibition on such loops precludes resolution theorem proving (one method for reasoning with logic statements). At first, this seems excessively restrictive, but radio and user interaction timetables preclude unbounded inference, so the constraint formalizes practical engineering properties of market-worthy systems. Bounded recursion allows incremental learning of <User/> preferences, <RF/> procedures, and serModels by incremental extension of prior knowledge.

2.5.2.2 Mass Customization

The introduction into markets of CRs that learn substantial new behaviors from their users will usher in an era of mass customization unlike anything previously experienced. Think of it. Your CR isn't a fixed-function information system. It is a clean sheet of paper. Unlike a laptop-style clean sheet of paper that you personalize by buying software packages and loading them into the system, this blank sheet of paper is an information sponge, constantly learning from you, about you, about who you are and what you do and like so that it can always know what you want and be fully prepared to get it done for you. Radar O'Riley from MASH? Not initially, but eventually no doubt.

2.5.2.3 Privacy is Important

If my CR knows that much about me, I want it to be pretty much unhackable, don't I? Don't you? Surely, we want it to destroy all personal data if it is ripped off by some thief so that the loss of the hardware doesn't also result in identity theft. Privacy support systems are developing. For example, some laptops now offer fingerprint readers. The CRA envisions the continual monitoring of the user by the iCR, continually confirming that the <User/> is still the <Owner/> and reacting appropriately to changes in the state of the Owner–CR situation. Privacy support has many open research issues that are merely suggested in this text.

2.6 SYNOPTIC iCR FUNCTIONAL DEFINITION

A CR, then, is a SDR that:

Delivers personalized wireless information services,

Through natural, cognition-level user interfaces,

Flexibly accessing and managing the node's RF communications bands and modes locally on behalf of the user, while

Learning user preferences and

Protecting the user's private information, typically

Cooperating with other CRs and cognitive networks (CNs) via an industry-standard Radio Knowledge Representation Language (RKRL, e.g., RXML) while

Resolving conflicting ontological views

In an open software-defined cognitive radio architecture.

2.7 EXERCISES

2.1. What are the seven core capabilities of a cognitive system?

2.2. What key capability clearly differentiates a cognitive system from other kinds of intelligent systems? Why is this important?

2.3. Define a platform and illustrate platform configurations from different trade organizations. [*Hint*: Cover at least OMG and Windows-Intel.]

2.4. Find at least six different video sensor systems on the Internet. Characterize the image processing provided by the system and its cost. Which ones provide video perception APIs? Which API features are needed for the CRA?

2.5. Complete Exercise 2.4 for speech-to-text software tools.

2.6. Find on the Internet research, freeware, or shareware that may be useful in the CR development process. [*Hint*: Visit Ptolemy II and the SDR Forum web sites.]

2.7. What parts of a CR could you get from a major electronics retailer like Best-Buy™ or Radio Shack™? What about a local computer-electronics shop? You could start with a laptop, add a wireless LAN card, add an amateur radio receiver (e.g., for easy access to AM/FM broadcast and the Weather Channel), and add a GPS card and related software. If you bought all this stuff, how self-describing is it? Each software tool loadable into the laptop describes itself in terms that the Windows Registry cares about, but not in terms that you necessarily care about. Write the XML for your collection of stuff: <AACR-platform><Laptop>???</Laptop> and so on. This is a lot of work. Wouldn't it be nicer if the manufacturer supplied a CD-ROM with metalevel description of each capability? That way, instead of reading about the IRDA port or Blue-Tooth on the motherboard, your AACR itself could ingest the metadata into its own self-description.

2.8. Suppose you started Exercise 2.7 with a wireless PDA. What are the advantages and disadvantages of such an approach? Let's call whatever hardware you pick your CR-zero, your starting point for cognitive radio.

2.9. Consider QoI. Write a set of equations in Excel or C that capture the trade-offs for a VTC. Express at least three different ways of setting up a VTC as information requests that characterize the VTC in the information dimensions of Quantity, Timeliness, Validity, Accuracy, and Detail. Implement one on CR-zero.

2.10. Extend the CRA introduced in this chapter to wireline connections.

CHAPTER 3

EVOLVING FROM AWARE AND ADAPTIVE TO COGNITIVE RADIO

This chapter develops a use case in detail to further characterize the role of machine perception and AML in AACR evolution. Technical vignettes introduce important second-level data structures underlying the CRA.

3.1 REVOLUTION OR EVOLUTION?

Is iCR the next great leap in radio engineering? Probably not. There is a sequence of steps that markets support, and a leap to iCR may be a bridge too far. Some iCR use cases are compelling and technically daunting but not impossible. Initial steps toward iCR include embedded GPS for location awareness. Some automobiles help users navigate via GPS. The next steps clearly include market development of the FCC's policy opening unused TV channels for ad hoc networking. This chapter examines just that specific next step in SDR evolution, but using machine perception and AML for AACR evolution. The chapter identifies data structures that integrate SDR with machine perception, particularly speech and natural language. Simple tasks like naming a cognitive assistant, ontological support for dialog, and understanding a wireless task given the brittleness of natural language processing technology all require a mix of computer science and radio engineering insights; so do tasks like inventing a waveform, observing etiquette among AACRs, and learning in the FCC use case. Having examined these features of AML in AACR, the final section reexamines the iCR value proposition

Cognitive Radio Architecture: The Engineering Foundations of Radio XML
By Joseph Mitola III

with lessons learned from the era of AI hype of the 1980s. Early in the introduction of a technology, opportunities for unrealistic expectations abound. If CR is a carrier of hype disease, the final section is a hype vaccine.

This chapter, then, is a microcosm of the text, examining the core technology challenges and market imperatives of iCR in the context of a radio-centric use case, "Moving Day."

3.2 MOVING DAY

My wife, Lynné, and I are taking two automobiles from Virginia to Florida on moving day, 2001. Since we are driving together, we want to talk from car to car. I have the Ford and she has the Honda. I'm in front, and she's in back. Do we want to accrue a thousand minutes of prime time on our cell phones during this leisurely three day trip? I don't think so. I have several CB radios, but they take too much time to install and require tedious voice protocols, so we purchase a pair of Motorola T5200 short-range VHF walkie-talkies packaged like cell phones. NexTel™ subscribers might not do this, but we weren't so we bought the radios. What are the AACR and iCR alternatives?

3.2.1 Aware–Adaptive Radio Solution

At some point soon, our cell phones will use bands other than the fixed cellular spectrum allocations. FCC advocacy of greater spectrum availability to secondary users via CR is gaining momentum [64]. The FCC CR learns of TV-band availability via an XG spectrum-use policy broadcast. XG policy broadcast expresses in a precise formal language "Here and now it is OK to create ad hoc ISM wireless networks on a given UHF TV channel." This behavior can be hard-coded, or it can be learned.

Suppose my SDR cell phone accesses VHF mobile and UHF TV bands but lacks an air interface for T5200 voice service that Lynné and I need on our trip south. The situation evolves as follows:

My RF aware–adaptive SDR cell phone detects the "unused TV channel" policy on the FCC's spectrum policy broadcast. The policy states the GPS-location zone for use of TV Channel 68.

My cell phone advises my cellular service provider and asks for a short-range download.

The TV-band walkie-talkie personality is downloaded. Preprogrammed behavior ensues. When my cell phone with the TV-band policy detects that it can't use Channel 68 anymore as we travel south, it is preprogrammed to search for another TV channel. It finds Channel 24 instead and so it goes.

Later, the walkie-talkie personality is written over by a downloaded map that takes up substantial memory.

If a pair of T5200's cost $50 in 2001 dollars, I might be willing to pay about that much for the downloads to both of our cell phones, at least the first time. That revenue would be shifted from the local cut-rate electronics merchant to my cell phone bill, which to me is a NO-OP. Fifty bucks is fifty bucks. Thinking about it for a minute, I realize that I didn't have to buy $7 worth of batteries for the trip south, either, since we keep our cell phone batteries charged up on the cigarette lighter. I hope my cell phone walkie-talkie works with my brother's T5200 when we are hunting next week. If the TV-band walkie-talkie personality can't talk to a T5200 in VHF I will have to get my money back.

3.2.2 AML Alternative

This section offers a cognitive version of the FCC unused TV-spectrum use case. The iCR is more autonomous than the TV-band AAR. The players in this use case are Joe, the <Owner/>, Lynné, the <Wife/>, Dan, his <Brother/>, and Genie, the <Self/>, a CWPDA embodying iCR technology.

1. Joe says, "Computer." (He sometimes calls his iCR "Computer" for the famous scene in that Star Trek movie.) The iCR says, "Good day, Master, your wish is my command." (Recall "I Dream of Genie"?) Joe asks the iCR for walkie-talkie link to wife/kids in the other car.
2. The iCR, who also knows herself[1] as Genie, asks, "How, Master?" If Genie were sold by a CWN service provider, she could "phone home" with Joe's request for a walkie-talkie SDR download, just an AAR with a nifty user interface. Not this CR. Genie is from Super-Soft, the world leader in cognitive information systems in this scenario. If Joe hadn't been a radio engineer, Genie would say, "Master, I have the Super-Soft Cognitive Radio Ontology so I will become a walkie-talkie for you." In this case, though, Joe suggests Genie look in the TV bands for empty spectrum.
3. Genie says, "What radio frequency band?" Since Joe is a radio engineer, he suggests: "Try an unused TV channel." Genie has the Super-Soft Cognitive Radio Ontology (CRO) and Cognitive SDR Subsystem. The ontology gives it the capability to talk to <Users/> informally about radio and to talk to radio peers in radio technical jargon.
4. Genie searches the TV bands: "TV Channel 68 is empty." Genie asks, "Will Lynné's car follow you within 500 meters?" To which Joe replies, "Yes." Genie reasons that for short range, she can close the link within FCC ISM Part 15 rules, generating an air interface that meets walkie-talkie constraints. Genie discovers by listening to <RF/> that there is a policy broadcast channel and confirms that Channel 68 is authorized,

[1] Over-personification emphasizes the artificial synthesis of an intelligent entity, Genie, who talks with a female voice.

advising the FCC network that she will transmit with 0.25 watts EIRP via FM voice. The FCC Cognitive Spectrum Network (FCC CSN) approves the plan. Genie can be trusted to verify its plans with the FCC before implementing them.

5. So Joe says, "OK, great. Did you check with the FCC?" Genie says, "Of course, Master. You know I would never do anything to get us in trouble with the FCC."
6. Genie sends a setup message via CDPD to CR2 in Lynné's auto: "Use UHF 68 in FM push-to-talk mode with 100 kHz modulation," Genie says (paraphrasing her RXML).
7. Genie and CR2 use UHF 68 for Lynné and Joe until they detect channel occupancy. These iCRs obey radio etiquette, constantly monitoring Channel 68 for TV, the primary user; they are secondary users, obligated to defer. They know what they are doing, and they are responsible to obey the rules.
8. When autonomously detecting the conflict for TV68, Genie infers that the situation closely matches step 3 above, except that from experience, Genie expects Master to advise, "Search for an unused TV channel." With CBR, Genie continuously compares all prior experience with the current situation, binding the most relevant experience to the current <Scene/>. (This is how CR1, the research prototype, works.) Genie's CBR binds steps 3–6 above to resolve the conflict, step 3 above to step (a) below, step 4 to step (b), and so forth.
   (a) Genie searches for an unused TV channel. She finds UHF 24 unoccupied.
   (b) Genie plans to switch to UHF 24 via binding <Unused>TV24 </Unused>.
   (c) Before Genie transmits anything, she tries to reach the FCC, but the FCC is not available. The local spectrum-use broadcast tells Genie that Channel 24 is OK for ISM use. Genie's core CRO tells her to wait for the FCC, but reasoning over uncertainty implies that the FCC would probably (90%) approve because of their recent approval of TV68.
   (d) To resolve the internal conflict, Genie says, "Master, I cannot reach the FCC, and we cannot use TV Channel 68 now because it is now occupied. Channel 24 is not occupied, and the FCC broadcast says it is OK in principle to use Channel 24. Since the FCC approved my prior request to use low power FM transmission that I synthesized for this occasion, I believe they would approve, but they do not reply. My trust instinct tells me that if you take responsibility, then I may use Channel 24."
   (e) "Thank you, Genie. Permission granted. Next time you can reach the FCC, tell them I gave you permission."

(f) Genie sends a message to CR2 advising it to switch to UHF 24 for the next transmission.

Genie knows a lot about radio, but she had no predefined walkie-talkie <Waveform/> for the TV bands. Genie synthesized it by finding walkie-talkie in the ontology <CRO/>, adapting its parameters to TV Channel 68. CBR then enabled Genie to maintain communications autonomously in spite of a temporary lack of access to the FCC. Genie not only provides connectivity but also is trustworthy. Genie is smart enough to discover Dan's new T6880 and link them together.

AARs emerging in 2003–2006 to use the TV bands are the foundation for iCRs like Genie. CR1, the research prototype, is not as capable as Genie, but the differences soon may be a matter of product engineering, not a lack of sensors, perception, or computational intelligence technology.

3.3 DEVELOPING AML FOR GENIE

While it is fun to write use cases, this book is about a CRA that realizes them in hardware and software *components* that work together via *design rules* to achieve the desired *functions* in an integrated product. That realization will take some time, but at present, one can sketch the structure of the software and experiment with CBR steps toward AML. This section therefore replays the Genie use case, explicating the CRA functions, components, and design rules.

3.3.1 Genie's Name

In step 1 above, <Owner/> addresses <Self/> with the speech segment "Computer." To interpret this properly, Genie must know "Computer" as a name. Genie needs to recognize "name" as a stimulus and to respond to one's name. The following data structures apply.

Expression 3-1 The <Self/> Grounded in the Universe

```
<Universe>
  <Self>
    <iCR-platform> . . . (capabilities from above) </iCR-platform> . . .
    <Name> "Computer" "Genie" </Name> </Self>
  <Scene> <Self/> <RF-environment/> <User-situation/>
    <Space-grounding> Location </Space-grounding>
    <Temporal-grounding/> </Scene> </Universe>
```

The <Self/> exists in the <Universe/>. There is a <Scene/> in which the <Self/> is located that also includes an <RF-environment/> and a <User-

situation/>, all grounded in space and time. The <Self/> has two names, each expressed in speech (indicated by quotation marks). Auditory perception correlates these two names with external speech stimuli. How did Genie learn her names?

3.3.2 Speak When Spoken To

In step 1 above, Genie responds, "Good day, Master, your wish is my command." The data structure relating <Name/> and <Response/> may be a serModel.

Expression 3-2 serModel of Salutation–Greeting Dialog

<serModel>
<Stimulus> "Computer" </Stimulus> <Experience></Experience>
<Response> "Good day, Master, your wish is my command."
</Response> </serModel>

The anonymous serModel is shallow. In an expert system, this serModel could be a named if–then rule. Since iCR uses AML, this serModel is learned not programmed. How did Genie acquire this serModel?

3.3.2.1 Acquiring serModels

To acquire the serModel of Expression 3-2, the perception system grounds the <User/> and the <Self/> in a <Training/> <Scene/>, a setting for the autonomous acquisition of serModels. The iCR <Self/> is grounded in space–time. Therefore, all the stimulus–response (SR) pairs are annotated with space–time. Each is learned in some place at some time and from an identifiable <Source/>. Each applies in a class of places and times expressed in a computational model of the equivalence of similar settings.

Expression 3-3 Training Setting Defined as Place of Purchase

<Training-setting>
<Place-of-purchase>
BestBuy™ Wal-Mart™ Circuit City™ . . . </Place-of-purchase>
<Owner's-home>
<Ask/> </Owner's-home> </Training-setting>

When purchased, the iCR knows its location from a trusted source like a secure WLAN or a salesperson (a somewhat trusted source). Suppose Genie learns her name when first purchased ("Right this way, Sir. You and Genie should spend a few minutes in the fitting room so she can get used to your voice and learn enough about you to know you as her owner.")

Expression 3-4 Use Settings Overlap Training Settings

```
<Use-setting>
  <Place-of-purchase> <Here> BestBuy™ Ocala, FL </Here>
    </Place-of-purchase>
  </Use-setting>
```

In addition, Genie knows from Expression 3-4 that unless otherwise specified, the space–time grounding for a behavior is <Now/>[2] and <Here/>. Settings for training and use are not identical but overlap. Genie may be trained at any of the places of purchase listed but she may perform place of purchase actions only at the Ocala, Florida store. Why should a radio engineer care about this?

Expression 3-5 Radio Training and Use Settings

```
<Training-setting>
  <Place-of-purchase> BestBuy™ Wal-Mart™ Circuit City™ . . .
    </Place-of-purchase>
  <FCC-spectrum-policy> <RF> 1600kHz </RF> <Trust> High </Trust>
    </FCC-spectrum-policy>
  <Owner's-home>
    <Ask/> </Owner's-home> </Training-setting>
<Use-setting>
  <Place-of-purchase> <Here = BestBuy™ Ocala, FL>
    </Place-of-purchase>
  </Use-setting>
```

The same data structures that mediate learning about the iCR's name readily mediate spectrum-use policy as in Expression 3-5. According to this RXML, the iCR learns spectrum use via a FCC broadcast on 1600kHz, and the degree of trust afforded to this broadcast is "high." The radio may use what it learns on the FCC policy broadcast anywhere it chooses. Thus, policy broadcasts bind policies in space–time for accurate interpretation, avoiding over- or undergeneralization of the policy. The <Training-setting/> and <Use-setting/> share semantics in <User/> and <RF/> domains.

Referring to Expression 3-2, such a serModel that does not specify a time and a place of applicability is not fully groundable. serModels may be grounded either with specific place and time, or with a class of location (e.g., <Home/>) and time (e.g., <Weekends/>). Although the quotes on the response "Good day . . ." in Expression 3-2 imply speech synthesis, the communications

[2] <Now/> refers to the ontological primitive. <Now> paired with </Now> delimits time while <Now/> refers to the current instant in the real world: <Now/> is a metalevel reference to the real world, not to the computer's clock which may be wrong.

medium may be explicated in the serModel. Adding <Here/><Now/> <Say/> to the serModel grounds it in space–time and an action capability. While it is relatively easy to express and to code such behavior as expert system rules, it is more flexible to learn such serModels through training. CR1, the research prototype, remembers everything (the CD-Rom illustrates this) and constantly attempts CBR on everything. It learns of a training setting through the flag "train" during which it acquires serModels. When it succeeds in applying a model, as CR1 often does, it asks permission to act, such as <Here/> and <Now/> to <Say/> something.

3.3.2.2 Stereotypical Learning

The salutation and FCC broadcast are examples of stereotypical situations. People learn these situations by exposure, observation, and participation. The CRA enables training by being exposed to structured stimulus–response pairs in reliably detectable situations so that "good" knowledge is acquired autonomously. Good knowledge may include off-nominal cases and negative reinforcement to learn what not to do as well as what to do. Since stereotypical situations have readily detectable features, they are good candidates for training.

3.3.3 Understanding a Request

The last part of step 1 in the two-car use case asserts that the user could explain the need for a walkie-talkie link to the other car. Speech recognition technology is powerful, but far from perfect, particularly in noisy acoustic settings (e.g., not broadcast news) and for conversational speech (not structured speech like the news). Speech recognition accuracy may be expressed in equal error rates, the error rate at which the probabilities of errors of different kinds are equal. Speech equal error rates are about 20% for untrained speakers and as low as 5% for trained dialog in acoustically moderate scenes (e.g., headphones in an office, but not a cocktail party or a busy airport concourse). The top 10 alternate text interpretations of a sufficiently long speech segment such as a phrase or a clearly articulated sentence contain the correct text about 95% of the time.

Suppose the request for walkie-talkie support for the trip to Florida is as follows:

1. "We are going to Florida by car."
2. "I will be in the front car."
3. "Lynné will be in her car behind me."
4. "We need to talk from car to car."
5. "Can you and her CR create a walkie-talkie between us?"

Suppose these five phrases are the exact words (ground truth) used to request walkie-talkie. The term "car" occurs five times, the most of any isolated word in the request. Suppose the speech recognizer got that right 80% of the time. In addition, walkie-talkie is a radio word, so further suppose Genie has a rich repertoire of speech models of such radio words and that it recognized that. Suppose, in addition, Genie has strong speech models for the fifty states, recognizing "Florida." The text from speech is not perfect, but Genie recognizes "Lynné," "car," "Florida," and "walkie-talkie." Even with brittle speech technology, this level of recognition is about 95% likely.

Genie checks her inventory of things that she can do with pairs of things she recognizes. "Car ⇔ walkie-talkie" evokes a mobile radio CBR case to create a point-to-point short-range voice link from one moving vehicle to another. The serModel for the vignette could be the following:

Expression 3-6 serModel Template of a Dialog to Form a Voice Link

```
<serModel>
  <Dialog><Phrase1>
  <Stimulus><Person/> <Vehicle/><Talk/> </Stimulus>
  <Response> <Here/><Now/>
    <Say> "Would you like me to create a <Voice-link>
    between <Person/> in <Vehicle/>?" </Say> </Response>
  </Phrase1>
<Phrase2><Stimulus> <Positive-response/> </Stimulus>
  <Response>
    <Here/><Now/> <AND> <Say> "OK, your wish is my command."
      </Say>
    <Initialize><Voice-link/> </Initialize> </AND>
    </Response></Phrase2> </Dialog>
</serModel>
```

The template consists of two phrase exchanges. The stimulus for <Phrase1/> is the recognition of <Person/>, <Vehicle/>, and <Talk/> from speech addressed to Genie. The second phrase instantiates in response to any positive response to Genie's question, yielding a verbal response <AND/> a task to initialize a <Voice-link/>, a SDR procedure. Therefore, Genie offers to create a walkie-talkie radio link between two cars. A question–answer system could embellish the dialog. The serModel not only asks permission but also implements the requested action. <Initialize/> instantiates a voice link by instantiating a SDR template.

3.3.4 Phone Home or Collaborate?

In step 2 of the walkie-talkie vignette, Genie wonders, "What RF?" (on which to build a walkie-talkie voice link). The SDR template lacked the details for <RF/>. Should Genie phone home, connecting to a home CWN, or introspect, trying to invent the missing pieces? With contemporary cell phone technology, the choices for new waveforms are limited and mostly up to the service provider. If the phone doesn't have a "walkie-talkie" button or it's not listed in the users' manual (Where did we put that thing?), then the consumer has little choice but to call the service provider. The consumer doesn't ask *the phone*. The phone has insufficient knowledge on-board. The iCR, on the other hand, may proactively search the semantic web, XG broadcasts, and peer iCR networks via knowledge discovery and data mining (KDD) [65] with voice interaction [66] and a semantic web ontology to automate interaction with the consumer.

In the near term, the user may type "walkie-talkie" into the service provider's Search window on the handset and get an offer to download that personality for a small fee. This scenario has many positive features, such as amortizing the cost of the "walkie-talkie" personality over many users and assuring that the personality is supported by a chain of responsibility in case it generates illegal interference. Although NexTel users sometimes refer to their VHF service as "walkie-talkie," the handset communicates with a tower, not with another handset like an actual "walkie-talkie." This service has a digital control channel, while walkie-talkies have no such adult supervision. How could an iCR synthesize such a capability instead of depending on a download?

3.3.5 Genetic Algorithm (GA) Introspection

Genie's alternative to the download could be the autonomous synthesis of the needed <Waveform/> via genetic algorithm (GA) [67–69]. The synthesis of a new air interface historically requires highly trained radio engineers and years of work, including ITU-R standardization. Computer-based alternatives include the GA that solves optimization problems in rugged fitness landscapes where the lack of smooth surfaces prevents conventional optimization. GAs generate a large population of individuals (waveforms) that are candidate point solutions, selecting the most fit among that generation to procreate a next generation, creating offspring by parenting with inheritance, mutation, and crossover, and iterating until a sufficiently good member of the population has been found or resources are depleted.

GA introspection to invent the walkie-talkie air interface could be based on a population of air interface components for machine-generated possibilities. Since there are many known approaches to physical layer radio connectivity, one could enumerate physical layer modes for the GA (such as AM, FM, BPSK, QPSK, MSK, or 16 QAM). Figure 3-1 illustrates how genes represent components and parameters of the SDR Forum's original Information Transfer Thread [70]. The gene string specifies the physical layer, such as (100000) indicating [analog] AM as the physical layer channel symbol, the product of that vector with the Channel Gene String (AM, FM, BPSK, QPSK, MSK, 16 QAM). A similar strategy applies to instantiation of the other parameters of waveform objects of the Forum's SCA. An additional

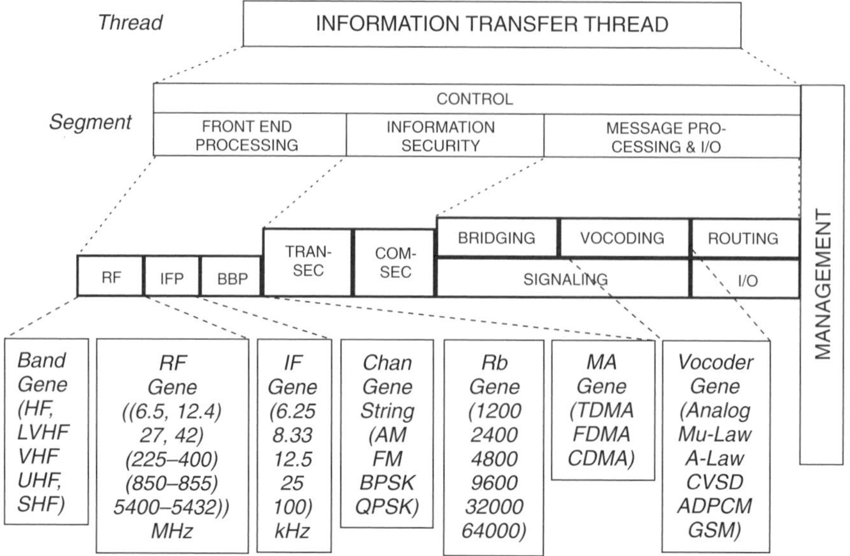

FIGURE 3-1 SDR thread parameterized for waveform genome.

chromosome string specifies source coding of voice (such as analog, A-law, Mu-law, CVSD, ADPCM, RPE-LTP) with (000001) selecting the GSM RPE-LTP vocoder. Other strings code the bit rate of digital source codec or bandwidth of analog source. Yet another string codes the control method (push-to-talk, control-channel/traffic-channel, etc.). Multiple access (collision, FDMA, CDMA), duplexing (simplex, half-duplex, full duplex), and every other aspect of the candidate air interface are so coded. Strings for data rate (1200, 2400, 4800, 9600, 32000, 64000 bits per second) and for bandwidth (6.25, 8.33, 12.5, 25, 50, 100 kHz) create further possibilities. All these vectors are gene strings to the population generator that creates an initial population of candidate waveform "individuals."

Given a candidate walkie-talkie <Waveform/> population, the GA needs a method of evaluating the merit of each individual. A Matlab® [71] script could interpret the strings to simulate each candidate. Fitness requires a metric space and a relatively quick way of evaluating each individual from hundreds to thousands of candidates. The GA simulates the behavior of several individuals on a standard task in a common RF test environment or in a radio propagation simulator [72, 73]. With a population of 100 individuals and 10 ms on a blade server for each evaluation, 1000 iterations require about 16.7 minutes. CWPDA time could be 10–100 times as long or three to seven Moore's Law cycles before GAs are practical on a stand-alone CWPDA. GAs introspect autonomously but do not avoid the Gödel paradox because one cannot specify a tight upper bound for a result to be generated. Thus, GAs are deprecated for stand-alone iCRs but not for CWNs with relatively unbounded resources.

For an overview of schemes for radio design by a GA see [74]. Ideas include shaping crossover and mutation based on templates that reflect good radio design, adjusting parameters of the search space, heuristically repairing failed members of the genome, representing the air interface with novel basis functions, and adjusting metalevel parameters genetically. Some of these techniques are combinatorially explosive.

GAs also manage the radio spectrum [75]. Parameters for multimode waveforms like IEEE 802.11 may be optimized by GAs selecting among QoS parameter sets like BER, bandwidth, spectrum efficiency, power, data rate, and interference generated by the waveform. Rondeau [76] distributes GAs across multiple platforms, highlighting the problem of terminating before convergence.

3.3.6 Waveform Templates

Genie could also synthesize waveforms by learning parameter settings of a generic air interface or waveform template. Jondral and Wiesser [77, 78] synthesize diverse air interfaces such as GSM, IS-136, and UTRA-FDD by parameterized Gaussian impulses for the channel symbols, polyphase filtering for sample rate optimization, and template parameters for Burst Length, Precoding (On/Off), NRZ (On/Off), Modulation Number, Spreading Factors (I&Q), Filter Number, and Sequence Length (I&Q). Such templates compactly encapsulate waveforms for autonomous instantiation through introspection.

Cognitive entities may be said to "wonder" about something if (1) they introspectively discover they have partial knowledge and (2) cannot find a plan to fill in the blanks, so they (3) autonomously allocate resources to plan generation. A waveform template without all parameters specified constitutes partial knowledge for template instantiation. Expression 3-7 enables Genie to recognize the association of a verbal expression "walkie-talkie" as a reference to a parameterized waveform template that can be <Initialized/>.

Expression 3-7 <Self/> Includes Walkie-Talkie Knowledge

```
<Self/> . . .
   <RF-capabilities> <Waveforms> <GSM> . . . </Waveforms>
   <Waveform-templates>
   <Jondral-Wiesser> GSM, IS-136, UTRA-FDD </Jondral-Wiesser>
   <Walkie-talkie> <Voice> <Range> 500m </Range> <Band> VHF UHF
      </Band>
   <Power> 100mW </Power> <RF> <Ask/> </RF> <Template#URI/>
   </Voice> </Walkie-talkie> </Waveform-templates>
. . .</RF-capabilities> . . . </Self>
```

Expression 3-7 offers a (partial illustrative) semantic template for the walkie-talkie function. The semantics are <Voice/>, <Range/>, <Band/>, <Power/>, and<RF/> ontological primitives known to the <Self/> and expressible to the <User/>, AACRs, and CWNs. Band, power, and range are constrained and the exact RF carrier frequency is unspecified and thus imply goals. In step 3 of the use case, Genie says, "What RF?" following the advice <Ask/> of Expression 3-7.

3.3.7 Ontological Support for Dialog

RXML could describe representative waveforms (e.g., APCO 25, CDMA, DECT, GSM, JTIDS, NTDR, PCS-1800, PCS-1900, SINCGARS, UWC-136, and Walkie-talkie) expressing features of air interfaces such as RF channels, R_b (chip rate, bit rate, or channel symbol rate), error control, framing, security, multiple access, multiplexing, source coding, and spectrum spreading. A fully developed RXML would enable Genie to share experience with other iCRs regarding existing air interfaces to learn band/mode-dependent QoI optimizations. Challenges to such AML for AACR evolution include radio domain semantics.

3.3.7.1 Radio Domain Semantic Inconsistencies

Ontology engineering for AACR to evolve toward Genie must overcome the domain-specific ascription of specialized meanings to common terms. Each major air interface (GSM, CDMA, etc.) is a unique ontology domain: across domains the same term has different meanings. For example, "Channel" means 100kHz of analog bandwidth in FM broadcast; one of eight logically associated frame to frame time slots in GSM; and an offset of the long code in CDMA. XML namespaces resolve such ambiguities syntactically but for effective management of radio resources, iCR must disambiguate contexts to use the correct namespaces appropriately. For example, suppose Joe asks Genie to "conference Dan into the walkie-talkie chat" and Dan has a CDMA cell phone. Genie must place a call to Dan linking his CDMA voice channel to Joe's TV-band transmit function. Conversion from CDMA to baseband vocoder to analog FM introduces coding noise, decrementing QoI. If Genie learns to format the Joe–Lynné ad hoc network to Dan's CDMA channel coder, QoI is enhanced in a generalized digital walkie-talkie.

The CDMA walkie-talkie is similar to the walkie-talkie modes of the Joint Tactical Radio System (JTRS) [79], for example, SINCGARS in single-channel analog voice. The Joint Tactical Information Distribution System, (JTIDS) [183] channel includes both time division slots and direct sequence coding. Each waveform thus has distinct semantics for "channel".

RXML assists in ontology resolution by always citing a source for knowledge. Sources (e.g., CYC, RDFS, OWL, or the FCC) [80] assist ontology resolution. Authoritative sources include the GSM MOU Committee and the ITU. If no authoritative source is specified, then when a <Self/> knowledge chunk is learned iCRs autonomously insert the place, time, and source of the update. This feature of RXML originated in RKRL 0.4 [145]. Although far from a universal solution, RXML namespaces and source semantics facilitate semantics alignment across radio domains. Synthesizing CR behaviors that relate radio ontologies to the <User/> requires the general world knowledge of the CRO.

3.3.7.2 Cognitive Radio Ontology (CRO)

In step 3 of the Genie use case, Genie is advised to "Search for a band with an unused TV channel." The use case asserts a cognitive radio ontology (CRO). Such ontology enables iCR to converse with users informally about radio services. In step 4 of the Genie use case, she searches the TV bands reporting "TV Channel 68 empty." The dialog requires the general world concepts of Expression 3-8.

Expression 3-8 CRO Contains Persons, Places, Things, and Actions

<CRO 0.1> <Universe>
<Persons> Owner Lynné FCC < Self> Lynné's-CR </Self> </Persons>
<Places> car </Places>
<Things> Band TV channel car meter range link ISM transmission radio signal constraints walkie-talkie order-wire CDPD broadcast URI Watt EIRP FM voice confirmation CN-One </Things>
<Attributes> empty 68500 Yes close unused local 0.25 </Attributes>
<Actions> Search follow generate listen transmit </Actions>
</Universe> </CRO 0.1>

With the intuition from the game "Twenty Questions," the embryonic CRO 0.1 includes persons, places, and things with attributes and actions. Linguists also categorize these by their linguistic role as nouns, adjectives, and verbs. Semantic role and linguistic role differ, particularly in less formal speech. Conversational speech is rich in anaphora and ellipsis, general references with words such as "that" and "this," which people interpret with relative ease. Language processing technology readily parses well formed sentences into parts of speech and word roles. Informal dialog is not as readily analyzed. In the CRO, a car is both a place and a thing: a thing in which a person can ride and a place from which to transmit defining time-dependant range of a radio transmission.

iCRs use the CRO to detect in speech the things that they can do, limiting the scope of dialogs in which they attempt to participate. Coding such general world knowledge for this purpose requires substantial technology development. Although language technology offers many domain-independent tools for parsing and interpreting well formed language, it offers only general tools for synthesizing speech plans and actions, leaving the analysis of semantic intent of radio jargon to the radio engineer. This text draws from language tools but adapts them to radio-related domains of discourse.

3.3.8 Authority and Trust

The iCR is "trusted" in step 4 of the Genie use case to verify its plan with the FCC. In GSM, a random challenge accesses the home location register (HLR) for call authorization. Similar secure protocols may contribute to trust relationships among AACR network authorities.

The trust relationship between an iCR and an authority includes shared semantics for dissimilar (learned) data structures of wireless domains: etiquette, user preferences, and QoI enhancing activities. The XG policy language [81] develops computational ontologies sharable among iCRs and trusted entities. XG may evolve toward autonomous coordination between a cognitive spectrum authority and a user's iCR for an autonomously synthesized air interface. Genie requires iCRs and their trusted authorities to evolve versions of a CRO like Expression 3-8. The AACR <Self/> defined in RXML in the companion CD-ROM offers an initial CRO.

3.3.9 Cooperation Among iCRs

In step 6 of the use case, Genie sends a setup message via cellular digital packet data (CDPD), one of its SDR modes, to CR2 in Lynné's car. Modern plug and play protocols enable electronic systems to exchange capability data, such as the Microsoft Registry, CORBA, and the SDR Forum's secure download [27]. The Genie-to-CR2 message would be structured more like Expression 3-9. The syntax CRO: . . . invokes the CRO namespace. CR2 learns by being told that it is to use a known template. Generally, iCRs learn by being told and by observing other radio communications: legacy radios, AARs, and iCRs in cooperative CWNs. Underlying these abilities is the grounding of message symbols to shared semantics, negotiating message symbols to communicate unambiguously.

Expression 3-9 CR-to-CR Communications Are Formal and Brief

CRO: <Request> <Air-interface>
<Walkie-talkie-template> <RF> 650 MHz </RF>
</Walkie-talkie-template>
<Acknowledge> CR1.Genie. 555.1212 </Acknowledge> </Air-interface>
</Request>

Techniques for formal cooperation illustrated in Expression 3-9 include the Knowledge Query and Manipulation Language (KQML) [82], CoABS [83], the DARPA Agent Markup Language (DAML) [127], OWL [84, 85], and the Java Agent Development Environment (JADE) message syntax. OWL and JADE have active user communities and tool sets focused on enabling infrastructure. Domain relevance to communications includes OWL for network management information bases [86, 87]. Ontology-based radio (OBR) research [88] points the way toward formalizing radio domains. Thus, the subsequent treatment doesn't reinvent these technologies, but introduces methods, tools, and lessons learned from these communities into the CRA.

3.4 LEARNING ETIQUETTE

Continuing to analyze the Genie use case, step 7 calls for the iCR's use of UHF TV Channel 68 until they detect channel occupancy. As two automobiles progress south on the Eastern seaboard, the RF environment changes. Microscale changes occur on millisecond time lines with the structure of multipath fading in dozens to hundreds of wavelengths. Power changes mesoscopically as the vehicles transit Fresnel zones a few meters in extent in SHF in urban settings and tens of kilometers in extent in the 1500 kHz AM broadcast band in the rolling hills of South Carolina. Macroscale changes occur as the vehicles traverse dozens to hundreds of kilometers entering and leaving the radiation patterns of low, medium, and high band AM, FM, and TV broadcast stations. Politeness of the level suggested by Genie requires prior knowledge, propagation modeling, and cooperation.

3.4.1 Prior Knowledge and Skill

To obey radio etiquette, Genie constantly monitors Channel 68 for primary TV broadcast as a persistent goal. Genie may be assisted in predicting changes to channel occupancy by prior knowledge like the <Spatial-knowledge-DB/> of Expression 3-10.

Expression 3-10 Prior Knowledge Assists in Adherence to Radio Etiquette

<Universe>
<Environment> <Spatial-knowledge-DB> < Schema/> <Content>
[1 (Washington, DC), 1500 kHz, WTOP, Strong] <Primary-user/>
[2 (Rocky Mount, NC), 794 MHz, TV68, Strong] <Primary-user/>
</Content> </Spatial-knowledge-DB> </Environment> </Universe>

Although there is no spatial knowledge that says [3 (Washington, DC), 794 MHz, TV68, UNK] in the exogenous <Spatial-knowledge-DB/>, there would be such an entry in its internally generated (endogenous) data base as the iCR searched for but observed no such signal in the DC area. With such knowledge, the iCR could share with another iCR that it hadn't observed TV68 in the DC area. Confirmation from several mobile iCRs about TV68 reinforces the further inference that TV68 is not in use anywhere in the Washington, DC area. The exogenous confirmation of endogenous predictions increases Genie's confidence in the plan to use that spectrum for the car-to-car walkie-talkie link. If, on the other hand, some iCRs observe a weak TV68, say, on the northern fringes of the DC area, Genie should pick a channel about which there were no such observations. Thus, etiquette entails not only backing off when a legacy user is detected, but proactively sharing knowledge among iCRs. Polite iCRs need RF sensors that detect weak legacy users like distant TV stations. Optimization through message exchange among intelligent agents is studied at length by the complex adaptive systems community [56], with implications for CWNs.

3.4.2 Case-Based Learning

In step 8 of the use case, Genie is acting as a walkie-talkie but doesn't have a channel because TV68 is active in the new location. As previously noted, step 8 closely matches step 3, but prior advice suggests "Search for an unused TV channel." Genie stored the prior experience as in Expression 3-11. Previously, in Expression 3-6, Genie had employed an existing serModel to instantiate a voice link, had applied the PTT-template, asking the user for advice, and recorded that advice as shown.

Expression 3-11 Walkie-Talkie Dialog Stored at 5 AM

<Dialog-walkie-talkie> <Phrasel>
<RF-capability> <Washington-DC> < 5 AM> <Voice-link/>
</5 AM> </Washington-DC> <Apply: PTT-template (RF = <Ask>)/>
</RF-capability >
<User-situation> <Washington-DC> <5 AM> <Asked> "What RF?"
<Response> <Told> <Search/> <TV> </Told> <Unused>
</5 AM> <Washington-DC> </Response>
</Phrasel> . . . </Dialog-walkie-talkie>

With today's brittle speech recognition algorithms, capturing even such a simple interaction this accurately is not easy. Advances in noise suppression, speaker isolation, and coarticulated word recognition render isolated phrases expressed to a known person in response to a known question as a credible mid-term research objective. Genie's analysis of the next phrase of step 3 results in Expression 3-12.

Expression 3-12 Phrase 2 of Walkie-Talkie Dialog with Spectrum Search

<Dialog-walkie-talkie> <Phrase2>
<Washington-DC> <5 AM>
<Search> (From: 614 MHz, To: 806 MHz, 6 MHz) </Search>

<Found> Unused-TV-Channel (RF: 794 MHz, 6 MHz) </Found>
<Lookup: Spectrum-DB RF = 794 MHz/> = "TV68"
<User-situation> <Asked> "TV68 OK?"</Asked>
<Response> <Told> <OK/> </Told> </Response> </User-situation> . . .
</Phrase2> </Dialog-walkie-talkie>

These two phrases must be integrated for a complete learning experience. Genie matches Phrase 2 to the attempt to instantiate the walkie-talkie air interface because the walkie-talkie goal has been achieved. Genie's procedure for achieving step 8 uses CBR to bind the <Scene/> to experience as follows:

(a) Genie binds Phrase 1 to step 8 to recall the advice to <Search/> for an unused TV channel. UHF 24 is not occupied, and
(b) Genie binds <Unused-TV-Channel:TV68/> from Phrase2 to <Unused-TV-Channel: TV24/> in the current setting.

The confirmation, "Thank you, Genie. Permission granted," provides reinforcement needed by CBR to complete the training experience.

The discussion of the data structures and computational procedures above reinforces a research agenda to overcome the acknowledged limitations of today's perception technologies toward the realization of iCRs like Genie, Bert, and Ernie.

3.5 VALUE PROPOSITION FOR AML IN AACR

In the last part of the use case, Genie acquires new knowledge via supervised machine learning. Genie had to bind current sensory perceptions in RF and user domains to a prior experience. The CR systems engineer's grasp of radio, wireless services, sensory perception, user ontology, planning technology, and machine learning is needed to synthesize Genie across these disciplines to autonomously enhance QoI for the mobile user. The cost of such technology versus the benefit is the cornerstone of the AACR value proposition. This last section therefore reviews the value proposition of AML for AACR. AML is fundamental to iCR, so the incorporation of AML into the CRA facilitates the migration of SDR toward iCR.

3.5.1 Why the Radio and Not the User?

Why not just let the programmers program the modes and the users control the radio? We don't need a cognitive agent in there, do we? This is a central question for AML in AACR. Encountering one case at a time, the well trained user can do it all without Genie. As the complexity of personal communications resources increases, there is a combinatorial explosion of rules about what to do and not to do in terms of network availability, power control, radio bands, and air interfaces that can be used or not used with each other or in specific places or at specific times. The service provider decreases this complexity for the cellular network user. But the complexity is more evident and growing in heterogeneous consumer markets like Home RF, WiFi, and BlueTooth where proliferation brings increasing complexity with interference, privacy, and consumer acceptance implications. Customers in those markets are increasingly reluctant to take on the complexity, so there may be a glass ceiling on wireless market penetration without AML to reduce complexity.

One could augment SDR with data structures like the <Self/> and <Environment/> and good-old expert system rules for flexibility. Such approaches no doubt will continue to propel wireless systems flexibility forward. However, the artificial intelligence (AI) approach of writing expert systems rules has proved flawed economically in complex domains in the past.

For some military applications in the 1980s, it made less sense to pay a team of PhDs to continually update rules than to just build a dumb system and train the relatively smart users to work out ways to overcome its limitations. High expectations and broken promises yielding the AI winter of the late 1980s resulted largely from the failed economics of expert systems compared to trainable people. To survive in the markets of the future, AACRs must overcome similar economics—consumers are pretty smart, so it will take breakthrough AML technology to win them over, and it must be affordable.

Good compromises for the average user can be shipped as embedded expert system rules. To synthesize Genie, algorithms must aggregate historical patterns from unstructured data, infer learning from unstructured situations, and bind current experience to prototypical situations both preprogrammed and learned in the past so that the system continually and gracefully learns (and learns to forget) from its experience. The decades-long process of evolving SDR to Genie along feasible AACR markets can be informed by considering the economics of iCR-class expert systems.

3.5.2 Economics of iCR-Class Expert Systems

In the 1980s Professor Ed Feigenbaum led the Mycin project, which diagnosed bacterial infections in the bloodstream using a rule-based control system. Here's a simplified Mycin Rule:

If (AND (EQ Blood:Gram-stain Neg) . . .) then (Assert
  (Causative-agent:ID "E-coli" EQ), 30)

which reads: *"If the Gram-stain of the Blood is Negative, and if . . . (other conditions are met), then assert that the Identity of the Causative-Agent is E-Coli, with a confidence of 30."*

A database of hundreds of such rules, of course, came to be called a knowledge base (KB) because it formalized knowledge in a given domain. Companion systems Dendral, meta-Dendral, and SU-X interpreted signal streams such as spectrograms to infer the identities of the sources of the signal streams with impressive results. Investment in expert systems technology and applications exploded. Core university texts like *Artificial Intelligence* [89], *Pattern-Directed Inference Systems* [90], and *Artificial Intelligence Programming* [91] showed how to construct and search "game" spaces such as checkers and chess by applying rules that represented legal moves and board states [12]. By applying this control structure plus some additional estimation and control theory, Mycin, Dendral, meta-Dendral, and SU-X, a military system, stimulated DoD funding of AI. Even automatic knowledge acquisition seemed within reach. Adding gas to the fire (meant in the most complimentary way), Professor Randy Davis wrote Tieresias, a program that autonomously acquired rules directly from expert medical doctors [318]. Rule bases incrementally acquired decision trees for diagnostic applications. Rules also applied arcane knowledge to configure minicomputers. In the middle of the hype, it seemed like there was nothing that couldn't be revolutionized using expert systems. Why that turned out not to be the case is a crucial insight for AACR. Eventually we (yes, I was a perpetrator) began to deploy these rule-based expert systems. By the end of the 1980s, they hadn't fared as well as we had hoped.

People were more affordable. In many applications, "three PhDs" were needed to write expert systems rules to capture the knowledge that one military enlisted person had regarding, for example, how to repair an engine. To the degree that knowledge of engines changes over time, the PhDs must continually update the rules. This was not cost-effective because the ability of the enlisted person to learn far outstripped the speed of rule synthesis. Thus, to continue to employ the team of PhDs to replace the one enlisted person was not cost-effective, and the lack of learning ability of the expert systems was not mission-effective.

3.5.2.1 Learning Through Perception

It wasn't just a lack of AML algorithms: the problem lay in the inability of the machines to learn from the physical environment, to perceive, to isolate entities in a scene, to observe, to converse, and thus to learn incrementally and naturally. In other words, it was not the lack of expertise in the initial expert system so much as the inability of the initial expert system to maintain its expertise on its own that led to many a disappointment after an impressive initial demonstration. So a majority of knowledge-based expert systems of the 1980s failed in deployment for lack of sustainability. The relevance is that if an iCR requires too much support from anybody, the user included, but especially from expert systems or worse AML or ontology maintenance programmers, then it may fail sustainability economics.

Some AI applications of the 1980s took three orders of magnitude more MIPS compared to less sophisticated customized code. The learning capability of even the most junior applications programmers far outstripped the autonomous learning ability of knowledge-based expert systems, which was close to zero. Thus, many potentially helpful AI algorithms of many types were not widely deployed. It was much more cost-effective for a manager to pay a conventional programmer to program a new application in an efficient language like C than to work within the limitations of an expert system shell. Fortunately, the economics of computing platforms have changed by a factor of 500 – 1000 in the intervening two decades (1984 to 2004). Today for a given level of complexity, the hardware resources are much less important than functionality and time to market. This motivates a closer look at AML economics versus both the expert system and the hard-coded wireless systems solutions.

3.5.2.2 Reinforcement Learning

Conventional AI technology like expert systems can make radios increasingly aware and adaptive, but not able to learn much from experience. There are three broad classes of machine learning: supervised, unsupervised, and reinforcement. A child instructed by a parent on how to light a fire exemplifies supervised learning. The oracle (the parent) provides ground truth. The learner's job is to recognize and to apply the skill learned from the oracle later, in a different setting, through a distorted perception system, with noise, and in other ways less than identical to the training setting.

An algorithm that ingests 1000 digitized written samples of the numerals from 0 through 9 and attempts to derive the ten classes of symbol without the associated ground truth is performing unsupervised machine learning. The algorithm employs some very general principles, such as a similarity metric with statistical tools to bin the samples into N bins. If it gets N right, then the algorithm has attained a degree of successful unsupervised learning. If k of the samples are associated with the wrong bin, then the error rate is k/1000 and one can populate a confusion matrix of the ten correct bins versus the incorrect bins.

A child playing with matches exemplifies an important variation on unsupervised learning called reinforcement learning. When the child burns a finger, the environment provides feedback to the learning process through the child's perception channels. Ouch! Reinforcement learning is neither completely supervised nor completely unsupervised, but behavior is either rewarded or discouraged by forces in the environment [92]. Completely unsupervised learning is not common in nature.

This text applies all three types of ML to AACR, showing how AML contributes to QoI enhancement goals of AACR. As a minimum AML in AACR could shape demand. With the deployment of 3G/4G and 802.11 hot-spot WLANs in wireless PDAs, there are dozens of modes of communication among which users and networks can choose. Why pay for the network operator's hot-spot LAN to upload a 10MB attachment of an email when you could use your corporate LAN for free in 10 minutes? PDA location awareness and AML (about the cost of connections versus the user's typical movement patterns) could shape demand, creating potentially disruptive forces in converged computer and communications markets. Opportunistic use of TV-band spectrum seems certain with adaptation approaching the Genie vision. But there are many DARPA-hard challenges in making progress toward Genie. A decade ago, there were many DARPA-hard challenges facing SDR, and several generations of SPEAK*easy*, JTRS, and commercial development later, there are SDR base stations and more. There may never be an ideal software radio in a handset, but the research agenda both identified a very important point on the horizon to navigate by and characterized the journey as a hard one. From the trade press one may get the impression that cognitive radio is "the answer" or is "almost here," but not from this text. The core technical chapters that follow present numerous technical challenges with suggestions and technical ideas, but with no argument that the way ahead is easy. Interesting and rewarding, yes; easy, no.

3.6 EXERCISES

3.1. Set up a computer speech system.

3.2. Dictate the five sentences of the explanation of Section 3.3.3 into the word processor. Dictate the explanation ten times, counting the errors of the errorful transcripts.

3.3. Set up CR1. Once you get used to the inference hierarchy and learning phases, link the speech recognizer into CR1.

3.4. In Exercises 3.2 and 3.3, what happened to the keywords Florida, Car, and Walkie-talkie?

- **(a)** Train your CR on those words.
- **(b)** What level of recognition is observed now?
- **(c)** The first two are very common words in talking about moving to Florida by automobile. Program your CR to recognize new topics (e.g., Florida and car) in informal speech. [*Hint*: Apply Zipf's Law. Use Zipf's Law to detect a new topic.] Google "bag of words."

3.5. What machine learning software is available on the Web? Could you just use any of that as a black box? Give it some data and it will learn? Which computer languages do you need to know in order to modify it for integration into a CR?

3.6. Find Java source code on the Web for an expert system shell, for uncertainty (use the Uncertainty in Artificial Intelligence (UAI) bulletin board if you need help; search for Bayes' inference if still not satisfied).

- **(a)** Does any of this look particularly applicable for integration into your CR-zero to help make it more intelligent? You can also try for source code in other languages such as C and C++ or C#.
- **(b)** What would be the benefits and potential drawbacks of this kind of development with freeware? With commercial tools?

3.7. What other lessons, issues, and technical ideas can you draw from Genie and the aspects of that use case developed in this chapter?

CHAPTER 4

AUTONOMOUS MACHINE LEARNING FOR AACR

This chapter develops AML as the core technology for AACR mass customization, focusing on reinforcement learning in which the iCR independently detects the learning opportunity, shapes the dialog and proactively verifies the acquisition of enhanced skill without burdening the user or violating radio etiquette.

4.1 MACHINE LEARNING FRAMEWORK

Machine learning texts typically develop ML strategies, data structures, algorithms, and parameter tuning for relatively simple problems so that the student may understand the ML method. Such simple problems include blocks-world, Rubik's Cube, and Towers of Hanoi [287]. More realistic examples include learning the structures of cells, inducing natural language grammars [93], balancing a rod, and exploring a maze [92]. Examples of ML applications to important problems include medicine, data mining, wireless channel coding [63], and cellular network admissions control [92] among others. Proceedings collect technical papers for emerging topics, such as agent technology in telecommunications [94] and ML in SDR [95].

This chapter applies relevant ML techniques to AACR as a self-contained overview of AML for AACR evolution, with knowledge objects (KOs) and domain heuristics (DHs) developed in subsequent chapters.

Cognitive Radio Architecture: The Engineering Foundations of Radio XML
By Joseph Mitola III

4.1.1 The AACR ML Framework

Algorithms that learn may be parametric, defined over continuous domains; or symbolic, defined over discrete domains; or both. Learning to recognize the difference between a speech utterance "Charlie" and the background noise of a crowded restaurant is parametric learning. Learning that the Owner's <Name/> is "Charlie" is symbolic learning. Systems like CLARION acquire both types of knowledge at once. The AACR ML framework for these consists of the following.

AML Framework for AACR

1. *Analysis* of the *problem space*, which includes:
   (a) *identification* of the problem space as a subset of the user's space–time–RF world, and
   (b) *autonomous detection of problem space interest items, to ground* the iCR's internal abstractions to sensory domains, external abstractions (e.g., protocols), and entities (e.g., people).
2. *Tailoring* the *learning algorithm* to the problem space.
3. *Acting to acquire knowledge and procedures* through experience:
   (a) this may include off-line *training* followed by use and
   (b) on-line *learning* to acquire additional knowledge
   (c) with positive and/or negative *reinforcement* from the environment.
4. *Internalization* of the data, knowledge, and procedures into *encapsulated skills*, for example, serModels (stimulus–experience–response models or the srModels of CR1), followed by
5. *Real-time enhancement and refinement* of the *encapsulated skills*, including validation, constraint discovery (negative reinforcement from novel situations), and the sharing of the newly acquired skill with other cognitive entities (AACRs and CWNs).
6. Off-line introspection to more fully internalize experience for future use.

This framework relates to AACR through steps performed by people (P) or algorithms (in a CWN, AAR, or iCR); through skill definition at design time (D), in real time (RT) "on-line," or through autonomous introspection ("sleep" mode) with CWN assistance ("prayer" mode).

Table 4-1 shows user interface and RF capabilities of SDRs defined by people at design time. New SDR capabilities are downloaded from a labor-intensive design-evolution infrastructure. AAR's skills may be developed interactively via preprogrammed ML adaptation within constraints. GAs in a CWN, for example, may digest experiences of a population of AARs to autonomously evolve skills, while people validate evolving skills for

TABLE 4-1 Skill Definition Frameworks

Steps	SDR	AAR With ML	iCR With AML
Analysis	People/Design	People/Design	P/D + **iCR sleep**
Selection	People/Design	People/Design; **CWN**	**CWN/ iCR RT sleep**
Acquisition	People/Design	**AAR/RT/CWN**	**iCR RT sleep/CWN**
Internalization	P/D + **Downloads**	**CWN download to AAR**	**iCR Sleep**
Refinement	**SDR/RT**	**AAR-CWN**	**iCR-iCR-CWN**

regulatory norms, QoS, QoI, and standards of AACR etiquette. With AML skill acquisition, CRs autonomously analyze problem sets, select ML algorithms, acquire new skills, and internalize new skills through reinforcement and introspection, a tall order indeed. This chapter does not present a specific solution to these cutting edge challenges of computer science and radio engineering, but instead characterizes research relevance within an architecture framework to facilitate the insertion of AML technologies as they mature.

4.1.2 Problem Space Analysis

The CRA AML problem space may be analyzed in terms of inference hierarchies (Figures 2-3 and 2-4), where stimuli aggregate from primitives to inferred entities in <Self/>, <RF/>, and <User/> domains. Problem subspaces occur horizontally and vertically in these inference hierarchies as illustrated in Figure 4-1. Entities interact in dialogs and scenes. Expectations of entity consistency over space–time assist in grounding sensory perceptions to the evolving operational ontology, the integrated internal view of the relevant aspects of the outside world. Vertical inference identifies known and novel abstractions. At the lower levels of sensory perception, algorithms *identify* objects in known classes where AML requires differentiating them from noisy backgrounds and tracking them in space–time, integrating RF, audio, and visual scenes. At the higher levels, the problem is to identify interactions among entities per stereotypical situations. RF-domain interactions include preprogrammed protocols from the physical layer through the applications layer of the ISO stack to be associated with user preferences autonomously. User-domain protocols are the social interactions among the <Self/>, the <User/>, and significant <Others/> for information services.

Object identification for AML succeeds only if the identity is *grounded* to internal ontological primitives in the <Scene/> with sufficient accuracy for the desired QoI. This degree of grounding falls short of general object identification. For example, the AACR that identifies the <User/> as the <Owner/> accurately but misclassifies the owner's eldest daughter, <Sue/>, as the younger daughter, <Ellen/>, for a few minutes may not make errors in its wireless

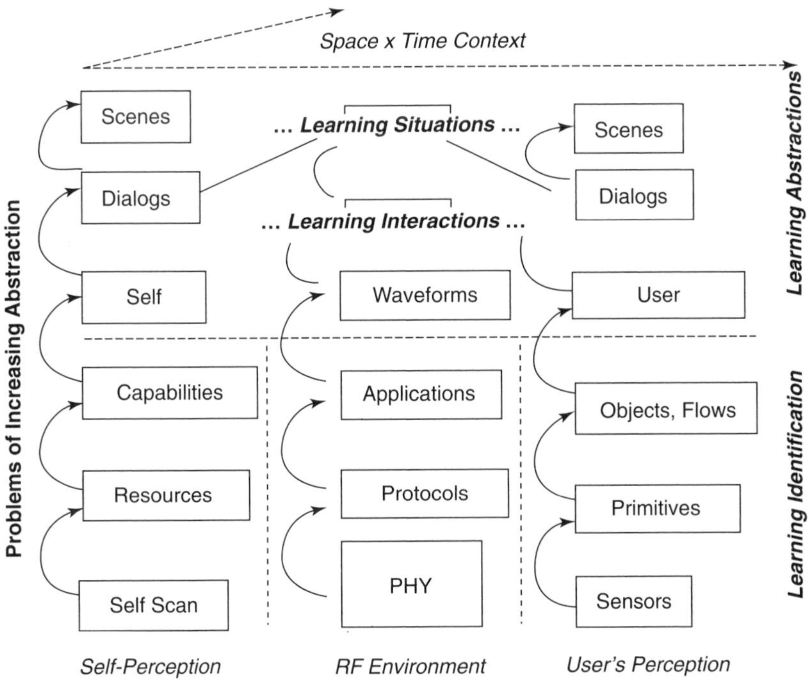

FIGURE 4-1 Identifying ML problem sets: identification, interactions, and situations.

access tasks. Computational models may degrade from model contamination as in the dialog shown in Figure 4-2.

A dialog management system, for example, based on VoiceXML [222], could readily generate such a dialog, and a ViaVoice [66] class speech recognizer could readily recognize the allowed responses "Sue," "Ellen," "Yes," and "No," although less reliably in noisy environments. The iCR structures its speech *actions* to limit the difficulty of interpreting the response to simplify grounding. The AACR proactively shapes the learning environment, enabling learning only if criteria for reliable AML are met. If the person replies "Yes" in step 6, and if the voice and face match <Ellen/> to some degree, then the iCR could update both speaker model and face model of <Ellen/>. The recognition of strong, variable acoustic noise may disable auditory learning while visual noise, such as a sea of faces in a busy airport, could also disable the learning of new faces. Both contribute to effective <Scene/> analysis by suppressing opportunities to learn erroneous associations.

Algorithmic analysis of CRA problem spaces may exploit generic properties of learning situations with little domain-specific knowledge. But even unsupervised ML requires a priori human tailoring, such as the identification of features of the problem space to be represented to the ML algorithm. The

1. AACR: "Are you Sue or Ellen?"
2. If the answer is not clearly (high probability/low uncertainty) "Sue" or "Ellen" then,
3. AACR: "Sorry but it is important for me to know who you are. Could you please answer 'yes' or 'no'? Are you Sue?"
4. (Response is clearly "No.")
5. "Are you Ellen?"
6. (Response is clearly "No.")
7. "Are you somebody else?"
8. (Response is clearly "No.")
9. The iCR jokes, "Are you kidding?" If laughter is clearly detected in the background, then the iCR may infer that it is in a no-win social situation where it cannot get a valid grounding, so it stops trying, disallowing use until <User/> can be grounded.
10. The iCR might warn those nearby, "OK, the joke is on me. But don't expect me to be helpful if I can't tell who you are."

FIGURE 4-2 ML dialog in which the iCR shapes entity identification.

multiple hierarchy of Figure 4-1 aggregates stimuli in time and space so that dialogs like that of Figure 4-2 autonomously tailor the ML opportunity to the capabilities and limitations of the iCR (e.g., shaping the conversation to learn about <Sue/> and <Ellen/>). Having analyzed the problem space(s) of a set of use cases, one is ready to tailor ML algorithms to AACR applications.

4.1.3 Tailoring ML Algorithms

Tailoring ML algorithms to AACR requires a strategic grasp of learning protocols. Tailoring includes defining data structures for learning abstractions (the vertical flow in the hierarchy) and interactions (the horizontal flows). Interactive learning requires metalevel algorithms to autonomously adjust the augmentation of applicable data structures. For example, the dialog of Figure 4-2 forces yes/no answers to tailor the information flow for entity identification in face and speaker perception spaces.

Tailoring includes describing AML algorithms in canonical forms as in Figure 4-3: a part that computes statistics, a part that identifies the features, and a part that labels significant features to acquire new knowledge. Labeling may be autonomous (unsupervised), interactive with user or CWN (supervised), or interactive with the environment via feedback (reinforcement learning).

In this canonical form, a general *statistical analysis* algorithm computes a feature set of the data presented to the algorithm through the tailored information flow. A domain-independent *feature identification* algorithm discovers potentially important features. A domain-dependent algorithm could then

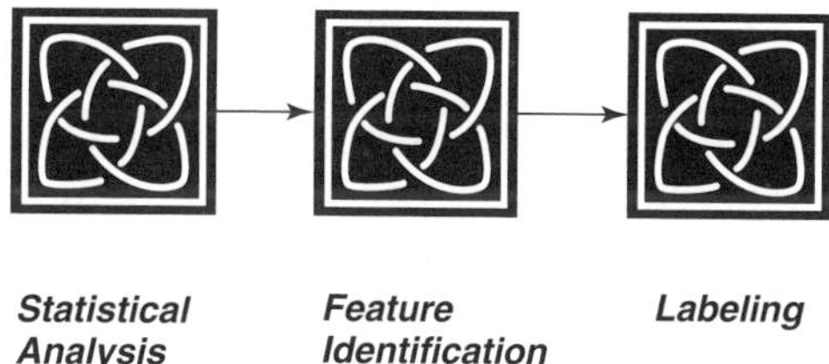

FIGURE 4-3 Typical structure of statistics-based ML algorithms.

label the important features. For example, a video algorithm might label intense pixels of a bimodal distribution as "Light" and the less intense pixels as "Dark." Similarly, a RF-domain algorithm might *label* a large number of low intensity spectrum points as "Noise," while the strongest bins are "Signals."

The canonical process for tailoring AML to AAR then consists of the following:

1. Structuring the iCR's data into parallel hierarchies to guide <Self/>, <RF/>, and <User/> domain grounding in space–time as in Figure 4-1.
2. Synthesizing behaviors that proactively shape the flow of information to AML components, guiding stimulus–response sequences.
3. Remembering stimulus–response experiences indexed to space, time, RF, and the user view of the <Scene/>.
4. Structuring ML components to perform:
   (a) Domain-independent analyses (e.g., statistics, sequence, entropy, etc.).
   (b) Identification of potentially important domain-independent features.
   (c) Learning by domain-specific labeling of the features either (i) autonomously (unsupervised), or (ii) interactively from an authority (supervised), or (iii) from positive and negative reinforcement.
5. Validating, amplifying, and refining the AML structure and parameters over time.

The balance of this chapter tailors a generic pattern recognition algorithm, the histogram for unsupervised, supervised, and reinforcement learning. Subsequent chapters develop steps 3–6 (page 81) of the AML Framework for AACR.

4.2 HISTOGRAM AS A DISCOVERY ALGORITHM

To learn autonomously, an iCR needs an unsupervised ML algorithm that applies domain-independent methods to identify potentially interesting

AACR data to learn to enhance QoI with that data. A histogram $H(\ )$ counts the number of occurrences of observations in a defined range, binning N observations into K bins, H_k. When $H(\ )$ estimates how well N samples fit a hypothetical probability distribution $P(x)$, the K bins are termed class intervals. For example, the chi-squared test compares H_k to $P_k(x)$ for Gaussian noise [96]. Histograms are standard tools for speech and text analysis; and for low level vision contrast enhancement, region splitting, and spatial clustering [97]. Histograms map data onto frequency-of-occurrence, enabling the detection of nonrandomness, entropy, and thus potentially useful information in a feature space [97]. Histogram-based AML typically includes pattern analysis tests for degree of interestingness.

4.2.1 The Mathematical Histogram

A histogram algorithm from introductory probability and statistics is defined in Equation 4-1. In this discrete formulation, Y, the phenomenon of interest, occurs in a metric space as a function of Z, the integer subset of the real line for N points.

$$
\begin{aligned}
H(K; Y\{y_i\,[i:1,\ldots,N]\}) \\
&= \textstyle\sum_i \{|y_i \in [\{\lambda(k) = Y_{\min} + k * ((Y_{\max} - Y_{\min})/K)\}, \lambda(k+1)]|\}, \\
&\quad 0 \le k < K \qquad (4\text{-}1)
\end{aligned}
$$

In general, $\{\lambda(k)\}$ is an anonymous function; $\lambda(k)$ is the lower bound of histogram bin k; and the floor $[*]$ has an open supremum; $|x| = 1$ if x is true, else zero. Since $\lambda(k + 1)$ is the upper bound of bin k, $|y_i \in [\lambda(k), \lambda(k + 1)]|$ counts y_i that are $\geq$ the lower and $<$ the upper bounds of bin k. If all values are equally likely, then for all k, $H(k) \equiv H_k \cong H_j$, for any $j \neq k$, but if one bin is larger than the others, then the values are not equally likely, so information may be implicit in $\{y_i\}$.

4.2.2 The Histogram Algorithm

In an AACR, $H(\ )$ could be implemented per the PDL of Algorithm 4-1. The first For loop finds the largest and smallest values in the one-dimensional array Y. The second For loop counts the number of values of Y that occur in K bins between $Y_{\min}$ and $Y_{\max}$. The last For loop labels significant deviations from an approximately equal distribution of values among bins as potentially <Interesting/>.

Algorithm 4-1 Interestingness Histogram of Array *Y* with *K* Bins

```
Program Design Language (PDL) pseudocode for H(K;Y)
Y( ) is an array-object of N real numbers from 1 to N;
Y.annotate(<Uninteresting/>);
```

```
K is a integer > 0; Bin( ) is an array-object of K integers
  from 1 to K
Ymax = Ymin = Y(1)
For i between 2 and N{
    If Y(i) < Ymin then Ymin = Y(i)
    If Y(i) > Ymax then Ymax = Y(i)} End For
Span = Ymax-Ymin; // may be zero
BinSpan = Span/ K
For i from 1 to N {
    Index = floor{(Y(i)-Ymin)/BinSpan} +1
    Bin(Index).Increment} End For
Significant = Alpha*N/ K //* Alpha is real > 1 which is the
  threshold of significance*//
For k between 1 and K{
    If (Bin(k) > Significant) then Y. and Bin. Annotate
      (<Interesting/>)} End For
```

This is a retrospective or batch histogram because all N values are known before $H(\ )$ is computable. $H(\ )$ is augmented with pattern detection to identify as potentially <Interesting/> any significant divergence from the uniform distribution. $H(\ )$ estimates the probability distribution of Y to the degree that relative frequency of occurrence informs probability, specifically if and only if (1) N is sufficiently *large*; (2) the random process of Y is *stationary* (having time-invariant statistics) and the random process is (3) *ergodic*, that is, the statistical parameters of a time series faithfully represent the statistical structure of the underlying random process(es).

In both RF and user domains, this is rarely the case, so design drivers for AACR learning include the detection of noiseless data, of conditions when stationary–ergodic assumptions approximate the <Scene/>, and of changes in the underlying processes such as statistical inconsistencies.

4.2.3 Histogram in the Ontology

For the augmented histogram to be available to an iCR, it is described to the <Self/> via ontological primitives, expressing the capabilities of the algorithm for autonomous use.

<Histogram/> may be described in terms of <Function> Finds largest and smallest elements of a vector, counting elements in a uniform range from smallest to largest and identifying <Interesting/> features. </Function>, summarized in RXML as the satisfaction of the <Goal/> to <Discover/> (Expression 4-1).

Expression 4-1 Histogram as a Goal-Satisfaction Function

<Histogram> <Domain> <Vector> Y <Index> [1 . . . N] </Index> </Vector> </Domain>

<Range> {<Interesting/> <Uninteresting/>) </Range>
<Goal> <Discover> Y </Discover> </Goal>
</Histogram>

The iCR recognizes, asserts, satisfies, and introspects over <Goals/>, including to <Discover/> <Interesting/> things to enhance QoI without knowing exactly what discovery may ensue. The CRA includes the principle that if something is novel, then <Self/> should try to <Discover/> it.

Expression 4-2 Novelty Implies Discovery

<If> <Novel> Y </Novel>
<Then> <Goal> <Discover> Y </Discover> </Goal> </Then>
</If>

Relating <Histogram/> to <Discovery/> mechanizes bottom-up knowledge bootstrapping. <Discovery/> implies related goals to <Validate/> new knowledge through <Reinforcement/> from <Use/>.

As an algorithm, <Histogram/> consumes resources as a function of both input space $|Y| = N$ and analytic complexity K, the number of bins in $H(\ )$, so one CRA design rule specifies that such algorithms be described at least in terms of finite, tightly bounded <Memory/> and <Processing/> resources measured in <MIPS/> (millions of instructions per second). The algorithm, goal alignment, and resource model establish <Histogram/> as an ontological primitive AML component via the CRA.

4.3 USER-DOMAIN LEARNING

Suppose a user who is "planning a trip" stores the "Map of the World" of Figure 4-4 on his iCR. The "Map" is novel, so the iCR has a <Goal/> to <Discover/> the "Map" to assist in "planning a trip." Without supervision <Histogram/> detects that "Map" has exactly two intensity states <Light/> and <Dark/> and is <Interesting/>. The light bits are "Land" and the dark bits are "Ocean."

The iCR pursues the <Histogram/> discovery that the "Map" is <Interesting/> via supervised learning as illustrated in Figure 4-4. Since "Map" has just two colors, the histogram of intensity for number of bins $K = 10$ yields a bipolar histogram where the light values are counted in the Y_{min} bin and the dark values in the Y_{max} bin, while eight bins between are empty. The figure shows this statistical structure by the two vertical bars and center y-axis. The large number of occurrences of a few discrete values concentrated in a few bins expresses high likelihood of learning compared to $H(\ )$ where two bins are only slightly above the mean. <Histogram/> could compute degree of

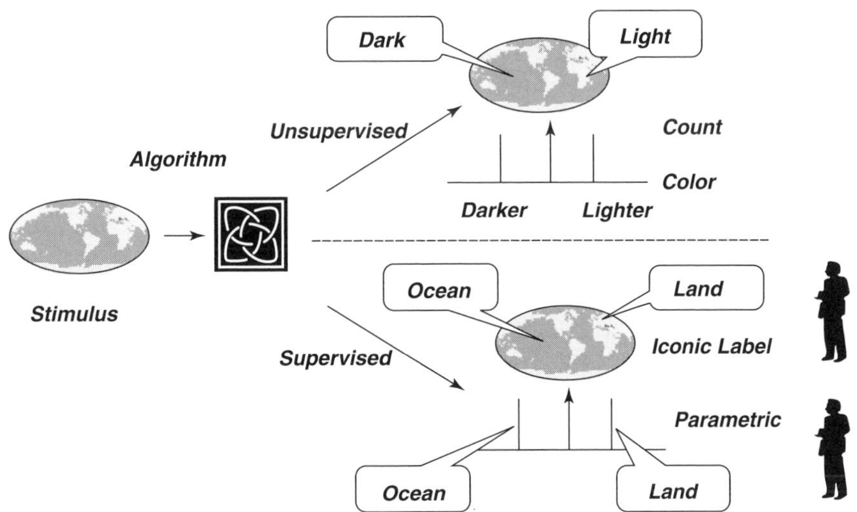

FIGURE 4-4 Unsupervised and supervised learning with iconic or parametric labeling.

interestingness of $H_{\max}$, the bin with largest count, and thus of Y as the peak to average ratio, $H_{\max}/((\Sigma_K H_i)/K) = K*H_{\max}/N$.

4.3.1 Autonomous Detection of Two Colors

Unsupervised ML employs general discovery methods [98, 321] with domain-independent features [321]. The <Histogram/> test for <Interesting/> begins the domain-independent labeling process. Suppose the map of Figure 4-4 has a million pixels. <Histogram/> discovers that 400,000 pixels are light blue while 600,000 are dark blue and none are any other color. It bifurcates "Map" into two classes, <Min-bin/> and <Max-bin/> or <Light/> and <Dark/> noting this discovery ontologically in Expression 4-3.

Expression 4-3 <Histogram/> Discovers <Interesting/> Intensity

<Histogram> <Novel> "Map of the World" <Index> [1 . . . 1000000]
</Index> </Novel> . . .
<Goal> <Discover> "Map of the World" </Discover> </Goal>
<Discovery> <Interesting> "Map of the World" </Interesting>
<Interesting> <Min-bin/> </Interesting>
<Interesting> <Max-bin/> </Interesting>
</Discovery> </Result> </Histogram>

<Histogram/> bifurcation dichotomizes the data [99] into two positive classes. Implicitly, there are negative classes, the conspicuous absence of pixels between the peaks. Strong negative classes imply an absence of noise, potentially <Interesting/> by itself, but not developed further in this introductory treatment.

4.3.2 Sensitivity of the Alpha Learning Parameter

Y is <Interesting/> if at least one of the H_i exceeds some fraction Alpha of the number of items expected in a given bin if the data were distributed uniformly over $[Y_{max}, Y_{min}]$. Methods for defining Alpha depend on the structure of random processes in Y. If the problem space is ergodic, stationary, or cyclostationary then probability distributions yield consistent statistics (e.g., mean, variance, Nth order moments). Even if the problem space is not perfect, finite-scope statistics model many problems accurately for some duration (a quasistationary interval), or for limited purposes (e.g., channel noise covariance). Hypothesis tests (e.g., chi-squared, Student's-t, gamma confidence intervals) inform Alpha for consistent random processes [100]. For introductory purposes, Alpha of 1.5 to 2 make Y <Interesting/>.

Alpha set too low categorizes uninteresting noise as <Interesting/>, but set too high misses something truly <Interesting/>. Lenat [320] reported similar problems with AM: with the interestingness thresholds too low, AM was lost in combinatorial explosion of apparently pointless hypotheses; with it too high, AM spent most of its resources on a few boring cases, rarely generating anything very interesting. Set just right, AM identified as interesting data structures reminiscent of Peano's axioms of arithmetic.

AACR therefore needs Alpha-parameter control loops such as <User/> interaction. With low Alpha, <Users/> tire of too many pointless questions. With Alpha too high, there are no user complaints, but little user-specific service customization. In a (risky) metacontrol law, iCRs adjust Alpha interactively for regular QoI customization but few complaints.

Alpha may adjust to an estimated noise covariance to ignore a fixed fraction of noise alarms for a constant false alarm rate (CFAR). The gamma distribution computes a threshold T such that for a specified distribution the probability of exceeding threshold T equals a specified rate, for example, 0.02 for 98% probability of $t < T$ [101]. Alpha equal to T yields a learning alert for statistically infrequent but meaningful AML opportunities.

4.3.3 Sensitivity of the *K* Learning Parameters

In the color histogram, K equal to 10 shapes the learning experience. Too many bins yields too few items per bin, so many samples are needed to populate the bins sufficiently for Alpha occurrences to be statistically represented if present. For example, if there are 200 items in the population (say, 200 countries in the world), then a uniform distribution with $K = 10$ puts 20 items in each bin. For Alpha of 2, greater than 40 countries must be in a bin for nonuniformity of potential interest. Increasing K increases the sensitivity of $H(\ )$ for fixed Alpha. This same change of K from 10 to 20 also reduces noise immunity, increasing the probability of <Interesting/> labels on less interesting features of the population.

Metalevel variation of K may determine the sensitivity of <Interestingness/> to K, for example, in sleep epochs or in CWNs. For the colored map of Figure 4-4, there is no difference in the number of bins declared <Interesting/> with as few as 3 bins and an Alpha of < 4/3.333 or with 1000 bins and any Alpha up to (400,000/1000 =) 400. Such insensitivity of <Interestingness/> parameters indicates both relatively noise-free data and strong learnable features. High sensitivity to learning parameters Alpha and K indicates subtle structure not easy to discover either autonomously or with mixed initiative.

4.3.4 Nonuniform a Priori Distributions

Alternatively, the values of Y may be distributed in random intervals, so $P(Y)$ is a Poisson distribution in the number of items per interval (e.g., like a queue). An interestingness test for Poisson distributions uses $N*$Poisson (kT) as the expected count in the bins appropriately normalized. The chi-squared test estimates the accuracy of the Poisson hypothesis:

$$\chi^2 = \sum_{k=1}^{K} (p_k - H_k(Y))^2 / H_k(Y) \qquad (4\text{-}2)$$

The chi-squared parameter χ^2 estimates the degree to which $H(Y)$ conforms to $P(\ )$ for K class intervals. An interestingness criteria Alpha$*\chi^2$ estimates <Interestingness/> with respect to the Poisson hypothesis. Fractals, for example, have heavy tails that do not conform to $P(\ )$. If bins exceed $(N/K)*E(H_k)*$Alpha, then $H(\ )$ is locally inconsistent with a Poisson distribution. The inhomogeneous Poisson (IHP) test maximizes the probability of a Jth order Poisson process [102]. A priori models of other sample processes yield a priori distributions, <Priors/>, that are otherwise distributed: normal, exponential, binomial, and so on. Historically, only scientists discovered priors and applicable contexts but GAs now sometimes yield discovery comparable to human experts with little human intervention [103, 104].

As mentioned earlier, Lenat's AM [320] was among the first to use <Interestingness/> to guide AML. AM and its successors Eurisko and CYC showed that AML is less viable as the conceptual distance from initially known data structures increases. Thus, the CRA emphasizes <User/> and CWN validation of incremental AML so AACRs operate on or near data points verified by an authority. Unsupervised AML alerts are opportunities for the interactive validation, clarification, or amplification of the <Interest/> item via mixed-initiative interaction.

4.3.5 Labels From the Supervisor

Since the "Map of the World" is <Interesting/> consistently over K and Alpha, the iCR asks the <User/> to act as AML supervisor, validating the finding and adjusting the labels. The iCR asks and the user labels the light blue "Land" and dark blue "Ocean." These labels establish links to <Land/> and <Ocean/> known to the <Self/> through the CRA's abstractions <Universe> <Earth> <Land/> <Ocean/> </Earth> </Universe>. Parametric labeling in the supervised ML research literature typically happens before the fact, with data offered to the ML algorithm in vectors (Index, Class, Value), such as (1, Land, 234434) or (4, Ocean, 343568).

In this case, the system initiated the label acquisition dialog that updates the internalized world model (per Expression 4-4).

Expression 4-4 Histogram Aggregates User Labels

<Histogram> <Domain> "Map of the World" [1 . . . 1000000]
</Domain> . . .
<Goal> <Discover> "Map of the World" <Discover/>
<Progress> "Map of the World"
<Interesting> <Intensity>
<Label> Y[1] <Light/> "Land" <Source> <User/> <Scene/>
</Source> </Label>
<Label> Y[10] <Dark/> "Ocean" <Source> <User/> <Scene/>
</Source> </Label>
</Intensity> </Interesting> </Progress> </Goal> </Histogram>

Internalization refines the autonomously ascribed properties <Light/> with the <User/> <Label/> "Land" that readily maps to the CRA ontological primitive <Land/>. Another <Source/> may offer another <Label/> and <Label/> may be time or location dependent. Verbose XML tags accurately internalize the dialog. Extracting such information by asking the right questions in the right way is an art form.

4.3.6 Operational Ontology

The operational ontology interates discovery with QoI enhancement. Consider Figure 4-5. Instead of asking the user to label two colors out of context, this graphical dialog points to regions corresponding to an <Interesting/> value labeling the centroid of the largest contiguous collection of <Dark/> pixels with a question intended to invoke one label.

If the user says, "It's the Pacific Ocean," all the <Dark/> gets <Label> Pacific Ocean </Label>. With minimal spatial reasoning, the iCR could ask of the next largest clump of dark blue, "This too?" The user now says, "No, that's the Atlantic Ocean." How can an iCR navigate through such semantics?

FIGURE 4-5 Semantic labeling request may seem more natural.

Dialog systems for extremely limited domains like 800 directory assistance are well in hand, but informal dialog navigation is on the cutting edge of language research [105].

Expression 4-5 A Priori Ontology Assists AACR in Focusing Questions and Interpreting Responses

<Universe> <Stars> <Constellations> <Milky Way/> </Constellations>
</Stars>
<Solar-system> <Sun/> <Planets> <Mercury/> <Venus/>
<Earth> <Land> <Continents> <Asia/> . . . </Continents> </Land>
<Oceans> <Antarctic/> <Atlantic/> <Pacific/> <Indian/> <Arctic/>
</Ocean> </Earth>
<Mars/> <Asteroids/> <Jupiter/> <Saturn/> <Uranus/> <Neptune/>
<Pluto/> </Planets>
</Milky Way> </Universe>

The a priori ontology of Expression 4-5 establishes <Ocean/> at a higher level of abstraction than <Ocean> <Pacific/> </Ocean> to match the speech segment "Pacific Ocean." Confronted with such an answer and aided by Expression 4-5, the AACR traverses the ontology to synthesize the clarifying question: "Is all the dark blue the Pacific Ocean? Please answer yes or no." The user answers "no." So iCR replies, "OK, then the dark blue covers all the world's oceans, correct?" By now, the astute radio engineer observes that such dialogs are tantamount to a message sequence chart (MSC) in a protocol stack (Table 4-2).

The MSC of Table 4-2 shows the dialog where the user functions as an "eighth layer" of the seven-layer ISO protocol stack. As protocols accommodate errors, so must dialogs, mediated in AACR by a mix of natural language voice, text, and/or pushing buttons. Variability of natural language argues that AACRs shape such dialogs toward simple negative (<No/>) and positive (<Yes/>) responses to patterns of ontological <Primitives/>. After a successful dialog, the iCR projects from its own GPS coordinate system to the user's "Map of the World" to further plan the trip on the <User/>'s <Map/> enhancing QoI. If the dialog fails, the iCR recovers with "Sorry, I was trying to learn

TABLE 4-2 Dialog as MSC: The User as the Eighth Layer of the Protocol Stack

<Self> <Goals>	<Message/> <Self/>-to-<User/> ⇒; Reply ⇐	<User> (thinks)
<Label><Dark/>	⇒ "What is all this stuff? (Pointer in <Dark/>)"	Listens/Says
<Listen/>	⇐ "<Pacific/> <Ocean/>"	
<Label> <Ocean/>	⇒ "Is all the <Dark/> blue the <Pacific/>? Please <Respond/> <Yes/> or <No/>."	(1 - oops) (2-dumb PDA)
<Listen/>	⇐ "<No/>"	Continue 1
1 <Verify/>	⇒ "OK, then the <Dark/> blue covers <All/> the <Earth/>'s <Ocean/>s, correct?"	1 (dumb PDA)
1 <Pop-goal>	⇐ "<Yes/>"	1 END

Note: AACR supplies ontological <Primitives/> in <Message/> to a dialog generator and the speech system recognizes them to select among predefined AACR templates for <Labeling/>.

about your Map of the World to help you, but I can't understand so let's just drop it for now." In other dialog failures, the AACR may ask the user to "take control of the radio" while the AACR acquires a training experience and learns.

The well-known set-theoretic idea [106], "Occam's Razor," underlies the <Histogram/> labeling strategy. The iCR searches for the *smallest set* of <Primitives/> known to <Self/> in an a priori ontology that *covers* the discovery. Since the ontology of Expression 4-5 bifurcates <Earth/> into <Land/> and <Ocean/> and <Histogram/> bifurcates "World" into <Dark/> and <Light/>, the alignment of <Earth/> with "World" covers both subsets. The <User/> response "Ocean" aligns <Dark/> to <Ocean/> and "Land" verifies the set cover. Such use of ontology operationalizes taxonomy and spatial containment, the fundamental set–subset relationships of spatial entities. The iCR reasoning about the subsets guides the <User/> to mutually grounded spatial abstractions.

Operational ontology structures both a priori and current world knowledge into a comprehensive computational taxonomy, a Dewey decimal system, agreed to among international standards bodies like the International Telecommunication Union (ITU) and World Radio Conference (WRC). Lenat's CYC [325] is comprehensive, but comparing CYC's informal models of radio to professional models like the GSM MOU Z.100 SDL models of GSM underscores the futility of a single universal taxonomy, far short of ontology. Moving toward universality, however, the semantic web community adopted OWL [84, 85] as a core semantics standard, defining web resources, fundamental abstractions like "Thing," and so on. The IEEE Standard Upper Ontology (SUO) held promise for abstractions like <Universe/> in SUO concepts like

TABLE 4-3 Comparing OWL, Radio XML, and Conversational Expressions of Knowledge

Conversational	OWL	Radio XML
Informal, unknown alignment of semantics among user, <User/>, <RF/> and use cases	Formal Web-based RDF, RDFS, DAML, and OIL semantics	Hierarchy of semantic <Primitives/> with text and graphical definitions
Most concise; uses contextual disambiguation; problematic if contexts are not accurately resolved	Most precise formal logic and set-theoretic treatment for iCR	Hierarchical; language avoids anaphora; may be somewhat circular
Fluid and natural for user interfaces	Readable by an expert with software tools	Readable with practice without software tools

#Physical-Universe$, but OWL's leverage of the Web-based Resource Definition Framework (RDF) RDF-Schema (RDFS), DAML, and OIL has greater expressive power and a growing community of users and research tools. OWL Communities of Interest (COI) define their own domain-specific ontologies resolvable against other COIs by interontology mappings [107].

In spite of these challenges, this book develops the AACR-specific operational ontology <Self/> in RXML, in simplified syntax but not inconsistent with OWL. To reduce tedium this book uses a form between OWL and conversational speech called Radio XML format (Table 4-3).

From <Histogram/> and dialog, the iCR knows about <Ocean/> and <Land/> on the <User's/> "map of the world." Mutual agreement that the <Dark/> regions are "oceans" must be fully reflected in the <Self/>'s ontological structures to complete the *grounding* of the user's term "ocean" to the iCR ontological <Primitive/> <Ocean/>, to invoke existing knowledge about <Oceans/>. The internalization of the AML dialog above *reinforces* via reinforcement learning (RL) [92] the discovery by <Histogram/> of two <Interesting/> classes. Fully grounded, the iCR can project cities of travel onto the <User/> "map"; identify wireless services during travel; and access travel and entertainment via existing travel-assistant skills for this <Trip/>. Knowledge thus internalized enhances QoI. Internalization of the <Histogram/> dialog associates prior knowledge of the <Earth/> to the "map" artifact of the discovery (Expresssion 4-6).

Expression 4-6 <Earth/> Internalizes AML of "Map"

```
<Self> . . . <Universe> <Milky Way> . . .
  <Earth> <Alignment>
  <Domain> <Earth> <Land/> <Ocean/> </Earth> </Domain>
```

<Range> <Earth> "Map of the World .jpg"
<Land> <Intensity> 234434 </Intensity>
<Source> <Histogram/> <Context/> </Source> </Land>
<Ocean> <Intensity> 343568 </Intensity>
<Source> <Histogram/> <Context/> </Source> </Ocean> </Range>
</Alignment> </Earth> . . . </Milky Way> <Constellations/>
</Universe> . . . </Self>

This mapping requires <Domain/> and <Range/> specifications for a correspondence that is 1:1, ONTO, and invertible. In this <Alignment/> the <Source/> is the <Histogram/> with <Context/> pointer to the dialog trace for introspection should future <Sources/> conflict with this internalization. <Histogram/> also internalizes success (see Expression 4-7).

Expression 4-7 <Histogram/> Internalizes the AML-Dialog Success

<Self> . . . <Histogram> <Domain> "Map of the World" [1 . . . 1000000]
</Domain> . . .
<Goal> <Discover/> <Label> <Alignment>
<Domain> "Map of the World. jpg" </Domain>
<Range> <Earth/> </Range>
<Domain> "Map of the World. jpg" <Intensity> 234434 </Intensity>
</Domain>
<Light> <Label = "Land" <Source = <Owner/> <Context/>/>
</Light>
<Range> <Land/> </Range>
<Domain> "Map of the World. jpg" <Intensity> 343568 </Intensity>
</Domain>
<Dark> <Label = "Ocean" <Source = <Owner/> <Context/>/>
</Dark>
<Range> <Ocean/> </Range> </Alignment> </Label> </Goal> . . .
</Histogram> . . . </Self>

There are many types of <Land/>, such as, for example, in the <RF/> domain of space communications, where the semantic primitive <Land/> refers to the ground station, antennas, and mobile handsets on <Earth/>. The above internalizes <Universe> . . . <Earth> <Land/> . . . while satellite communications expresses <RF> . . . <SATCOM> <Land/> . . . </RF>. Ship-to-shore radio links internalize land as <RF> . . . <Maritime><Land/> . . . </RF>. Thus, the terminal ontological primitive <Land/> does not stand alone, but with context strings <RF/>, <SATCOM/>, <Maritime/>, and so on. The results of the histogram are no longer <Interesting/> as such because <Label/> accomplishes the <Goal/> that <Interesting/> initiated. They may be relabeled <Informative/> if QoI enhancement is reinforced through subsequent

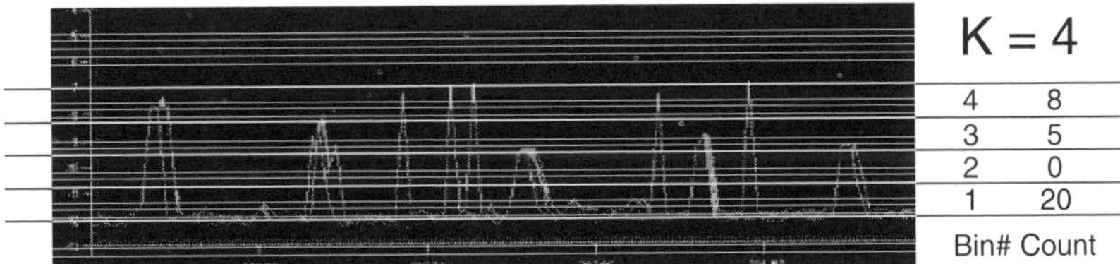

FIGURE 4-6 PSD histogram in UHF (notional counts). (Courtesy of Bruce Fette, General Dynamics, Fall Church, VA. Used with permission.)

events. Operational ontology thus achieves the systematic evolution of ontology by context-sensitive assimilation of experience.

4.4 RADIO-DOMAIN LEARNING

The <Histogram/> readily discovers <Interesting/> features in the RF domain. It characterizes spectrum occupancy from signal strength estimated from power spectral density (PSD). With space–time grounding and autonomous modeling of the radio propagation, iCR signal discovery enables spectrum-use etiquettes. It may ask a spectrum <Authority/> like the FCC or a home CWN about the <RF/> domain. The replies may be generated by human experts or algorithms.

Suppose an AACR has little <RF/> domain understanding, but can compute the PSD via the fast Fourier transform (FFT). Since PSD represents energy, PSD-<Histogram/> of midbands that are neither as crowded as HF nor as sparse as EHF cluster into two groups: (1) a large number of relatively low power noise samples with low power interference and (2) a smaller number of relatively high power adjacent channel signals or co-channel interference (Figure 4-6). The number of items in each of four PSD amplitude bins of <Histogram/> H(4; PSD) is shown to the right in the figure.

With $K = 4$ and Alpha$*(N/K) = 7$, the Fette data [155] from the FCC use case fills up the low bin and the high bin with statistically significant numbers of hits compared to the middle bins. Both low power and high power bins are <Interesting/>.

4.4.1 Unwrapping the RF Histogram

Pointers from the PSD to the raw data may identify data points contributing to bin H_k. Define <Histogram*> $H^*(K;$ PSD) </Histogram*>with an audit trail linking H_k to PSD$_i$ so that from H_k the iCR can retrieve just those PSD points contributing to that bin. H^* could use a companion matrix $H^*[1 \ldots K]$ to PSD $[1 \ldots N]$ such that $H^*[i] - k$ for PSD$[i]$ in H_k. The retrieval of the raw PSD data points "unwraps" the bin.

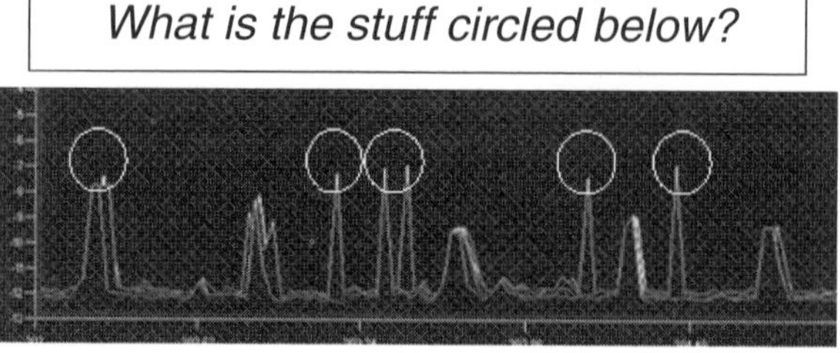

FIGURE 4-7 Iconic labels of the PSD mediate supervised ML.

The AML dialog unwraps <Interesting/> <Histogram*/> bins to identify the signal peaks and ask "What does this stuff mean?" (Figure 4-7). There is little a priori semantic grounding in the question posed. The <Authority/> knows about PSD. Does the iCR know that these items are <Signals/> versus <Noise/>? The <Authority/> could answer "Peaks," "Signals," or something else. Answers like "Strong signal" help iCRs to honor radio etiquette. Alternatives include "interference," "push-to-talk signals," and "FM signals." Better semantic alignment promotes knowledge transfer. <Histogram*/> identifies entropy-related opportunities to acquire knowledge, and semantic alignment creates the knowledge bridge <Label/>ing these <Signals/> as such for flexible spectrum etiquette.

4.4.2 Noise Versus Signal

The low-intensity PSD values (in Figure 4-7) contribute a large number of counts in the lowest <Histogram*/> bin H^* [1]. Suppose the <Authority/> classifies these as "Noise." For AML, the <Label/> should <Align/> with the conceptual primitive <Noise/> in the iCR's a priori ontology <Self/>. That ontology also must describe signals and interference to bootstrap spectrum occupancy knowledge for spectrum etiquette.

Expression 4-8 Naïve RF Ontology: RF Consists of Signals and Noise

```
<RF>
  <Signal> <Definition> Information-bearing electromagnetic (EM)
    wave
    </Definition> </Signal>
  <Noise> <Definition> Non information-bearing electromagnetic (EM)
    wave
    </Definition> </Noise>
</RF>
```

Expression 4-8 divides the RF-domain into signals and noise, both forms of electromagnetic (EM) wave propagation. Such definitions suffice for basic coursework and simpler ontologies like open CYC™ but fall short of AACR needs. Noise includes Brownian motion [108], a thermal property of materials. The low noise amplifier (LNA) establishes a noise temperature for the rest of the radio, but the LNA is an electronics component, not a radio wave. For reasoning about noise sources, iCRs need richer definitions. Dictionary definitions illuminate the uninformed but also fall short, such as the following from the *Merriam–Webster Collegiate Dictionary*:

> **noise** **1**: loud, confused, or senseless shouting or outcry **2 a**: sound; *esp*: one that lacks agreeable musical quality or is noticeably unpleasant **b**: any sound that is undesired or interferes with one's hearing of something **c**: an unwanted signal or a disturbance (as static or a variation of voltage) in an electronic device or instrument (as radio or television); *broadly*: a disturbance interfering with the operation of a usually mechanical device or system **d**: electromagnetic radiation (as light or radio waves) that is composed of several frequencies and that involves random changes in frequency or amplitude **e**: irrelevant or meaningless data or output occurring along with desired information **3**: common talk : rumor; *esp*: slander **4** : something that attracts attention "the play T will make little noise in the world—Brendan Gill" **5** : something spoken or uttered

Definition 2 conveys an intuition for noise, but it mixes the concepts of noise and interference without differentiating them sufficiently for AACR.

The <RF/> ontology of Expression 4-9 redefines <Noise/> as a measurable quantity that fluctuates because of Brownian motion.

Expression 4-9 RF Ontology: RF Includes Energy That Consists of Either Noise or Signal

<RF>. . .<Energy>
<Noise> <Definition> Brownian random fluctuations of a measurable quantity
</Definition> </Noise>
<Signal> <Definition> Information-bearing fluctuations of a measurable quantity
</Definition> </Signal>
</Energy></RF>

Here, <Noise/> is a measurable process changing with time ("fluctuations") but providing no information. <Signal/>, on the other hand, provides <Information/>. Tailoring <Information/> to a specific <User/> via <RF/> is a canonical <Goal/> supportable via these definitions. From Brownian motion, iCR may relate Boltzmann's equation for temperature, bandwidth, and noise power (Equation 4-3) to the thermal noise power of an electrical circuit:

$$p = kTB \tag{4-3}$$

T is absolute temperature in kelvin units (K); B is bandwidth in hertz (Hz); and Boltzmann's constant k is 1.380658E-23 joules/K. Since 1 watt is 1 joule/second and hertz are inverse seconds, the units of power are watts. In radio engineering, if T_0 is 290 K, then kT_0 is −172.977 dBm/Hz, or about −170 dBm per kHz of thermal noise power.

Expression 4-10, an operational ontology, augments the textual definitions with equations to compute expected <Noise/> power. With an ability to measure <Energy/> in the circuits of the <Self/>, a noise-aware AACR can inspect <Energy/> sources—including RF—for <Noise/> power.

Expression 4-10 Noise Semantics Internalized Via the Boltzmann Equation and Histogram

```
<RF>. . .<Energy>
<Noise> <Definition>Brownian random fluctuations bearing no
  <Information/> </Definition>
  <Power> <Estimate><*>
     <Boltzmann> <k> 1.380658E-23 </k> </Boltzmann>
     <Temperature><Standard> 290 <Units> K </Units> </Standard>
        </Temperature>
     <Bandwidth> <W> ? </W> </Bandwidth> </*> </Estimate>
        </Power>
  <Detection> <Domain> <PSD> Y </PSD> </Domain>
     <Histogram*> Y <NOT> <Interesting/> </NOT> </Histogram>
     </Detection></Noise>
<Signal> Information-bearing fluctuations of a measurable quantity
  </Signal>
</Energy></RF>
```

A PSD without an <Interesting/> <Histogram/> probably does not contain <Signal/> because it lacks sufficient statistical differentiation to convey <Information/>. Some apparently-random signals may be transformed to signals that are clearly differentiated, for example, via an adaptive equalizer or via despreading. There are many ways of differentiating signal from noise, but <Histogram/> illustrates the internalization in the <Self/> of a general-purpose tool autonomously employed to assess a sample sequence for <Information/>.

4.4.3 Signals Bear Information

Venerated models formulate radio communications as information-bearing signals passing through channels from source to sink. Human communications occur between one or more originators and one or more recipients. Cognitive radio enhances SDR by explicitly formulating the originator, the need to communicate, and the intended recipient as <Entities/> and aspects of an enhanced communications model. <Originators/> are sources with intent, so reception should satisfy that intent. Signals satisfy communications needs of <Originators/> and <Recipients/>, while noise and interference limit the QoS hence QoI and thus the degree of satisfaction of those <Needs/>.

Expression 4-11 <Communications/> Includes <Needs/>

<Communications> <Path> <Signal> <Source> <Entity>
<Originator> <Need> <Recipient> <Information/> </Recipient>
</Need>
</Originator> </Entity> </Source>
<Channel/>
<Sink> <Entity> <Recipient> <Information/> </Recipient> </Entity>
</Sink>
</Signal> </Path> </Communications>

Expression 4-11 reads, "Communications is a path for a signal between a source (an entity that needs to send information to a recipient) through a channel to a sink, the entity that is the recipient of the information." <Need/> to convey information creates a clear <Goal/> for iCR: to <Discover/> and <Satisfy/> the <Needs/>. A <Recipient/> is an <Entity/> with a <Receiver/>, which could be the recipient's own eyes or ears. <Recipients/> consist of the <Self/>, the <User/>, and other <Entities/> including nonhumans (e.g., a FCC iCR monitoring for spectrum abuses). For epistemological balance, <Nature/> is the postulated <Entity/> that originates and receives natural <Signals/> like <Lightning/>. If a tree falls in the forest and nobody (human) hears it, it definitely makes a sound, for <Nature/> is the universal listener.

Expression 4-12 <Receiver/> Defined

<Receiver> <Definition> <Thing/> that <Transforms/> <Signal/> to
<Information/>
</Definition> </Receiver>

The functional definition of receiver is that which transforms signals into information. By this definition, a human language translator also is an audio <Receiver/>.

Expression 4-13 <Interference/>

<Interference> <Definition> {<Signal/> <Noise/>} that degrades
<Receiver/> function
</Definition> </Interference>

The functional definition of interference implies explicit measurement of the performance of a <Receiver/> (e.g., data rate) that may be physical, functional, or hypothetical.

Expression 4-14 Generic Model of Communications as Signal in Space

<RF>. . .<Energy> . . . <Noise/> . . .
<Signal> <Definition/> <Power/> <Detection/> </Signal>
<! – Generic Model of Communication –>[1]
<Entity> <Need> <Information> <Source>
<Transmission> <Time> <Space>
<Propagates> <Signal/> <Channel/> <Recipient/>
<Interferent/> </Propagates> </Space> </Time> </Transmission>
</Signal>
</Source></Information> </Need> </Entity> . . . </Energy> . . . </RF>

The RXML expression adds the act of <Transmission/> in space × time that <Propagates/> via a channel to recipients and <Interferents/>, <Entities/> with whom the signal interferes. This refined model explicates the dual role of any communications <Transmission/> as a possible generator of <Interference/>, a relationship over which AACRs must reason for spectrum etiquette.

Expression 4-15 Interference From <Nature/>

<! -- Natural Interference -->
<Entity> <Nature> <Need> Cloud-motion <Information> Lightning
<Source> Static-discharge
<Transmission> RF-EMP <Time> <Space> <Propagates/>
<Signal > High-power-fractal </Signal> <Channel> HF </Channel>
<Recipient> <Nature/> </Recipient> <Interferent/> </Space> </Time>
</Transmission>
</Source> </Information> </Need> </Nature> </Entity>

In Expression 4-15, natural cloud motion creates a need for a static discharge, which generates a broadband electromagnetic pulse (EMP), propagating in space × time through the HF channel to nature as the recipient and to no particular <Interferent/>. The <Information/> content is the fact conveyed by nature that a lightning strike has occurred at a specific time and place.

[1] <! – begins a comment while –> ends that comment. Comments help people but may be read by a NL-capable algorithm. Comments are not normative while the RXML ontological <Primitives/> are normative.

Expression 4-16 GSM in the Generic Communications Model

```
<! -- GSM Voice Communication -->
<Entity> <User/> <Need> <Talk/> <Information> <Voice-call/> <Source>
  RPE-LTP <Transmission> GSM <Time> <Space>
  <Signal> Mobile-IMSI# <RF/> <Time-slot/> </Signal > <Channel>
    N-VA </Channel>
  <Recipient> GSM-BTS </Recipient> Burst-N1-M </Space> </Time>
    </Transmission>
  Call-N2 PSTN#N3 </Source> Call-complete </Information>
  Conversation-complete </Need> <Satisfy/> </Entity>
```

In Expression 4-16, the generic voice-communications model supplies RXML tags to a notional GSM voice call placed in Northern Virginia (N-VA). The tag placement scopes the ontological primitives for the signal-in-space, while the larger transmission includes the Public Switched Telephone Network (PSTN). The completion of the call satisfies the information transfer need of the <Entity/>.

Expression 4-17 Ontological Perspective on Signal and Interference

```
<! -- Detecting Signal in Noisy Communication -->
  <Detect> <Signal/> <Known/> </Detect>
  <Detect> <Domain> Y </Domain>
    <Histogram*> Y <Interesting/>
    </Histogram*></Detect>
. . . </Energy></RF>
```

The <RF/> part of the <Self/> ontology reflects this enhanced model of <Communications/> highlighted in Expressions 4-11 through 4-17. Communications are <Signals/> that satisfy a <Need/> of an <Entity/> to exchange <Information/> with an intended <Recipient/>. An <Interferent/> receives a <Transmission/> in <Time/> and <Space/> that decrements QoI. The AACR as <Originator/> adheres to etiquette if it trades off its <Needs/> against those of <Interferents/> in a way that reflects social norms.

This ontological perspective is illustrated in Figure 4-8 [145]. Any air interface occupies <Space/>–<Time/> and <RF/>. Multiple input multiple output (MIMO) exploits distinct physical paths in the channel (n, kn) via the aggregate channel [109–113]. The <Originator/> chooses a communications path that need not be wireless. The path could pass through WLAN to the <Home/> Internet service provider (<ISP/>) and thence to the desk-side PC of the <Recipient/>. Cellular paths generally pass through the PSTN, while WLAN paths traverse <ISPs/>. Point-to-point UHF links among sports enthusiasts illustrate increasingly rare purely RF paths between <Originator/> and <Recipient/>. Expressions 4-11 to 4-17 and Figure 4-9 thus outline

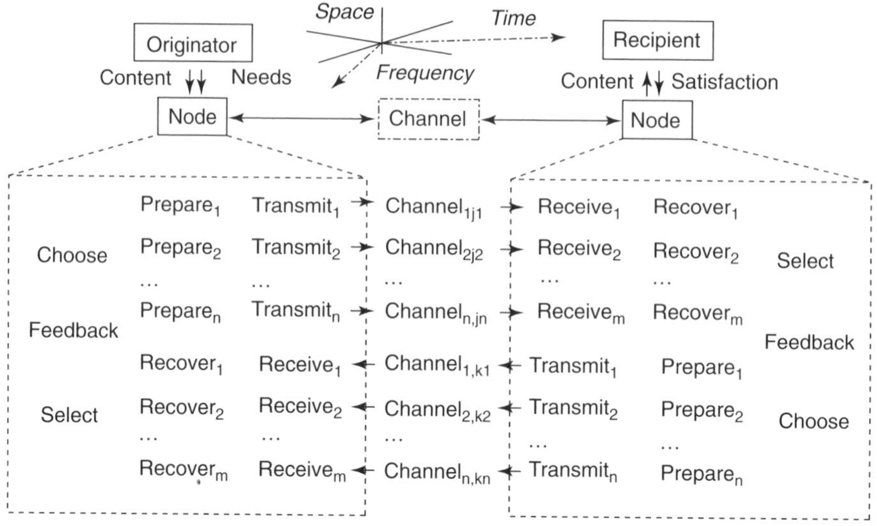

FIGURE 4-8 Communications defined in space–time–RF as satisfaction of originator need.

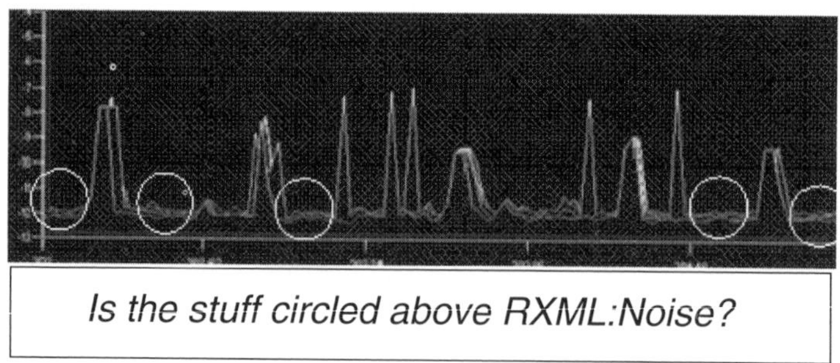

FIGURE 4-9 Query to <Authority> to confirm noise <Label/>.

the principle of signals as means for <Originators/> to share information with <Recipients/> to satisfy a <Need/>, the ontological stance of the CRA.

4.4.4 Understanding Noise and Signals

The ontological treatment of <Noise/>, <Signal/>, and <Interference/> above expands the significance of the fact that <Histogram/> detects <Signal/> as <Interesting/>. The items circled in Figure 4-9 are <Interesting/>, so the <Self/> may infer that $H[1]$ contains <Signal/>, while this is actually <Noise/> learned when the iCR asks an <Authority/> whether bin $H[1]$ is <Signal/>.

Low power signals may contaminate the <Noise/> bin, so an <Authority/> labels $H[1]$ "Noise and low power signals," augmenting the <RF/> ontology with <Power/> as a numeric quantity, <Power> ?Number <Units> dB </Units> </Power>.

The high powered bins designated by <Histogram/> as <Interesting/> are labeled <Signal/>. High powered noise may contaminate them, such as lightning strikes at HF. PSD labeling works well in SHF and above, but not at HF because distant interference and natural noise, especially lightning strikes, dominate the noise floor. Similarly, <Labels/> are not accurate in densely occupied bands (e.g., cellular) in peak traffic loading, heavily occupied with signals and co-channel interference. The iCR could overgeneralize, learning that points with high PSD values are <Signals/> and points with low PSD values are <Noise/>, but that does not work in HF, an overgeneralization. Band, mode, and other constraints appropriately limit the applicability of PSD occupancy knowledge. The companion CD-ROM address many of these in RXML, but an open source industrial strength radio ontology does not yet exist.

With the augmented PSD <Histogram/>, the iCR can learn about signals and noise through experience with a skilled <Authority/> supervising the ML as in Figure 4-10.

A response like <Adjacent-channel/> from the <Authority/> could be met with a request to download additional RXML knowledge about <Adjacent-channel/> signals should the <Goal/> warrant. In UHF, the <Authority/> might designate signals as <TV/>. If the <Self/> is looking for <Sports/> on <TV/>, then it could refine the query by asking an <Authority/> for the <TV/> <Channel/> broadcasting <Program> "The Superbowl" </Program>. While turning CWN into the TV Guide™ may not seem cost-effective, if that is what makes the <User/> happy, then that may be "where the money is."

The PSD <Histogram/> discovery process plus <Labeling/> taught the <Self/> to use <Histogram/> as a signal/noise dichotomizer where the background is primarily thermal noise. Once the <Self/> learns <Signals/> and

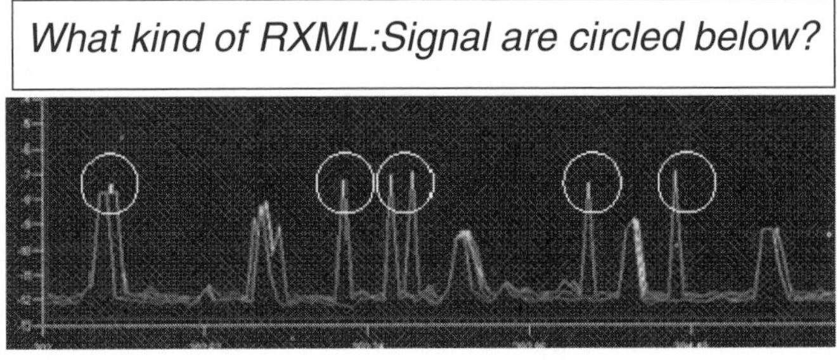

What kind of RXML:Signal are circled below?

FIGURE 4-10 <Histogram/> discovery + RXML knowledge yield informed query.

<Noise/>, it could request an algorithm from the <Authority/> to differentiate signals from noise. The <Authority/> could download a <Squelch/> object in response. A <Self/>-referential <Goal/> to <Minimize/> the computational resources could motivate the query, substituting <Squelch/> for <Histogram/>, replacing AML with an encapsulated skill.

Expression 4-18 To <Optimize/> <Resources/> <Query/> an <Authority/> for a <Skill/>

<Self> <Procedure>
<Definition> "A <List/> of <Goals/> in a <Plan/>" </Definition>
</ Procedure >
<Optimize> <Domain> < Procedure /> </Domain>
<Range> <Skill/> </Range> </Optimize>
<Procedure> <Goal> <Optimize> <Sequence/> </Optimize>
</Goal>
<Procedure> <Query> <Authority/> <Skill> <Resource/> </Skill>
</Query>
</Procedure> </Procedure> </Self>

The iCR downloads <Squelch/>, the optimized computing <Resource/> offered. The CRA incorporates explicit and detailed representations of <Space/>, <Time/>, and <RF/> to associate knowledge with <Scene/> to acquire skills and to share knowledge.

The <Histogram/> discovery process applies to signal phase–space, the complex plane as well as the frequency domain of the PSD. The two classes discoverable by the <Histogram/> would be the two states of a BPSK channel symbol, for example, with zero and π radians the most commonly observed values of the BPSK phase plane. Researchers have published approaches to classifying relatively large collections of such channel symbols [114]. A library of <Signal-type/> functions may be structured into an ontological collection <Signal-types>. <Signal-type/> might express degree of belief in the class. PSD-related recognizers might estimate signal bandwidth and the number of peaks in the spectrum. Although a specialized algorithm may perform better than AML using <Histogram/>, AML will deal with previously unknown cases. Subsequently, an interference recognizer synthesized by a GA from samples of previously unknown interference, could outperform a preprogrammed recognizer [115].

General AML techniques like <Histogram/> raise a strategic question about the level of learning. How much of what should be learned or attempted? How much should be preprogrammed? What should the iCR ask of a cognitive network in a prayer cycle, versus introspection versus interaction? Each of these questions implies multidisciplinary research issues identified and addressed but not yet solved.

4.4.5 The Ontological RF <Histogram/>

The <Self> assimilates <Histogram/> by description as an ontological primitive in terms of other primitives (Expression 4-19).

Expression 4-19 Evolving Operational Ontology of the iCR <Self/>

<Self>
  <Name />
  <Owner> KTH </Owner>
<iCR-platform>
  <RF-environment>
    <RF-capabilities> <Waveforms/>
    RF-sensors </RF-capabilities>
  <RF-knowledge>
  <Spatial-knowledge-DB/> </RF-Knowledge> </RF-environment>
  <User> <Current-user/> <Authorized-user/> <User-situation/> </User>
  <Autonomous-control> … methods …</Autonomous-control>
  <Environment> <Space-time-grounding/> </Environment>
<Histogram> <Domain>Y[1 … N] </Domain> <Memory/>
  <MIPS><Domain>Y[1 … N] </Domain> 38*N </MIPS>
  <Goal> <Scene> <User>/> <RF>/>Learn(Y) </Goal>
  <Reinforcement> <Positive>
  <Learn> <Domain> <User>/> <Scene/> "Map-of-the-World .jpg"
    </Domain>
  <Align> "Ocean" <Earth> <Ocean/> </Earth> </Align>
  <Align> "Land" <Earth> <Land/> </Earth> </Align> </Learn>
  <Learn> <Domain/> <RF/> <Scene/> "PSD" </Domain>
  <Align> "Signal" <RF> <Signal/> </RF> </Align>
  <Align> "Noise" <RF> <Noise/> </RF> </Align>
  <Align> <Histogram> <PSD> <Signal/> <Noise/> </PSD>
    </Histogram>
  <Detect> <Signal/> </Detect> </Align> </Learn>
  </Positive> <Negative/> </Reinforcement> </Histogram>
</iCR-platform>
</Self>

This internalization shows that <Histogram/> learned the semantic alignment of <User/> speech segments "Ocean" and "Earth" to prior <Self/> ontological primitives. "Ocean" refers to <User> <Speech> <wav> "Ocean" </wav> </Speech> </User>. <Histogram/> positive learning experiences are internalized in terms of the <Domain/> of learning in case the knowledge applies only to that <Scene/> or place and time.

The <Histogram/> with the related <Interestingness/> detector is a general skill. From the ability to acquire <User/>, knowledge of oceans and land, it

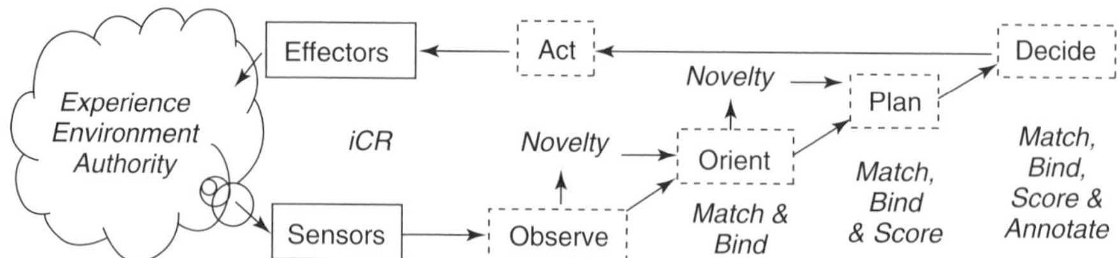

FIGURE 4-11 Reinforcement learning among iCR components.

supports <User>-domain skills. From the ability to learn about <Signals/> and <Noise/>, it supports the <RF> domain. Learning also is a property of the <iCR-platform/>, a top-level capability of the <Self>, where <Histogram> constitutes a general capability to <Learn/>. <Alignment/> defines equivalence classes for reactive responses, deliberative planning, and CBR. The RXML above expresses dynamic knowledge as well as repository knowledge. Real-time performance requires efficient application of these associations.

4.5 REINFORCEMENT, EXTENSION, AND CONSTRAINT DISCOVERY

This section further develops AML for AACR with methods for refinement, extension, and validation of discoveries.

4.5.1 Reinforcement Learning

The many types of reinforcement learning modify existing behavior by suppressing behavior that yields negative reinforcement and facilitating behavior that yields positive reinforcement. Figure 4-11 shows a flow of reinforcement learning among iCR cognition components.

The iCR acquires experience through its sensors. In the CRA, the iCR remembers everything, constantly comparing new sensory stimuli to prior experience, identifying new stimuli, sensory primitives, and stimulus sequences via *novelty* detection. Hierarchical novelty detection realizes a hierarchical multidimensional novelty vector (a tensor) of newness of current experience. To perceive positive and negative reinforcements, the iCR recognizes and isolates from the <Scene/> specific cues to actions, perceiving reinforcement via matching, binding, scoring, and annotation. *Matching* aligns current stimuli (sensory stimuli, perceived objects, and related abstractions) with stimulus memory. *Binding* associates specific stimuli in the <Scene/> with related internalized stimulus–experience-response sets that are abstractions of prior scenes. When identical items (stimuli or responses) are bound in a scene, they form conceptual anchors ("Islands of understanding" [116]). Dissimilar items may participate in a variable–value relationship where one acts as a label for the other (for corresponding but dissimilar-match binding). If a specific action is accurately matched and bound to the associated feedback, *scoring* yields reinforcement learning (RL). Successful reinforcement autonomously *annotated* (symbolically or with additional types of scores) facilitates metalevel retrieval, adaptation, and bootstrapped learning. The retrieval, binding, using, and scoring of relevant experience generalizes CBR [117] for the CRA.

Established reinforcement learning (RL) algorithms produce an adaptation policy from actual or simulated experiences. RL methods include time difference (TD) [92], dynamic programming (DP) [118], Monte Carlo methods [92], and Q-learning [119]. All RL methods associate rewards, values, or quality with a state or state–action pair, creating a policy that specifies a preferred action for each possible state. For example, Q-learning estimates the quality of candidate action **a** in state **s** via **Q**(**s**, **a**):

$$\mathbf{Q}_{k+1} = (1-\gamma(\mathbf{s},\mathbf{a}))\mathbf{Q}_k(\mathbf{s},\mathbf{a}) + \gamma(\mathbf{s},\mathbf{a})\mathbf{f}(\mathbf{s},\mathbf{a},(\max|_b Q_k(\mathbf{s},\mathbf{b}))) \quad (4\text{-}4)$$

The weight (1 − γ(**s**, **a**)) determines the degree of exploitation of current knowledge, while **f** shapes the search for new knowledge at the rate γ(**s**, **a**). As the number of iterations approaches infinity, Q approaches the optimal dynamic-programming policy with probability 1 [120]. Q-learning applies to call-admissions control in simulated networks [121] and robot control [122], among others. Call-admissions control has well-known mathematical structure and the possible states and actions are known in advance. Similar algorithms for such well structured domains appear in state–space automatic control [123], fixed-point maps [124], Kuhn–Tucker optimality, and GAs [74]. These methods may not apply readily to open domains like <User/>-specific jargon or a change in daily commuting pattern. Open domains are relatively unstructured, constantly admitting novelty, and thus somewhat out of reach of classical RL, automatic control, and optimization. In the AACR <User/> domain, the primary measure of goodness is whether the inconsistent and fickle user thinks the CR is "good" or not. Computational ontologies and structured dialogs assist in adapting RL to AACR. The technology is brittle, so the architecture must accommodate apparent suboptimality and contradiction, tracking in space–time the <User's/> changing needs and QoI patterns. Such AML applications are embryonic, so the text characterizes candidate technologies, approaches, and research issues, not pretending to offer closed-form solutions.

The plan–decide–act components of iCR in Figure 4-12 naturally align to RL, and the CRA does not preclude classical RL methods. However, although the properties of RL suffice for closed domains like games [125] and avoiding undersea mines [53], they are not well understood for open domains. For example, Q-learning often falls off cliffs in the cliff-walking problem using greedy methods to discover penalties associated with moves. One AACR

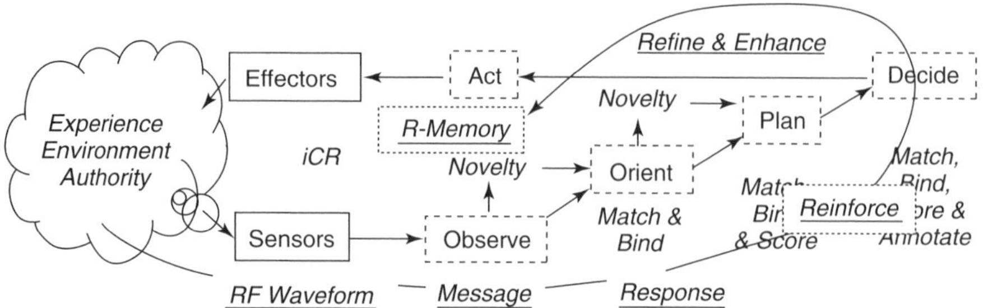

FIGURE 4-12 Formal messages are interpreted in the CRA.

equivalent of falling off a cliff is the violation of regulatory policy, so the FCC may not appreciate classical Q-learning. Other methods like SARSA [92] that do not violate policy converge less rapidly, perhaps causing user frustration over the time spent learning simple preferences. Therefore, the sequel tailors classical RL methods to iCR towards AML without unacceptable penalties in <User/> acceptance or <RF/> regulatory viability.

If the iCR structures its behavior to obtain feedback, then it is engaging in *active* reinforcement learning. Game playing programs that look ahead to reachable states from a given board state actively learn the benefit of the situation–action pair. Otherwise, a RL algorithm may learn the relationship between action and benefit without modeling the environment, instead using the actual environment *as* the model, for *passive* RL. Formal, informal, and CBR feedback assists RL in <User/> and <RF/> domains to develop, refine, and apply acquired knowledge.

4.5.2 Formal Reinforcement

One may envision in Figure 4-12 a path of reinforcement that minimizes learning errors. Formal reinforcement employs formal language for reinforcement from a validating authority. Formal languages avoid error sources of natural language (e.g., ambiguity) particularly with computational complexity of Chomsky's Level 2, context-free language parsed by push-down automata (PDA), or Level 1 finite state language parsed by finite state machine (FSM).

Formal languages for RL feedback include KQML [126], DAML/OIL/OWL [127], JADE [128], and RKRL [145]. A KQML-like query to validate a <Signal/> learned by <Histogram/> might be as follows:

Expression 4-20 KQML-like Request to Validate New Knowledge

(:ask-one (:content (:Validate (:Signal (121.5, –102 dBm) (:Here Chantilly, VA) (:Now 1258)))
(:receiver CWN-1) (:language RKRL) (:context :New-Knowledge)))

KQML has simple performatives (":ask-one") and structure (:receiver, :language, and :content among others) for iCR formal reinforcement.[2] In Expression 4-20, the AACR validates carrier frequency and signal strength of one of an observed signal. Responses to such validation messages from a CWN or other <Authority/> constructively constrain dynamic behaviors. A CWN response that validates the observation includes the identity of the primary user of this frequency (Expression 4-21).

Expression 4-21 Illustrative KQML Response From a CWN Identifying Signal

```
(:tell-one (:content (:Validated (:Signal (:Identity "Dulles Tower Air Traffic
  Control Primary")
   (:Place (:Latitude 242200 :Longitude 758833)
      (:Range (:Ground (20 :mi)) (:Air (:AGL (35 :kft) (220 :mi)) )
      (:Place-name "Chantilly, VA"))
      (:Time :Indefinite)))
   (:receiver CR1) (:language RKRL) (:context Knowledge-Store) . . .)))
```

The response identifies the inference at 121.5 MHz in Chantilly, VA with etiquette-enabling information: <Location/> (:Place) and <Identity/> "Dulles Tower Air Traffic Control." This refinement enables the iCR to defer to air traffic control and to differentiate "Dulles Tower" in "Chantilly, VA" from other signals on 121.5 MHz via location, range, and radio propagation.

In the CRA, such reinforcement from an <Authority/> has indices for retrieval by place, time, <Authority/>, and other <Scene/> features. In CR1 the indices are Java HashMaps.

The response of Expression 4-21 illustrates the role of <RF/> spatial reasoning in RL. An iCR interprets messages about space, time, and RF as follows. <Range/> represents distance from a reference point, a function of altitude above ground level, <AGL/>. Such specific <Range/> values can calibrate an iCR's model(s) of terrestrial and ground-to-air radio propagation. Even in the simplest scenarios, elevated entities like tall buildings and aircraft receive radio communications at ranges longer than ground-based receivers. Radio propagation modeling is neither difficult nor excessively resource intensive for CWNs. Models of terrestrial and ground-to-air radio propagation abound in Matlab, MathCAD, Excel, Analytica, RF-CAD, and ARRL. Propagation models inform <Self/> management of radio emissions, predicting and reasoning about conformance to published (e.g., XG) and accepted (iCR–iCR peer group) norms. Ontological primitives for <RF>

[2] Although KQML has lost its role to OWL in the semantic web, the radio engineering community is driven more by simplicity, compactness, and computational efficiency than the Internet community, placing simpler languages like KQML in a strong position to contribute early to AACR evolution.

<Propagation/> </RF> in the RXML <Self/> enable the autonomous use of relevant radio propagation models; without the RXML, the AACR does not know that it knows how to model radio propagation.

Semantic alignment of formal languages like KQML and OWL goes beyond isolated computational ontologies to social norms like RA agreement on XG protocols, industry agreement on SDR standards, and grounded agreement among peers, for example, a group of iCR reasoning from <Propagation/> models. Validation involves an <Authority/> with an opportunity to mutually align semantics and to augment and constrain iCR behavior. Validation assists the iCR in <Self/> diagnosis and metalevel improvement of learning, not just knowledge.

Formal languages in wide use for <RF/> include SS7 and the ISO protocol stacks. These often employ finite state languages, state machines, message sequence charts, and formalized semantics, making them relatively easy to define, implement, and use, but few are suited to the validation of new knowledge as in the KQML example. Industry-standard tools like ITU Z.100 Specification and Description Language (SDL) [44] and the Unified Modeling Language (UML) [51] support the definition, implementation, and deployment of formal languages. Although such formal languages enable AACR dialogs, they typically are not used for AML or real-time communications with an <Authority/>.

Agent development environments, on the other hand, promote the use of computational ontology in such communications to the degree that the <Self/> and the <Authority/> both use consistent software agents. The Java Agent Development Environment (JADE), for example, includes the JADE-Management-Ontology, with Query-Agents-On-Location and Where-Is-Agent actions. OWL object communication languages include graphical ontology development software tools like Network Inference's Construct, a Visio-based tool with export to OWL [129]. JADE and OWL both support inference. The Cerebra OWL inference engine is akin to JESS, the Java Expert System Shell. The sequel develops <Self/>–<Authority/> communications via KQML for its tutorial value, progressing to the more cumbersome but industrial strength OWL as need dictates. XML is the baseline metalanguage foundation for incorporating the best of KQML, RKRL, JADE, and OWL into RXML as a domain-specific language for AACR evolution.

Formal messages flow through CRA sensing, perception, interpretation, and use as illustrated in Figure 4-12. The <RF/> waveform and protocol stack are both effector and sensor for formal message exchanges. The <RF/> sensory domain presents messages from the air interface to the iCR cognition functions as they are <Observed/>. In these exchanges, the iCR knows message format but the content is *novel*. The novelty detector identifies novel content. The response is matched to the query in the <Orient/> process. Content interpretation consists of recognizing the relevance of the message to the query, generating a <Plan/> to Reinforce Memory (R-Memory) to refine and enhance knowledge (e.g., of 121.5 MHz).

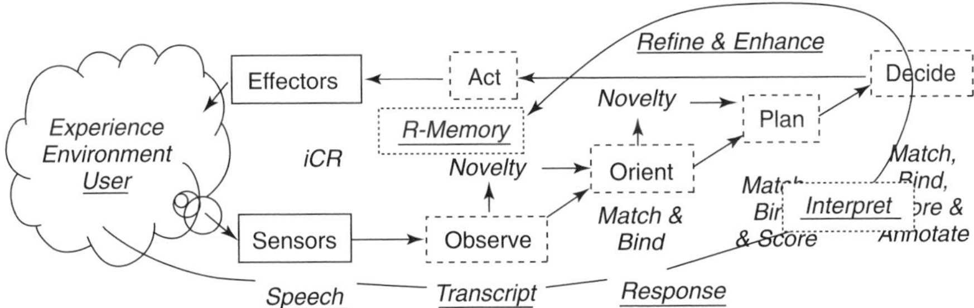

FIGURE 4-13 User reinforcement entails interpreting an errorful transcript.

Thus formal reinforcement to Q-learning may proceed from iCR to an <Authority/> by a mutually agreed formal language to be assimilated as illustrated in Figure 4-13.

4.5.3 Reinforcement Via Natural Language

Validation from a user requires skill with natural language (NL). Unlike formal languages, NL challenges radio engineering and AACR technology development. Most NL systems embed AML to some degree. As AACR evolves, the mix of preprogrammed versus autonomously acquired NL skills may shift toward AML for reduced cost of tailoring language to users.

During the last ten years, NL technology has matured significantly. Commercial products like IBM's ViaVoice [66] recognize spoken language, creating errorful transcript hypotheses with as high as 75% to as low as 5% word error rates. Commercial language translation systems convert text among English, Japanese, Chinese, Russian, Arabic, and the Romance Languages with high reliability. VoiceXML [222] generates spoken language dialogs, driving any of a variety of speech-synthesis software tools. This section introduces a strategy for adapting these tools to AACR evolution that focuses on (1) overcoming the error rates inherent in current NL technology and (2) facilitating the insertion of NL components as the technology matures.

To develop the role of NL in AACR, let us return to the <User/> domain and the map of the world. From a dictionary of English words <Self/> knows the two labels "Land" and "Ocean" as parts of Earth. An iCR can use the dictionary for reinforcement from the user as well, for example, substituting the dictionary definition for a new word in a verification dialog such as:

"So the light blue means the *solid part of the surface of the earth?*"

The user reinforces this expression for <Land/>. If the <User/> asks, "How did you know that?" the CR responds, "From the dictionary [Britannica 2003, noun]." For such a dialog, the errorful transcript must be sufficiently accurate to invoke <Land/>. If the CR infers the verb form of "land" instead, it could ask:

"So the light blue means *to set or put on shore from a ship?*"

The <User/> negatively reinforces that definition of "land." The dialog designer is on thin ice at this point. While a few users may enjoy training an AACR, many will want to shut it off. Failures like this might steer the AACR toward a metalevel strategy to infer less deeply, ask fewer questions, and focus on simpler tasks.

In the RF example, the <Self/> could ask a radio-aware user (e.g., a Ham radio operator) whether "signal" means "indication" or one of the more radio-specific definitions (all from the *Britannica* on-line dictionary):

> **Signal** **a**: an object used to transmit or convey information beyond the range of human voice **b**: the sound or image conveyed in telegraphy, telephony, radio, radar, or television **c**: a detectable physical quantity or impulse (as a voltage, current, or magnetic field strength) by which messages or information can be transmitted

Some users might not care, while others like Hams might prefer their own definitions at odds with the dictionary. Such is the nature of technical domains. Therefore, iCRs may assert prior knowledge of <RF> <Signal/> </RF> in their interactions with users, expert or otherwise, to avoid being taught inappropriately. VoiceXML could mediate the following dialog:

CR: "You said that high values of the power spectral density represent 'signal' correct?"

User: (Says or types) "Yes."

CR: "The term 'signal' is a technical term in radio. To me, it means 'a detectable electromagnetic quantity by which messages or information can be transmitted.' When using this term, I will refer to <Signal/> as s.i.g.n.a.l in written form and as 'signal-in-space' in verbal form, OK?"

If the user says anything but "Yes," then the iCR may explain that there really are few alternatives for dictionary conflicts on concepts with substantial a priori knowledge. Early adopters like Ham radio operators may enjoy such dialogs, while many users are confused, annoyed, and otherwise disenchanted with such dialogs. Thus, AML must continually detect <User/> attitude toward interaction. Positive reinforcement of training opportunities reinforces a strategy of interactive knowledge refinement, while negative reinforcement steers it away from aggressive user-domain learning with greater focus on simpler RF-domain tasks with more formal CWN interaction and less user NL interaction. The dynamic adaptation of strategies remains a research challenge in applied cognitive science.

As illustrated in Figure 4-13, reinforcement from a user via NL flows from the microphone sensor through *speech* interpretation to yield the errorful *transcript*.

In this NL flow, the interpretation of the *response* is much more difficult than the corresponding flow for formal languages. Background noise and conversations distract the human respondents, presenting irrelevant information, missed words, added words, and coarticulation in the errorful *transcript*. This makes it much more difficult to *interpret* the reinforcement accurately. The dialog may guide the user toward verbose yes/no answers. "Yes, I agree" typically is detected with greater reliability than "Yes" alone. Dialogs that offer choices more reliably acquire correct answers via speech than dialogs that ask open questions. "Do you like soup, nuts, or something else?" acquires <User/> <Preference/> more reliably than "What do you like?"

The NL knowledge accumulated by an iCR increases with time. For example, the number of names known grows with experience such as exposure to news stories [130]. Therefore, as the iCR's user encounters new experiences, iCR exposure to new names grows. iCR's ability to recognize new names and to learn their relevance in assisting the <User/> is therefore a key technology issue. With linear name growth, there may be a quadratic growth in interactions. The AACR designer therefore has to determine how to learn names that assist the user.

For example, if the user knows 100 people ("communicants"), to learn names via NL it must index speech accurately. The cell-phone user speaks directly into the microphone and delimits the name spoken into the voice phone book. Evolving AACRs may leverage such voice phone books to assist in extracting names from natural conversation, a much more difficult task that introduces errors in spoken name recognition. To connect wirelessly to an intended communicant, AACR must minimize errors in retrieving names from the spoken phone book. Errors may increase like the square or cube of the number of names. If the error rate for the $N = 100$ person phone book is $R_e = 0.001$, then the error rate for $N = 1000$ (10×) could be 100× or $R_e = 10\%$. Methods for mitigating errors include shaping the entropy of the database, for example, via the following dialog:

AACR: "The name 'Stan Smith' sounds a lot like an existing entry 'Sam Schmidt'. Are these the same people?"

User: "No, they are different people." (AACR detects "No" but the suffix is confusing.)

AACR: "You said that these are not the same person, correct?"

User: "Yes."

AACR: "If you have another name for 'Stan Smith' that doesn't sound so similar to 'Sam Schmidt', then it will be easier for me to help you connect to these two when needed. Would you like to try that?"

User: "Cut the crap." (The AACR ceases the dialog based on the negative response.)

Dialogs of this type appear in VoiceXML voice response systems with numeric keypad or limited vocabularies (e.g., numbers-only). As the size of the prior vocabulary increases, the likelihood that the AACR correlates speech to inappropriate prior knowledge also increases. WordNet [131] and WordSense [132] illustrate the linguistic challenges. WordNet is an unabridged on-line dictionary, originally developed by Princeton University, but now is a multilingual web resource with many contributors. WordSense software tools index WordNet to compute metrics of the semantic similarity of argument word senses interactively. Names, definitions, and word senses (e.g., verb versus noun sense for "rock") are core language technology increasingly relevant for iCR systems engineering as AACRs evolve toward truly transparent NL user interfaces.

4.5.4 Incremental Reinforcement With <Scene/> CBR

Speech could mediate reinforcement or the <User/> could press the buttons "Approve" or "Disapprove." Either way the iCR must know exactly what the user approves or disapproves. How does an iCR know that it generated interference? In both <User/> and <RF/> domains, the architecture must focus the attention and structure the relationship to the world so that reinforcement is received and accurately interpreted. CRA enables continuous CBR, parallel matching of the current <Scene/> to all prior experience to find similar cases. CBR accesses prior experience, matches it with the current <Scene/>, and adapts and applies the best prior experience to solve an immediate problem, for example, synthesizing a <Plan/>. There are many CBR applications like designing layouts for drying parts, developing dinner menus, adversarial argumentation, and supporting a help desk (Help-desk3), many of which are relevant [14], some with enterprise knowledge-management software tools [133–135]. In many CBR applications, the cases are vectors of attribute–value pairs from databases or GUIs. Each new <Scene/> offers a new vector consistent with the stored case base. In AACR, CBR addresses the current <Scene/> (attribute–value vector), current state of the <Self/> (a complex attribute–value vector), relevant priors (potentially dissimilar attribute–value vectors), and candidate state–action relationships (which are maps in high-dimensionality vector spaces). A <Case/> that matches a <Plan/> positively reinforces that <Plan/>.

Consider the <Histogram/> again. <Histogram> identifies <Interesting/> features of "Map of the World .gif," later labeled "Land" that aligns to <Self> <Earth> <Land/> </Earth> </Self>. The CR learns that its <Location/> can be shown on the "Map." When <On/> "Land" it should access terrestrial wireless services, but when crossing "Water" via <Aircraft/> it should not use the <Cellphone/>.

4.5.5 CBR <Anchor/>

To orient itself, it needs a <Scene/> description as in Expression 4-22. A <Scene/> is the multidimensional parameter space in which experience occurs. Each <Scene/> includes place, time, and inferred features of the environment, such as the user's state of mind (e.g., lost, in need). Computationally, <Scene/> consists of those features of those sensory perceptions that index and constrain experience. Thus, space, time, sound, vision, and RF are the dimensions of the sensory <Scenes/> for CBR.

Expression 4-22 <Scene/> Consists of Space, Time, <RF/>, and User-Domain Features

<Scene>
<Space> <Anchor/> <Zone/> </Space>
<Time> <Anchor/> <Epoch/> </Time>
<RF> <Anchor/> <Band/> </RF>
<Sensors> <Audio/> <Video/> <Controls/> </Sensors>
<User/> <Need/> <Plan/> </Scene>

Each CBR dimension needs one or more conceptual <Anchors/>, features that identify stereotypical <Scenes/> like <Home/>, <Work/>, and <Leisure/>. The anchor is the distinct, easily recognized, or unique reference point. A <Scene/> could be {<Home> <Dinner/>, <WiFi/>, <People> <Owner/> <Woman/> <Child/> </People>, <Conversation> <Plan> "trip vacation Paris" </Plan> </Conversation> </Home>}. Sensory perceptions estimate the people and the topic of the conversation. Speech perception notes repeated words like "trip," "vacation," and cities like "Paris." Although accurate topic spotting stretches current technology, the CRA includes speech topic-spotting interfaces for technology insertion.

An <Anchored/> <Scene/> presents an opportunity to <Learn>. Suppose the <Histogram/> learns <Land/> from "Map" in the <Home/>. <Home/> refers to the latitude–longitude and spatial extent of the <Owner/>'s <House/>. The <House/> is a learned subset of space, a <Zone/> distinguished by the GPS coordinates of learnable components: <Door/>s, <Driveway/>, and the <Yard/>. CRA levels also structure inferences hierarchically with <Scene/> indexing. This enables differentiation of sensory experience from vicarious experience: simulations that analyze experience and plan actions.

4.5.6 CBR Binding

CBR retains sets of problems and associated solutions, and to enable CBR, the CRA is data-intensive, storing the raw data characterizing the cases. Feature-space memory retains only a cluster center and covariance matrix instead of, say, the 4000 points underlying these parameters. CBR on the other hand retains the 4000 data points, retrieving the most relevant point and applying that solution to the current problem. CBR may adapt the solution to fit the <Scene/> by *matching*, *binding*, and *adjusting*. Successful new solutions contain the details of the problem for use in the future, adding another point to the database. CBR reinforcement differentiates the more successful solutions from the less successful, storing the successful solution to complete the CBR cycle: retrieve, reuse, revise, and retain. Commercial retrieval processes use nearest-neighbor algorithms, while others use decision trees (e.g., based on ID3) [117]. If the retrieval process is based on decision trees, then the retention process digests the new cases into updated decision trees.

4.5.7 CBR Reinforcement

CBR assists iCR with reinforcement. For example, in an earlier use case, an AACR reused knowledge about UHF TV to create a walkie-talkie between two automobiles.

Table 4-4 shows how an unused TV channel, UHF-13, used "earlier today" for a "walkie-talkie" service becomes a <Variable/> in the <Scene/> "<Earlier-Today/>" for the <Value/> UHF-25 by matching the <Unused-TV/> frame of the <Now/> <Scene/>. The <Value> UHF-25 </Value> substituted into the <Binding/> yields a walkie-talkie service in UHF Channel 25 <Now/>. Two kinds of binding apply. "Earlier today," the walkie-talkie service binds to UHF Channel 13, a TV channel at that time and in that place <Unused/> and thus available for the AACR. The CBR problem is <Need> walkie-talkie </Need>. It retrieves the most recent similar <Scene/> for <Reuse/> <Now/>. The reuse process may define islands of confidence at exact matches, using nonmatching values as variable–value pairs. While *<Need> Walkie-talkie </Need>* matches exactly in the prior and present scenes, the *<Process>* . . . *<Bind>* and *<Unused-TV>* aspects of the two scenes are only partial matches. The postulated CBR algorithm detects (via simple matching) that **UHF-13** from <Earlier-today/> occupies the same slot as <u>**UHF-25**</u> <Now/>, so **UHF-13** becomes the <Scene/> <Variable/> while <u>**UHF-25**</u> becomes the <Value/> <Now>. Established CBR algorithms adapt and apply prior experience reinforcing decisions this way [117] as does Bind ( ) in CR1.

The AML techniques of this chapter motivate the hardware and sensory-perception architecture. This chapter necessarily ignored many critical issues, such as reasoning with uncertainty given the errorful nature of computer vision and natural language technology. Subsequent chapters address these challenges.

4.6 LEARNING STRATEGIES

This chapter developed methods for discovery, unsupervised, supervised, and reinforcement learning for AACR. The <Histogram/> initiated the discovery process in unsupervised learning, marking statistically significant items as <Interesting/>. It identified potentially <Interesting/> features of a <Scene/> or <Domain/> without the aid of a teacher. The <Interesting/> features

TABLE 4-4 CBR Matching Corresponding Features of RF Scenes

<Earlier-today>	<Now>
<Unused-TV> **UHF-13** *</Unused-TV>* *<Need> Walkie-talkie </Need>* *<Process>. . . <Bind> Walkie-talkie* **UHF-13** *</Bind> </Process>*	*<Unused-TV>* **UHF-25** *<Unused-TV>* *<Need> Walkie-talkie </Need>* *<Process>. . . <Bind> Walkie-talkie* **?UHF-25?** *</Bind> </Process>*

enabled the AACR to proactively seek <Labels/>, formally from a CWN or informally via NL from the <User/> to develop hypotheses, to verify knowledge, and to obtain additional related knowledge, thus autonomously obtaining interactive supervision. Shaping the dialog to overcome the brittleness of natural language technology helps AACR to learn about a user.

The chapter also developed reinforcement learning, RL. AACRs may use RL to adapt to the environment. Unsupervised discovery via <Histogram/> initiates CBR matching of the current <Scene/> to prior experience, for example, to fill in the blanks of previously successful plans, obtaining reinforcement at key junctures. Contemporary RL methods like Q-learning measure positive reinforcement for appropriate behavior and negative reinforcement for inappropriate behavior. Since effective RL requires high correlation of a given <Scene/> to prior experience, the chapter developed an AML strategy for relating the current <Scene/> to the most relevant prior experience whether formalized a priori or learned recently. The methods were derivatives of CBR, introduced informally in the first chapter and developed in the use cases of Chapters 2 and 3.

The AML examples showed how to learn via interactions with CWNs and people using a priori knowledge. Radio XML structured both a priori and current knowledge. The <Self/> RXML represents a priori knowledge needed to bootstrap AML. RXML internalizes knowledge acquired through AML. RXML therefore works both as the external language for spectrum-use constraints and as an internal language for cognition. RXML expresses knowledge about the <Self/>, the outside world or <Universe/>, and the classes of entities of most relevance in the <RF/> and <User/> domains. Supervised, unsupervised, and reinforcement learning set the stage for the more complete definition of the CRA. Although SDRs may be controlled using conventional networks buttons and displays, the wireless evolution toward more sophisticated information services entails expanding user interfaces to include AACRs passively observing users in speech, language, and visual domains to identify preferences and needs without annoying GUIs. Speech, NL, vision, and cognition component integration requires formal semantics, necessitating industry agreement (e.g., on extending OWL to radio in the not too distant future). For AACR situation awareness, iCRs must detect and track relevant aspects of a <Scene/> identifying <User/> information states. Scene perception via rapidly maturing speech recognition and machine vision technologies assists with symbol grounding. Substantial hardware implications of such an approach are evident. In addition, the ubiquity of AML in AACR behavior argues for ubiquitous AML in the CRA, distinguishing CR1 from others where AML occupies a specialized subsystem if it is present at all.

Management of conceptual primitives motivated the use of namespaces. Although the majority of the ontological primitives developed so far originated in RKRL [145], best commercial practice favors XML as a standard metalanguage. Radio XML (RXML) merges an open framework for new conceptual primitives, a radio-domain core <RF/>, and a <User/>-domain core of stereotypical knowledge. A RXML namespace formalizes the status of the ontological primitives. Thus, <Signal/> becomes RXML:Signal, dropping the angle brackets. AACR evolution may employ RXML:RF ontology to varying degrees over time. Initially, the ontology enables industry to communicate more clearly with each other in a multisupplier market, much as the SCA has facilitated teamwork for SDR. As the AACRs move toward the iCR, degrees of reasoning over the ontology will increase. For example, as XG emerges, AACRs may exchange radio technical information with regulatory authorities, spurring the use of RXML.

There are many technology challenges for AML in AACR. Some are addressed further in the sequel but many are not. Intrinsic to AML is the further development of reliable algorithms for the grounding of internal symbols to external entities and actions for reliable AML. The subsequent treatment of sensory perception begins to equip one to ground symbols for the level of wireless services for evolution toward the use cases.

Real-world applications exhibit uncertainty at every level of perception from the raw stimuli where sensors produce noisy data to the perception of occluded entities (obstructions, auditory noise, radio interference, etc.), leaving the perception system to identify and track entities. This chapter meticulously avoided mentioning uncertainty that lurked just out of reach everywhere. Subsequent chapters introduce methods for handling sensory and perception uncertainty in AACR.

Some of the many other relevant ML technologies appear in Figure 4-14 [145]. Feature space methods like support vector machines (SVMs) and instance-based learning (IBL) with particular relevance to AACR are developed further, as are knowledge-based knowledge acquisition, neural networks, and genetic algorithms.

The breadth of ML techniques in this book suffices for initial development of AML in AACR. AML is a core technology of the subsequently defined CRA. Ubiquitous CBR in the CRA developed here could be implemented in an autonomous agent architecture or an increasingly autonomous robotics architecture like CLARION [53], SAIL [136], or RCS [264], the NASA/NIST standard robot architecture [137].

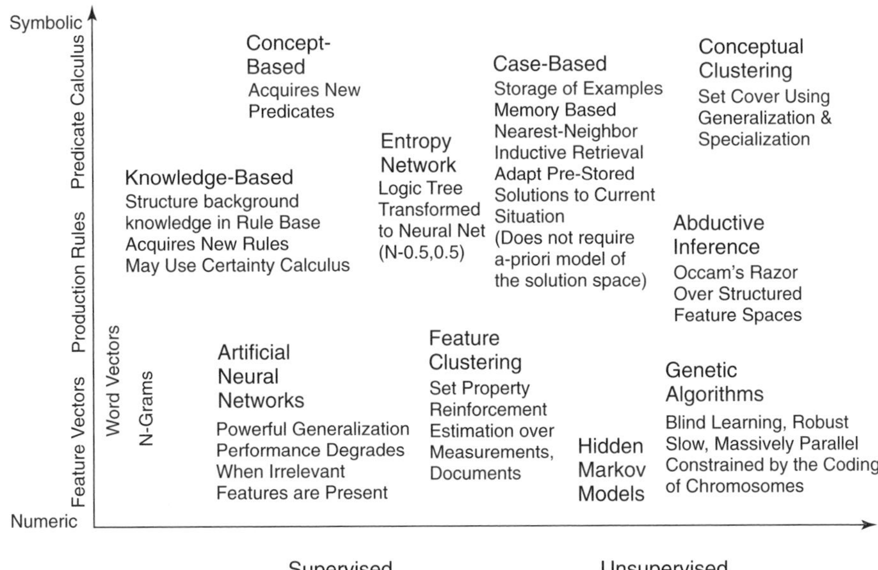

FIGURE 4-14 Relevant machine learning technologies.

4.7 EXERCISES

4.1. Enhance your favorite use case of a prior chapter with AML.

(a) How useful are design tools like UML?

(b) How does AML change the value proposition of the use case or product?

4.2. Computer-aided software engineering (CASE) tools can help you analyze the application domain to define an AML approach, but may not help you track the evolution through its training, performance, and reinforcement. How could CASE tools "fix" this? Consider & Builder.

4.3. Develop a consolidated RXML ontology of all the ontological primitives of the chapter. Identify the holes and fill them sufficiently to enable the use case of Exercise 4.1. Check yourself against RXML:Self from the companion CD-ROM/web site. Complete the RXML for the walkie-talkie use case. Complete the RXML for the Bert–Ernie child-protector use case.

4.4. Find on the Web two or three XML reference repositories of knowledge relevant to the use cases of Exercise 4.3. Identify specific trade-offs between the RXML repository and "OJT." How can the size of a repository be reduced by relying more on OJT? Explain with CBR OJT.

4.5. What would you do to <Histogram/> to address uncertainty? How would you train it to set uncertainty parameters autonomously? How would you gather sufficient examples to set learning and performance parameters a priori?

4.6. Write a histogram-based discovery algorithm in your favorite computer language. What is the best way to save the results of an entire learning episode or <Scene/>? How would you save the <Domain/> and <Interesting/> results for later use?

4.7. Write a CBR algorithm that accesses <Scenes/> that you saved in Exercise 4.6.

4.8. Does your CBR algorithm from Exercise 4.7 deal well with uncertainty? If not, then what would you do to improve it? If so, then present it with a case for which there is no good answer because of the need for a judgment based on experience.

4.9. Revise your algorithm so the CBR can explain how two different outcomes are both candidates and why it chose one over the other?

4.10. Aggressive cheating by leaning on the <User/> may be built into an AACR. Outline acceptable forms of cheating (e.g., how to cheat by asking <User/>, CWN, etc.). How can cheating help it become smart enough to succeed in the marketplace? What cheating is unacceptable?

CHAPTER 5

COGNITIVE RADIO ARCHITECTURE

Architecture is a comprehensive, consistent set of *design rules* by which a specified set of *components* achieves a specified set of *functions* in products and services that evolve through multiple design points over time [144]. This chapter develops the CRA by which SDR, sensors, perception, and AML may be integrated to create AACRs with better QoI through capabilities to observe (sense, perceive), orient, plan, decide, act, and learn in RF and user domains, transitioning from merely aware or adaptive to demonstrably cognitive radio.

This chapter develops five complementary perspectives of architecture. CRA I defines six functional components, black boxes to which are ascribed first level functions common to AACR design points from SDR to iCR and among which critical interfaces are defined. CRA II examines the flow of inference through a cognition cycle that arranges the core capabilities of iCR in temporal sequence for both logical flow and circadian rhythm for the CRA. CRA III examines the related levels of abstraction for AACR to sense elementary sensory stimuli and to perceive QoI-related aspects of a <Scene/> consisting of the <User/> in an <Environment/> that includes <RF/>. CRA IV examines the mathematical structure of this architecture, identifying mappings among topological spaces represented and manipulated to preserve set-theoretic properties. Finally, CRA V briefly reviews SDR architecture, sketching an evolutionary path from the SCA/SRA to the CRA. The CRA <Self/> provided in CRA Self .xml of the companion CD-ROM expresses the CRA in RXML.

5.1 CRA I: FUNCTIONS, COMPONENTS, AND DESIGN RULES

The *functions* of AACR exceed those of SDR. Reformulating the SDR network node to the AACR <Self/> asserts a *peer* capable of creating networks and needing functions by which the <Self/> accurately perceives the local scene including <RF/> and the <User/> and autonomously learns to tailor QoI to the specific <User/> in the current <RF/> <Scene/> and situation.

5.1.1 AACR Functional Component Architecture

The SDR components appear with the related cognitive components in Figure 5-1. The cognition components describe the SDR in Radio XML so that the resulting <Self/> knows that it is a radio and that its goal is to achieve high QoI tailored to its own users. RXML asserts a priori radio background and user stereotypes as well as dynamic knowledge of <RF/> and space–time <Scenes/> perceived and experienced. This knowledge enables both structured reasoning with iCR peers and CWNs and ad hoc reasoning with users while learning from experience.

The detailed allocation of functions to components with interfaces among the components requires closer consideration of the SDR component as the foundation of CRA. SDRs include a hardware platform with RF access and computational resources, with more than one software-defined personality. The SDR Forum has defined its software communications architecture (SCA) [27] and the Object Management Group (OMG) has defined its software radio architecture (SRA) [61], similar fine-grain architecture constructs for next-generation plug and play. These SDR architectures are defined in Unified Modeling Language (UML) object models [138], CORBA Interface Defini-

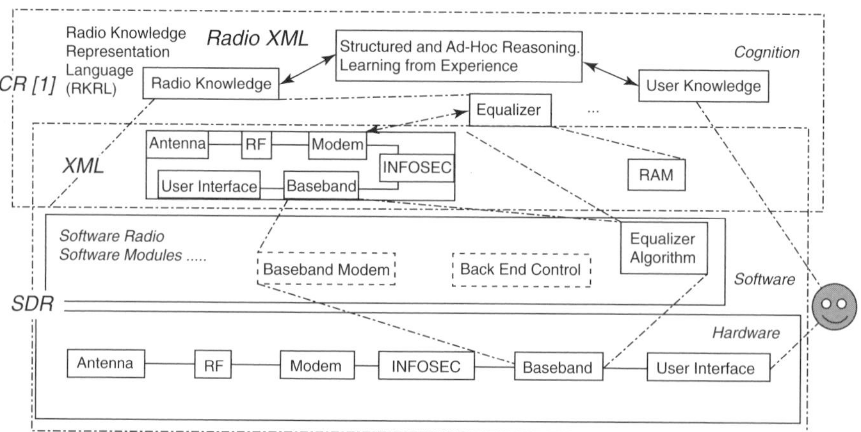

FIGURE 5-1 The CRA augments SDR with computational intelligence.

tion Language (IDL) [139], and XML descriptions of the UML models. The SCA emphasizes plug-and-play wireless personalities on computationally capable mobile nodes where network connectivity is often intermittent at best, while the SRA focuses on making the Web wireless.

The commercial cell phone community [140], on the other hand, led by Ericsson and Nokia, envisions a much simpler architecture for mobile wireless devices, consisting of two APIs, one for the service provider and another for the network operator. They define a knowledge plane in the future intelligent wireless networks that is not dissimilar from a distributed CWN. Their Wireless World Research Forum (WWRF) [328] promotes the business model of the user → service provider → network operator → large manufacturer → device, where the user buys mobile devices consistent with services from a service provider, and the technical emphasis is on *intelligence in the network*. This strategy no doubt will yield computationally intelligent networks in the near- to mid-term.

The CRA developed in this text, however, envisions the computational intelligence to create low cost ad hoc networks with the *intelligence in the mobile device*. This technical perspective enables the business model of user → device → heterogeneous networks, typical of the Internet model where the user buys a device (e.g., a wireless laptop) that can connect to the World Wide Web wirelessly via any available Internet service provider (ISP). The CRA builds on both the SCA/SRA and the commercial API model but integrates semantic web intelligence in Radio XML for mobile devices to enable more of an Internet business model. SDR, AACR, and iCR form a continuum facilitated by RXML.

The AACR node CRA consists of the minimalist set of six functional components of Figure 5-2. A functional component is a black box to which functions have been allocated, but for which implementing components are not specified. Thus, while the Applications component is likely to be primarily software, the details of those software components are unspecified.

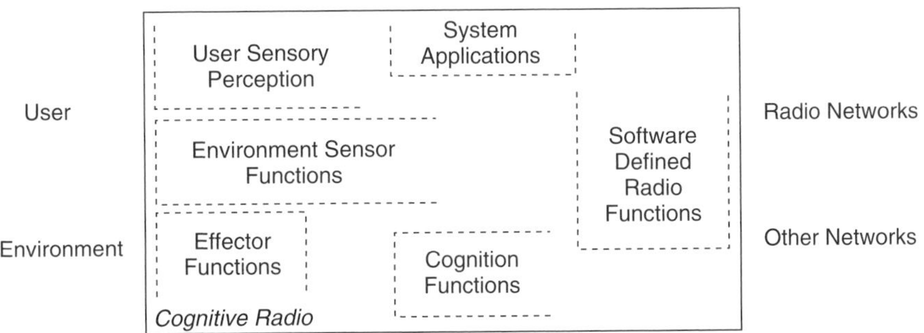

FIGURE 5-2 Minimal AACR node architecture.

The six CRA functional components are:

1. The *user sensory perception* (*User SP*) interface includes haptic, acoustic, and video sensing and perception functions. User SP functions may include optimized hardware, for example, for computing video flow vectors in real time to assist scene perception.
2. The local *environment* sensors (location, temperature, accelerometer, compass, etc.).
3. The *system applications* (media-independent services like playing a network game).
4. The *SDR* functions (which include RF sensing and SDR radio applications).
5. The *cognition* functions (symbol grounding for system control, planning, learning).
6. The *local effector* functions (speech synthesis, text, graphics, and multimedia displays).

These functional components are embodied on an iCR-platform, a hardware–software infrastructure supporting the six functions. For the capabilities described in the prior chapters, these components go beyond SDR in critical ways. First, the traditional user interface is partitioned into a substantial user sensory subsystem and a distinct set of local effectors. The user sensory interface includes buttons (the haptic interface) and microphones (the audio interface) to include acoustic sensing that is directional, capable of handling multiple speakers simultaneously and including full motion video with visual scene perception. In addition, the audio subsystem does not just encode audio for (possible) transmission; it also parses and interprets the audio from designated speakers such as the <User/> for a high performance spoken natural language interface. Similarly, the text subsystem parses and interprets the language to track the user's information states, detecting plans and potential communications and information needs unobtrusively, trusted to protect private information as the user conducts normal activities. The local effectors synthesize speech along with traditional text, graphics, and multimedia display as tasked by the cognition component.

Systems applications are those *information services* that synthesize QoI value for the user. Typically, voice communications with a phone book, text messaging, and the exchange of images or video clips comprise the core systems applications for SDR. Usually these services are integral to the SDR application, such as text messaging via GPRS. AACR systems applications break the service out of the SDR network for greater personal flexibility and choice of wireless connectivity without additional user tedium. The typical user might care if the AACR wants to switch to 3G at $5 per minute, but a particularly affluent user might not care and would leave all that up to the AACR.

The cognition component provides all the cognition functions from the semantic grounding of entities from the perception system to controlling the overall system through planning and actions, learning user preferences and RF situations in the process.

Each of these subsystems may contain its own processing, local memory, integral power conversion, built-in-test (BIT), and related technical features. This functional architecture is described to the <Self/> in RXML for external communications about the <Self/> and for introspection in Expression 5-1.

Expression 5-1 The AACR Has Six Functional Components

<Self> <iCR-platform/> <Functional-components>
<User SP/> <Environment/> <Effectors/> <SDR/> <Sys Apps/>
<Cognition/>
</Functional-components> </Self>

The hardware–software platform and the functional components of the AACR are independent. The architecture design principle is that the (software) functional components adapt to whatever RF–hardware–OS platform might be available. Platform-independent computer languages like Java apply.

5.1.2 Design Rules Include Functional Component Interfaces

These functional components of Figure 5-2 imply critical functional interfaces. The AACR *N*-squared diagram of Table 5-1 characterizes these interfaces. They imply an initial set of AACR applications–programmer interfaces (CRA APIs). In some ways these APIs augment the established SDR APIs. For example, the Cognition API adds a planning capability to SDR. This is almost entirely new and will be helpful for AARs to fully support XG. In other ways, these APIs supersede the existing SDR APIs. In particular, the SDR user interface becomes the User SP and Effector APIs. User Sensory APIs encapsulate perception, while the Effector API encapsulates actions like speech synthesis to give the AACR <Self/> its own voice. User SP and SDR status flow perceptions toward the cognition component from which Effectors and SDR accept tasks. These interface changes enable the AACR to sense the situation represented in the environment and to access radio networks on behalf of the user in a situation-aware way.

Interfaces 13–18, 21, 27, and 33 may be aggregated into an information services API (ISAPI) by which an information service accesses the other five components. Interfaces 25–30, 5, 11, 23, and 35 would define a cognition API (CAPI) by which the cognition system obtains status and exerts control over the rest of the system.

5.1.3 The Cognition Components

Figure 5-1 shows relationships among the three computational-intelligence aspects of CR—radio knowledge, user knowledge, and the capacity to learn.

TABLE 5-1 AACR *N*-Squared Diagram Characterizes CRA Internal Interfaces

From–To	User SP	Environment	Sys Apps[a]	SDR	Cognition[b]	Effectors
User SP	1	7	13 PA	19	25 PA	31
Environment	2	8	14 SA	20	26 PA	32
Sys Apps	3	9	15 SCM	21 SD	27 PDC	33 PEM
SDR[a]	4	10	16 PD	22 SD	28 PC	34 SD
Cognition[b]	5 PEC	11 PEC	17 PC	23 PAE	29 SC	35 PE
Effectors	6 SC	12	18	24	30 PCD	36

Key: P, primary; A, afferent; E, efferent; C, control; M, multimedia; D, data; S, secondary; others not designated P or S are ancillary.

[a] Information Services API consists of interfaces 13–18, 21, 27, and 33.

[b] Cognition API consists of interfaces 25–30, 5, 11, 23, and 35.

Interface Notes: Follow the numbers of the table:

1. *User SP–User SP*: Cross-media correlation interfaces (video–acoustic, haptic–speech, etc.) reduce uncertainty (e.g., if video indicates user is not talking, acoustics may be ignored for commands reducing errors and enhancing QoI.)
2. *Environment–User SP*: Environment sensors parameterize user sensor perception. Temperature and humidity extremes that limit video would be detected via this interface.
3. *Sys Apps–User SP*: Systems applications may focus scene perception by identifying entities, range, expected sounds via speech, and spatial perception interfaces.
4. *SDR–User SP*: SDR applications may provide expectations of user input to user SP perception to improve probability of detection and correct classification of perceived inputs.
5. *Cognition–User SP*: This is the *primary control efferent* path from cognition to the control of the user-SP component, controlling speech recognition, acoustic signal processing, video processing, and related sensory perception. Plans from the cognition component may set expectations for user scene perception, improving perception.
6. *Effectors–User SP*: Effectors may supply a replica of the effect to user perception so that self-generated effects (e.g., synthesized speech) may be accurately attributed to the <Self/>, validated as having been expressed, and/or cancelled from scene perception.
7. *User SP–Environment*: Perception of rain, buildings, indoor/outdoor can help set SDR parameters.
8. *Environment–Environment*: Environment sensors would consist of location sensing such as GPS or Glonass; temperature of the ambient; light level to detect inside versus outside locations; possibly smell sensors to detect spoiled food; and others that may surprise one even more. There seems to be little benefit to enabling direct interfaces among these elements.
9. *Sys Apps–Environment*: Data from the systems applications directly to environment sensors would be minimal.
10. *SDR–Environment*: Data from the SDR personalities directly to the environment sensors would be minimal.
11. *Cognition–Environment*: (Primary control path) Data from the cognition system to the environment sensors controls those sensors, turning them on and off, setting control parameters, and establishing internal paths from the environment sensors.
12. *Effectors–Environment*: Data from effectors directly to environment sensors would be minimal.
13. *User SP–Sys Apps*: Data from the user sensory-perception system to systems applications is a *primary afferent path* for multimedia streams and entity states that affect information services implemented as systems applications. Speech, images, and video to be transmitted move along this path for delivery by the relevant systems application or information service to the relevant wired or SDR communications path. Sys Apps overcomes the limitations of individual paths by maintaining continuity of conversations, data integrity, and application coherence (e.g., for multimedia games).
14. *Environment–Sys Apps*: Data on this path assists systems applications with physical environment and location awareness.
15. *Sys Apps–Sys Apps*: Different information services interoperate by passing control information and domain multimedia flows to each other through this interface.

Table footnote continued

16. *SDR–Sys Apps*: This is the primary afferent path from external communications. It includes control and multimedia information flows for all information services, with wired and wireless interfaces.
17. *Cognition–Sys Apps*: Through this path the AACR <Self/> exerts control over information services.
18. *Effectors–Sys Apps*: Effectors provide incidental feedback to information services through this afferent path.
19. *User SP–SDR*: The sensory-perception system may send limited raw data directly to the SDR subsystem via this path, for example, in order to satisfy security rules for biometrics.
20. *Environment–SDR*: Environment sensors like GPS historically have accessed SDR waveforms directly, such as providing timing data for air interface signal generation. The cognition system may establish such paths if cognition provides no value added. The use of this path is deprecated because all environment sensors including GPS are unreliable. Cognition has the capability to deglitch GPS, for example, recognizing from video that the <Self/> is in an urban canyon and therefore reporting location estimates based on landmark correlation.
21. *Sys Apps–SDR*: This is the primary efferent path from information services to SDR through the services API.
22. *SDR–SDR*: The linking of different wireless services directly to each other via this interface is deprecated. If voice services need to be connected to each other there should be a bridging service in Sys Apps.
23. *Cognition–SDR*: This is the primary radio control interface, replacing the control interface of the SDR SCA and the OMG SRA.
24. *Effectors–SDR*: Effectors such as speech synthesis and displays could provide state information directly to SDR waveforms via this interface.
25. *User SP–Cognition*: This is the primary afferent flow for perceptions especially the states of <Entities/> in the scene, landmarks, known vehicles, furniture, and the like.
26. *Environment–Cognition*: This is the primary afferent flow for environment sensors.
27. *Sys Apps–Cognition*: This is the interface through which information services request and receive support from the AACR <Self/>. This is also the control interface by which cognition sets up, monitors, and tears down information services.
28. *SDR–Cognition*: This is the primary afferent interface by which the state of waveforms is made known to cognition. Via this interface cognition can establish primary and backup waveforms for information services enabling the services to select paths in real time for low latency services. Those behaviors are monitored for quality and validity (e.g., obeying XG rules) by the cognition system.
29. *Cognition–Cognition*: The cognition system (1) orients the <Self/> to information from <RF/> via SDR and from scene perceptions, (2) makes plans, (3) makes decisions, and (4) initiates actions, including the control of resources. The <User/> may directly control anything via transparent paths through the cognition system that enable it to monitor the user to learn from the user's direct actions.
30. *Effectors–Cognition*: This is the primary afferent flow for effector status information.
31. *User SP–Effectors*: The user SP should not interface directly to the effectors, but should be routed through cognition for self observation.
32. *Environment–Effectors*: The environment component typically should not interface directly to the effectors.
33. *Sys Apps–Effectors*: Systems applications may display streams, generate speech, and otherwise directly control effectors once cognition establishes the paths and constraints.
34. *SDR–Effectors*: This path may link a SDR's voice track to a headphone, but typically SDR should provide streams to Sys Apps. This path may be necessary for legacy compatibility during migration but is deprecated.
35. *Cognition–Effectors*: This is the primary efferent path for the control of effectors. Information services provide the streams to the effectors, but cognition sets them up, establishes constraints, and monitors the information flows.
36. *Effectors–Effectors*: These paths are deprecated.

The minimalist architecture of Figure 5-2 and the functional interfaces of Table 5-1 do not assist the radio engineer in structuring QoI-related knowledge, nor do they assist much in integrating machine learning into the system. The fine-grained CRA of this chapter more fully addresses these three core cognition components.

First, radio knowledge has to be translated from the classroom and engineering practice into a body of computationally accessible, structured technical knowledge about radio. Radio XML is the primary means developed in this text for the formalization of radio knowledge. This starts a process of RXML definition and development that can only be brought to fruition by industry and over time, a process similar to the evolution of the SCA in the SDR Forum. RXML enables the plug-and-play of RF and user world knowledge for enhanced QoI as the SCA enables the plug-and-play of radio components.

The World Wide Web is now sprouting computational ontologies, some of which are nontechnical but include radio, like the open CYC ontology. They bring the radio domain into the semantic web, which helps people and algorithms know about radio. This informal knowledge lacks the technical scope, precision, and accuracy of authoritative radio references like the ETSI GSM Mou and ITU 3GPP. Not only must radio knowledge be precise, it must be stated at a useful level of abstraction, yet with the level of detail appropriate to the use case. Thus, ETSI GSM in most cases would overkill radio level of detail yet lack sufficient knowledge of the user-perspective functionality of GSM. In addition, AACR is multiband, multimode radio (MBMMR), so the knowledge must be comprehensive, addressing the majority of radio bands and modes. Therefore this text captures radio knowledge needed for competent CR in the MBMMR bands from HF through millimeter wave at a level of abstraction appropriate to internal reasoning, formal dialog with a CWN, and informal dialog with users. To begin this process, Table 5-2 relates ITU standards to CRA capabilities in an agenda for extracting content from formal documents that bear substantial authority, encapsulating that knowledge in approximate form that can be reasoned with on AACR nodes.

The table is illustrative, not comprehensive, but it characterizes the technical issues that drive cognition component architecture. ITU, ETSI, other regional and local standards bodies, and CWN supply source knowledge to the AACR node as the local repository for authoritative knowledge. The initial corpus of formalized radio knowledge is provided in <Self> <RF/> </Self> in the companion CD-ROM/web site.

Next, user knowledge should be formalized at the level of abstraction and degree of detail necessary for the CR to bootstrap user knowledge for QoI enhancement. Incremental knowledge acquisition was motivated in the introduction to AML by describing how <Histogram/> can identify learning opportunities. Effective use cases clearly identify the classes of user and the specific knowledge needed to customize envisioned services to enhance QoI. Use cases may also supply sufficient initial knowledge to render incremental AML not only effective but also enjoyable to the user.

TABLE 5-2 Radio Knowledge in the CRA

Need	Source Knowledge	AACR Internalization
Sense RF	RF platform	Calibration of RF, noise floor, antennas, direction
Perceive RF	ITU, ETSI, ARIB, RAs	Location-based table of radio spectrum allocation
Observe RF (sense and perceive)	Unknown RF Known RF conforms to ITU, ETSI, etc.	RF sensor measurements and knowledge of basic types (AM, FM, simple digital channel symbols, typical TDMA, FDMA, CDMA signal structures)
Orient	XG-like policy	Receive, parse, and interpret policy language
	Known waveform	Measure parameters in RF, space, and time; interact per protocol stack
Plan	Known waveform Restrictive policy	Enable SDR for which licensing is current Optimize transmitted waveform, space–time plan
Decide	Legacy waveform, policy	Defer spectrum use to legacy users per policy
Act	Applications layer	Query for available services (white/ yellow pages)
	ITU, ETSI, . . . , CWN	Obtain new skills encapsulated as download
	Air Interface	Operate waveform
Learn	Unknown RF	Remember space–time–RF signatures; discover spectrum-use norms and exceptions
	ITU, ETSI, . . . , CWN	Extract relevant aspects such as new feature

To relate a use case to the seven iCR capabilities, one extracts specific and easily recognizable <Anchors/> for stereotypical situations observable in diverse times, places, and situations. One expresses the anchor knowledge using RXML. Table 5-3 illustrates this process for the SINCGARS–Sparky and FCC unused-TV channel use cases.

Speech, language, and visual cues are constantly generated and tracked to discern user intent. Wearability of Charlie and Genie CWPDAs assists with continuous tracking of the user's state, and with the acquisition of visual cues. Cognitive meeting rooms [141] are being developed. The personalization of those technologies in AACR should propel this vision forward.

Staying better connected requires the normalization of knowledge between <User/> and <RF/> domains. If, for example, the <User/> says, "What's on one oh seven-seven," while in the car on the way to work in the Washington, DC area, then the dynamic <User/> ontology should enable the AACR to infer that the user is talking about the current FM radio broadcast, the units

TABLE 5-3 Use-Case Knowledge in the CRA Node Architecture

Need	Source Knowledge	AACR Internalization
Observe User	Sparky use case	Sense voice and face; perceive "Sparky" <User> <Name> "Sparky" </Name> <Speaker-model> <Face-model></User>
Observe Scene	TV use case	Sense and perceive Joe, Lynné, and Dan's voice and face <User> <Name> "Joe"</Name> <Speaker-model/> <Face-model/> </User>
Orient	Sparky use case	Recognize request for existing SINCGARS SDR waveform
	TV use case	Recognize request to create a walkie-talkie
Plan	Sparky use case	Extract SINCGARS training setup from the <SOI/> via database query and retrieval from network server
	TV use case	Create Part 15 ad hoc wireless network
Decide	Sparky use case	Get permission from Sparky via <Speech-recognizer/>
	TV use case	Get permission from Joe via <Speech-recognizer/>
Act	Sparky use case	Instantiate SINCGARS and enable training password
	TV use case	Instantiate walkie-talkie waveform template for UHF TV Channel 68
Learn	Sparky use case	Acquire training password "Second Guessing"
	TV use case	Adapt walkie-talkie waveform template to UHF 24 when legacy use is detected on UHF 68

are in megahertz (MHz), and the user wants to know what is on WTOP. If it can't infer this question, then it should ask the user to do the task, observing user selection of 107.7 FM, learning from observing. Tailoring to the user requires continually adapting the <User/> ontology with repeated regrounding of terms in the <User/> domain to conceptual primitives and actions in the <RF/> domain.

The process of linking user expressions of interest or <Need/> to the appropriate radio technical operations sometimes may be extremely difficult. Military radios, for example, have many technical parameters. A "channel" in SINCGARS may consist of dehopped digital voice in one context (voice communications) or a 25 kHz band of spectrum in another context. If the user says, "I need the Commander's channel," the SINCGARS user is talking about a "dehopped CVSD voice stream." If the same user a few seconds later says, "This sounds awful. Who else is in this channel?" the user is referring to interference with a collection of hop sets. If the CR observes, "There is strong interference in almost half of the available channels," then the CR is referring to a related set of 25 kHz channels. If the user then says, "OK, notch the strongest 3 interference channels," he is talking about a different subset of the channels. The question, "Is anything on our emergency channel?"

switches context from SINCGARS to <Self/>, asking about a physical RF access channel. Such exchanges eliminate the radio operator but demand cross-domain grounding. Candidate methods of cross-domain grounding call for associated architecture features:

1. *<RF/> to <User/> shaping the dialog* to express precise <RF/> concepts to nonexpert users in an intuitive way, such as:
   (a) Dialog: "If you rotate the remote speaker box it will make a big difference in reception from the wireless transmitter on the TV."
   (b) CRA implications: Include a *rich set of synonyms* for radio technical terms (<Antenna> ≅ <Wireless-remote-speaker> ≅ "Speaker box").
2. *<RF/> to <User/> learning jargon* to express <RF> connectivity opportunities in <User> terms.
   (a) Dialog: "tee oh pee" for WTOP, "Hot ninety two" for FM 92.7, "Guppy" for "E2C Echo Grand on 422.1 MHz."
   (b) CRA implications: *Facility for single-instance update of user jargon.*
3. *<User/> to <RF/> relating values to actions:* Relate <User/> expression of values ("low cost") to features of situations (<Home/>) that are computable (<CONTAINS> <Situation> <Home/> </Situation> </CONTAINS>) and that relate directly to <RF/> domain decisions.
   (a) <Situation/>: Normally wait for free WLAN for big attachment; if <AND> <Home/> <NOT> WLAN </NOT> </AND>, ask if user wants to pay for 3G.
   (b) CRA implications: *Associative inference hierarchy* that relates observable features of a <Scene/> to user sensitivities, such as <Cost/>.

These highlights of CRA cognition component design considerations are more fully developed in the CRA <Self/> companion CD-ROM.

5.1.4 Self-Referential Components

The cognition component must assess and manage all of its own resources, including validating downloads. Thus, in addition to <RF/> and <User/> domains, RXML must describe the <Self/> to the <Self/> for self-referential reasoning. This class of reasoning is well known in the theory of computing to be a potential black hole for computational resources. Specifically, any Turing-capable (TC) computational entity that reasons about itself can enter a Gödel–Turing loop from which it cannot recover. Thus, TC systems are known to be "partial"—only partially defined because the attempt to execute certain classes of procedure will never terminate. To avoid this paradox, the

CRA mandates the use of only "total" functions, typically restricted to bounded minimalization [142]. Watchdog "step-counting" functions [143] or timers must be in place for all reasoning and radio functions. The timer and related computationally indivisible control construct is equivalent to the computer-theoretic construct of a step-counting function over "finite minimalization." It has been proved that computations that are limited with reliable watchdog timers can avoid the Gödel–Turing paradox to the reliability of the timer. This proof is a fundamental theorem for practical self-modifying systems.

Briefly, if a system can compute in advance the amount of time or the number of instructions that any given computation should take, then if that time or step count is exceeded, the procedure returns a fixed result such as "Unreachable in Time *T*." As long as the algorithm does not explicitly or implicitly restart itself on the same problem the paradox is avoided. Although not Turing capable, such an AACR is sufficiently computationally capable to perform real-time communications tasks such as transmitting and receiving data and bounded user interface functions. Otherwise, the AACR eventually will crash, consuming unbounded resources in a self-referential loop. This is not a general result, but is highly radio-domain-specific, established only for isochronous communications. Specifically, for every situation, there is a default action that consumes $O(1)$ resources enforced by a reliable watchdog timer or other step-counting function. Since radio air interfaces transmit and receive data, there are always defaults such as "repeat the last packet" or "clear the buffer" that may degrade the performance of the overall communications system but that have $O(1)$ complexity. Since there are planning problems that can't be solved with algorithms so constrained, either an unbounded community of CRs must cooperatively work on the more general problems or the CN must employ a Turing-capable algorithm to solve the harder problems (e.g., NP-hard with large N) off-line.

Thus the CRA structures systems that not only can modify themselves, but can do it in such a way that they will not induce nonrecoverable crashes from self-referential computing.

5.1.5 Flexible Component Architecture

Although this chapter develops the six-component CRA and a particular information architecture, there are many possible cognitive radio architectures. The purpose is not to try to sell the six components, but to develop the architecture principles. The CRA and research prototype, CR1, therefore offer open-source licensing for noncommercial educational purposes.

5.2 CRA II: THE COGNITION CYCLE

The cognition component of the CRA includes a temporal organization and flow of inferences and control states, the cognition cycle.

5.2.1 The Cognition Cycle

The cognition cycle implemented in Java in CR1 [145] is illustrated in Figure 5-3. This cycle synthesizes the CRA cognition component in an obvious way. Stimuli enter the cognitive radio as sensory interrupts, dispatched to the cognition cycle for a response. Such an iCR sequentially observes (senses and perceives) the environment, orients itself, creates plans, decides, and then acts. In a single-processor inference system like a mote™ [281], the CR's flow of control also moves in the cycle from observation to action. In a multiprocessor system, temporal structures of sensing, preprocessing, reasoning, and acting may be more parallel. The process of Figure 5-3 is called the wake epoch because reasoning during this epoch of time is reactive to the environment. There also may be sleep epochs for introspective reasoning or prayer epochs for asking for help from a higher authority.

During the wake epoch, the receipt of a new stimulus on any of a CR's sensors or the completion of a prior cognition cycle initiates a new cognition cycle.

5.2.2 Observe (Sense and Perceive)

The CR observes its environment by parsing incoming stimulus streams. These can include monitoring speech-to-text conversion of radio broadcasts (e.g., the weather channel). In the observation phase, the CR associates

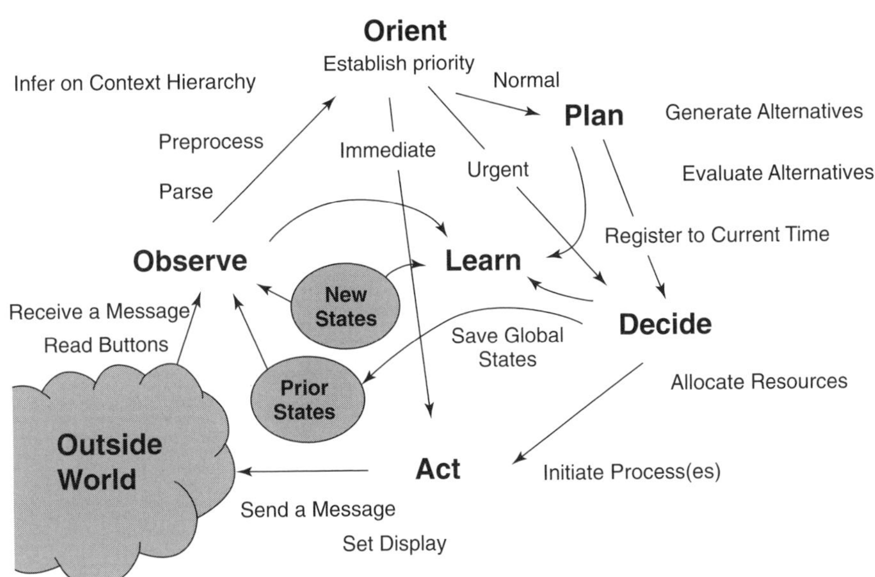

FIGURE 5-3 Simplified cognition cycle.

location, temperature, light level sensors, and so on to infer the communications context. This phase binds these stimuli to prior experience to detect patterns over time. CR1 aggregates experience by remembering everything. All the audio, all the emails, and all the radio situations that one might experience in a year occupies a few hundred gigabytes of space, depending on the detail retained. So the computational architecture for remembering and rapidly correlating current experience against everything known previously is a core capability of the CRA. The Observe phase embraces the User SP, the Environment, and SDR RF sensor processing of the CRA.

5.2.3 Orient

The Orient phase determines the significance of an observation by binding the observation to a previously known set of stimuli or <Scene/>. This phase operates on the internal data structures that are analogous to the short-term memory (STM) that people employ to engage in a dialog without necessarily remembering everything to the same degree in long-term memory. The natural environment supplies the redundancy needed to instigate transfer from STM to long-term memory (LTM). In the CRA, the transfer from STM to LTM is mediated by the sleep cycle in which the contents of STM since the last sleep cycle are analyzed respect to each other and to existing LTM. How to do this robustly remains an important CR research topic, but the process is identified in the CRA. Matching of current stimuli to stored experience may be achieved by stimulus recognition or by binding.

Stimulus recognition occurs when there is an exact match between a current stimulus and a prior experience. Reaction may be appropriate or in error. Each stimulus is set in a larger context, which includes additional stimuli and relevant internal states, including time. Sometimes, the Orient phase causes an action to be initiated immediately as a reactive stimulus–response behavior. A power failure, for example, might directly invoke an act that saves the data (the "Immediate" path to the Act phase in the figure). A nonrecoverable loss of signal on a network might invoke reallocation of resources, for example, from analyzing speech to searching for alternative RF channels. This may be accomplished via the path labeled "Urgent" in the figure.

The binding occurs when there is a nearly exact match between a current set of stimuli and a prior experience and very general criteria for applying the prior experience to the current situation are met. One such criterion is the number of unmatched features of the current scene. If only one feature is unmatched then binding may be the first step in generating a plan for behaving similarly in the current scene as in the last comparable scene. In addition to number of features that match exactly, instance-based learning (IBL) supports inexact matching and binding. Binding also determines the priority associated with the stimuli. Better binding yields higher expectation of autonomous learning, while less effective binding yields lower priority for the incipient plan.

5.2.4 Plan

Most stimuli are dealt with deliberatively rather than reactively. An incoming network message would normally be dealt with by generating a plan (in the Plan phase, the normal path). In research-quality and industrial-strength CRs, formal models of causality [296] would be embedded into planning tools. The Plan phase should also include reasoning over time. Typically, reactive responses are preprogrammed or learned by being told, while other deliberative responses are planned. Open source planning tools like OPRS [60] may be embedded into the Plan phase. Planning tools enable the synthesis of RF and information access behaviors in a goal-oriented way based on perceptions, RA rules, and previously learned user preferences.

5.2.5 Decide

The Decide phase selects among the candidate plans. The radio might alert the user to an incoming message (e.g., behaving like a pager) or defer the interruption until later (e.g., behaving like a secretary who is screening calls during an important meeting) depending on per-<Scene/> QoI metrics adjudicated in this phase.

5.2.6 Act

"Acting" initiates the selected processes using effectors that access the external world or the CR's internal states.

Access to the external world consists primarily of composing messages to be spoken in the local environment or expressed in text locally or to another CR or CN in KQML, RKRL, OWL, RXML, or some other appropriate knowledge interchange standard.

Actions on internal states include controlling resources such as radio channels. CR action can also update existing internal models, for example, by adding a new serModel to an existing internal set of models such as adding a word to the word sense set. Such new models may be asserted by an action of the <Self/> to encapsulate experience. Experience may be actively integrated into RXML knowledge structures as well. Knowledge acquisition may be achieved by an action that creates the appropriate data structures.

5.2.7 Learning

Learning depends on perception, observations, decisions, and actions. Initial learning is mediated by the Observe phase in which all sensory perceptions are continuously matched against all prior experience to continually count occurrences and to remember time since last occurrence of the stimuli from primitives to aggregates.

Learning may occur when a new type of serModel is created in response to an Action to instantiate an internally generated serModel. For example, prior and current internal states may be compared with expectations to learn about the effectiveness of a communications mode, instantiating a new RF mode-specific reactive serModel.

5.2.8 Retrospection

Since the assimilation of knowledge by machine learning can be computationally intensive, cognitive radio has “sleep” and “prayer” epochs for additional machine learning. A sleep epoch is a relatively long period of time (e.g., minutes to hours) during which the radio will not be in use, but has sufficient electrical power for processing. During the sleep epoch, the radio can run machine learning algorithms without detracting from its ability to support its user’s needs. Machine learning algorithms may integrate experience by analyzing statistical parameters running GAs and examining exceptions. The sleep epoch may rerun stimulus–response sequences with new learning parameters reminiscent of the way that people dream. The sleep cycle could be less anthropomorphic, employing a genetic algorithm to explore a rugged fitness landscape, potentially improving the decision parameters from recent experience. Learning opportunities not resolved in the sleep epoch can be brought to the attention of the user, the host network, or a designer during a prayer epoch so named for bringing a problem that the <Self/> cannot solve to a higher authority.

5.3 CRA III: THE INFERENCE HIERARCHY

The phases of inference from observation to action show the flow of inference, while the inference hierarchy organizes the related data structures. Inference hierarchies have been in use since Hearsay II in the 1970s, but the CR hierarchy is unique in its method of integrating machine learning with real-time performance during the wake epochs. An illustrative inference hierarchy includes layers from atomic stimuli at the bottom to information clusters that define action contexts as in Figure 5-4.

The pattern of accumulating elements into sequences begins at the bottom of the hierarchy. Atomic stimuli originate in the external environment including RF, acoustic, image, and location domains among others. The atomic symbols are the most primitive symbolic units in the domain. In speech, the most primitive elements are the phonemes. In the exchange of textual data (e.g., email), the symbols are the typed characters. In images, the atomic symbols may be small groups of pixels (“blobs”) with similar hue, intensity, texture, and so on.

A related set of atomic symbols forms a primitive sequence. Words in text, tokens from a speech tokenizer, and objects in images [146] are the primitive

Sequence	Level of Abstraction
Context cluster	***Scenes*** in a play, Session
Sequence clusters	***Dialogs***, paragraphs, protocol
Basic sequences	***Phrases***, video clip, message
Primitive sequences	***Words***, token, image
Atomic symbols	***Raw data***, phoneme, pixel
Atomic stimuli	External phenomena

FIGURE 5-4 Standard inference hierarchy.

sequences. Primitive sequences have spatial and/or temporal coincidence, standing out against the background (or noise). Basic sequences communicate discrete messages. These discrete messages (e.g., phrases) may be defined with respect to an ontology of the primitive sequences (e.g., definitions of words). Sequences cluster together because of shared properties. For example, phrases that include words like "hit," "pitch," "ball," and "out" may be associated with baseball. Knowledge discovery and data mining (KDD) and the semantic web offer approaches for defining or inferring the presence of such clusters from primitive and basic sequences.

A scene is a context cluster, a multidimensional space–time–frequency association, such as a discussion of a baseball game in the living room on a Sunday afternoon. Such clusters may be inferred from unsupervised machine learning, for example, using statistical methods or nonlinear methods like support vector machines (SVMs) [98]. The progression from stimuli to clusters generalizes data structure across sensory perception domains.

5.3.1 Vertical Cognition Components

Cognition components may be integrated vertically into this hierarchical data structure framework. For example, Natural Language Processing (NLP) tool sets may be embedded into the CRA inference hierarchy as illustrated in Figure 5-5. Speech channels may be processed via NLP facilities with substantial a priori models of language and discourse. AACRs need to access those models via mappings between the word, phrase, dialog, and scene levels of the observation phase hierarchy and the encapsulated speech component(s).

Illustrative NLP components include IBM's ViaVoice NLP research tools like SNePS [147], AGFL [148], or XTAG [149] and morphological analyzers like PC-KIMMO [150]. These tools go both too far and not far enough in the direction needed for CRA. One might like to employ existing tools using the errorful transcript to interface between the domain of radio engineering and such tool sets. At present, one cannot just express a radio ontology in Interlingua and plug it neatly into XTAG to get a working cognitive radio. The internal data structures needed to mediate the performance of radio tasks

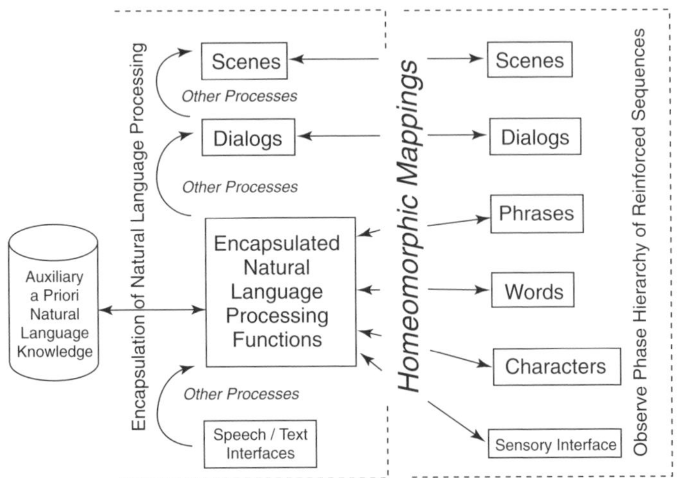

FIGURE 5-5 Natural language encapsulation in the observation hierarchy.

(e.g., "transmit a waveform") differ from the data structures that mediate the conversion of language from one form to another. Thus, XTAG wants to know that "transmit" is a verb and "waveform" is a noun. The CR needs to know that if the user says "transmit" and a message has been defined, then the CR should call the SDR function *transmit( )*. NLP systems also need scoping rules for transformations on the linguistic data structures. The way in which domain knowledge is integrated in linguistic structures of these tools may obscure the radio engineering aspects. Although experts skilled with language tools can create domain-specific dialogs, at present no tool can automatically synthesize the dialogs from a radio domain ontology. Integrating speech, vision, and data exchanges together to control a SDR is in its infancy and presents substantial technology challenges that motivated the inclusion of such vertical NLP tools in the CRA.

5.3.2 Horizontal Cognition Components

Radio skills may be embodied in horizontal cognition components. Some radio knowledge is static, requiring interpretation by an algorithm such as an inference engine to synthesize skills. Alternatively, radio skills may be embedded in active data structures like serModels through the process of training or sleeping. Organized as horizontal maps primarily among wake-cycle phases observe and orient, the horizontal radio procedure skill sets (SSs) control radio personalities as illustrated in Figure 5-6. With horizontal serModels there are no logical dependencies among components that delay the application of the knowledge. With First Order Predicate Calculus (FOPC), the theorem prover must reach a defined state in the combinatorially explosive

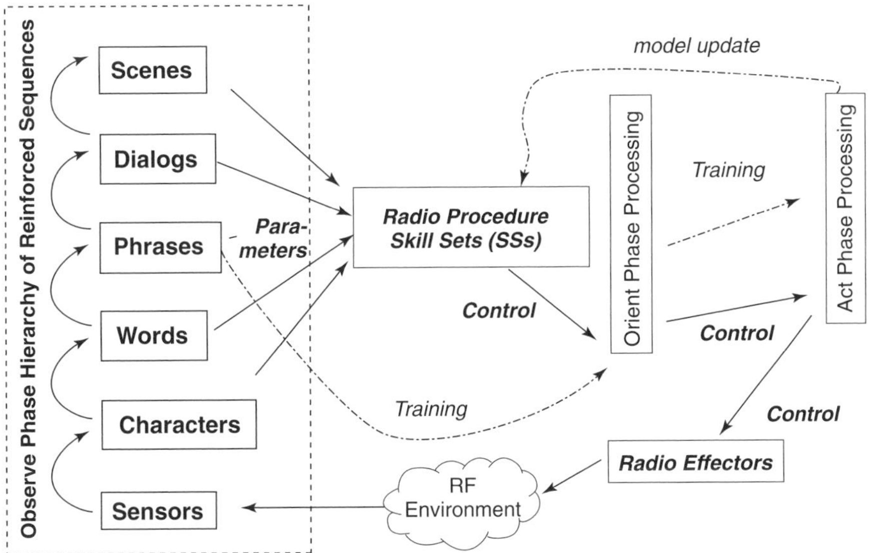

FIGURE 5-6 Radio skills respond to observations.

resolution of multiple axioms in order to initiate action. In contrast, serModels are continually compared to the level of the hierarchy to which they are attached, so their immediate responses are always cascading toward action.

Nothing precludes speech, text, vision, or other senory-perception domains from using horizontal cognition components to synthesize such reactive behaviors within one domain or across domains.

5.3.3 General World Knowledge

An AACR needs substantial knowledge embedded in the inference hierarchies. It needs both external RF knowledge and internal radio knowledge. Internal knowledge enables it to reason about itself as a radio. External radio knowledge enables it to reason about the role of the <Self/> in the world, such as respecting rights of other cognitive and legacy radios.

Figure 5-7 illustrates the classes of prior and dynamic knowledge an AACR needs to employ in the inference hierarchies and cognition cycle. It is one thing to write down that the Universe includes a Physical World (there could also be a spiritual world, and that might be very important in some cultures). Examples abound in the semantic web. It is quite another thing to express that knowledge in a way that the AACR can effectively employ. Symbols like "Universe" take on meaning by their relationships to other symbols and to external stimuli. In the CRA <Universe/> ontology, metalevel *abstractions* are distinct from existential knowledge of the physical Universe. In RXML,

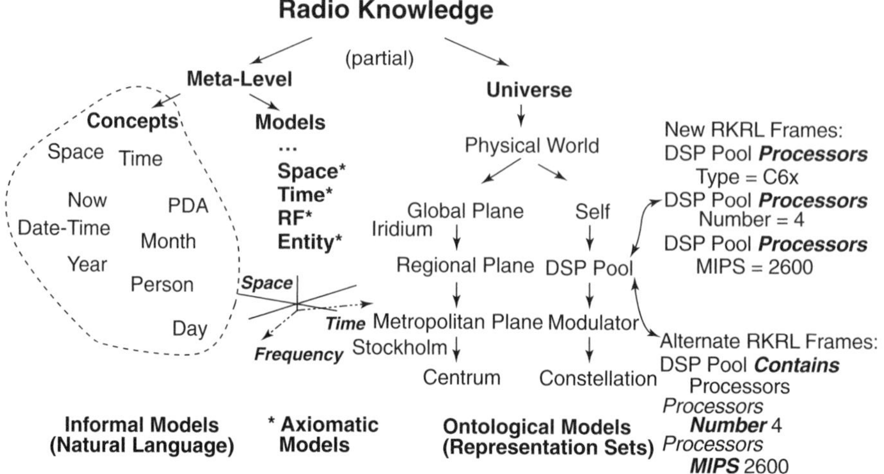

FIGURE 5-7 External radio knowledge includes concrete and abstract knowledge.

this ontological perspective includes all in a universe of discourse, <Universe/> (Expression 5-2).

Expression 5-2 The Universe of Discourse of AACR Consists of Abstractions Plus the Physical Universe

<Universe>
<Abstractions> <Time> <Now/> </Time> <Space> <Here/> </Space> . . . <RF/> . . .
<Intelligent-entities/> . . . </Abstractions>
<Physical-universe> . . . <Instances/> of Abstractions . . .
</Physical-universe>
</Universe>

Abstractions include informal and formal metalevel knowledge from unstructured concepts to the mathematically structured models of space, time and RF. To differentiate "now" as a temporal concept from "Now" as the Chinese name of a plant, the CRA includes both the a priori knowledge of "now" as a space–time locus, <Now/>, as well as associated functions ("methods") that access and manipulate instances of the concept <Now/>. Definition-by-method permits the cognition component to reason about whether a given event is in the past, present, or future.

Given the complexity of a system that includes both a multitiered inference hierarchy and the cognition cycle's observe–orient–plan–decide–act sequence with AML throughout, it is helpful to consider the mathematical structure of these information elements, processes, and flows.

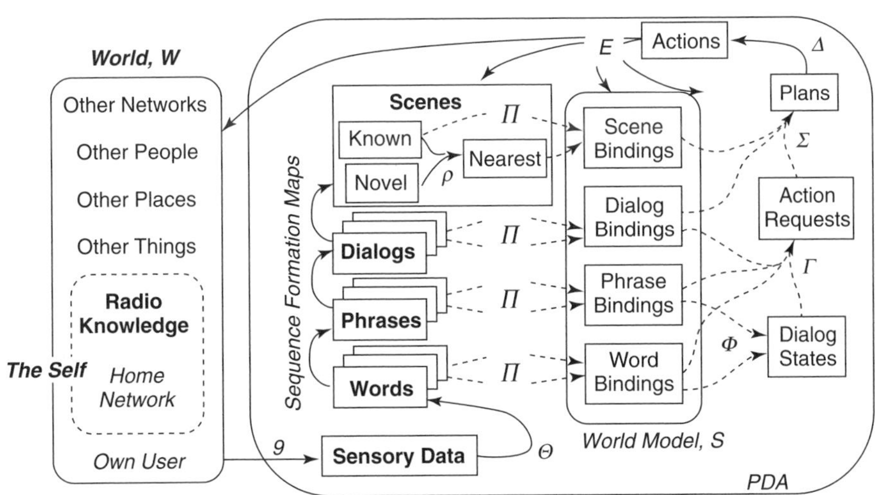

FIGURE 5-8 Architecture based on the cognition cycle.

5.4 CRA IV: ARCHITECTURE MAPS

Cognition components are implemented via data structures, processes, and flows that may be modeled as topological maps over the abstract domains identified in Figure 5-8.

The <Self/> is an entity in the world, while the internal organization of the <Self/> (annotated PDA in the figure) is an abstraction that models the <Self/>. The model data structures are generalized words, phrases, dialogs, and scenes that may be acoustic, visual, or perceived in other sensory domains (e.g., infrared). These structures refer to set-theoretic spaces consisting of a set X and a family of subsets Ox that contain $\{X\}$ and $\{\,\}$ the null set and that are closed under union and countable intersection, a topological space induced over the domain. The dissertation [145] and companion CD-ROM and web site develop these more theoretical considerations that helped shape the CRA.

Although the CRA provides a framework for APIs, it doesn't specify the details of the data structures or of the maps. Other theoretical issues for industrial strength CRA include properties of the architecture maps that reflect the following:

1. *Noise*, in utterances, images, objects, location estimates, and the like. Noise sources include thermal noise, conversion error introduced by the process of converting analog signals (audio, video, accelerometers, temperature, etc.) to digital form, error in converting from digital to analog form, preprocessing algorithm biases, and random errors, such as the accumulation of error in a digital filter, or the truncation of a low energy

signal by threshold logic. Dealing effectively with noise differentiates tutorial examples of cognition from the real thing.

2. *Hypothesis management*, keeping track of more than one possible binding of stimuli to response, dialog sense, scene, and so on. Hypotheses may be managed by keeping the N-best hypotheses (with an associated degree of belief), by estimating the prior probability or other degree of belief in a hypothesis, and keeping a sufficient number of hypotheses to exceed a threshold (e.g., 99% of all the possibilities), or keeping hypotheses until the probability for the next most likely hypothesis is less than some threshold. The estimation of probability requires a measurable space, a sigma-algebra that defines how to accumulate probability on that space, proof that the space obeys the axioms of probability, and a certainty calculus that defines how to combine degrees of belief in events as a function of the measures assigned to the probability of the event.
3. *Trustable training interfaces*, the reverse flow of knowledge from the inference hierarchy back down to the perception components. The recognition of the user by a combination of face and voice could be more reliable than single-domain recognition either by voice or by vision, so the user training the CR must be able to synthesize such cross-domain associations painlessly. Visual recognition of the Owner outdoors in a snowstorm, for example, is more difficult than indoors in an office. While the CR might learn to recognize the user based on weaker cues outdoors, access to private data might be constrained until the quality of the recognition exceeds some learned threshold balancing ease of use against trust.
4. *Nonlinear flows*. Although the cognition cycle emphasizes the forward flow of perception enabling action the CRA must accommodate reverse flows (e.g., from perception to training) and different flow rates among horizontal and vertical cognition components.

5.5 CRA V: BUILDING THE CRA ON SDR ARCHITECTURES

Cognitive radio may be realized via software-defined radio (SDR) with sensory perception, RF autonomy, and integrated machine learning of the self, the user, the environment, and the "situation." This section reviews SDR and the SDR Forum's SCA as a model of the SDR <Self/> and as a starting point for AACR evolution.

5.5.1 SDR Principles

Hardware-defined radios such as the typical AM/FM broadcast receiver convert radio to audio using radio hardware, such as antennas, filters, analog

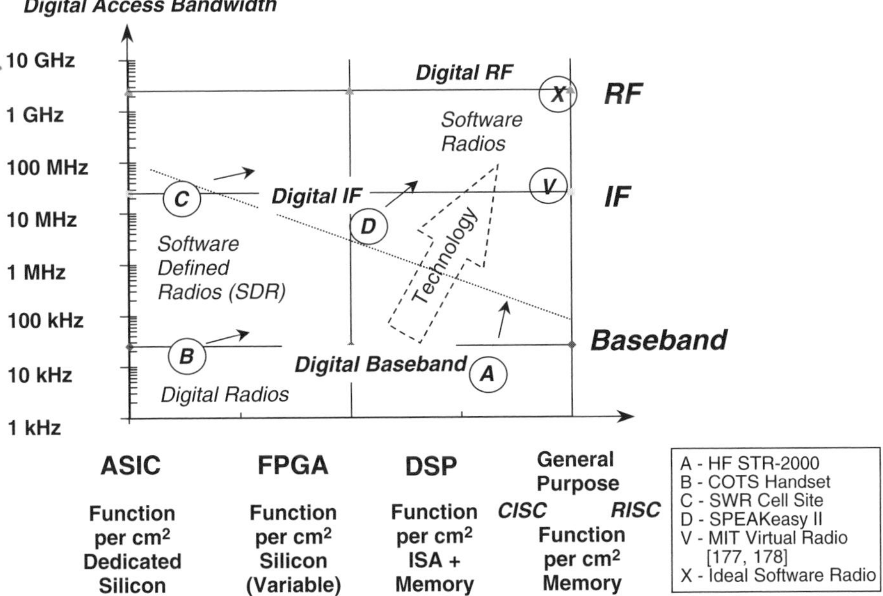

FIGURE 5-9 SDR design space shows how designs approach the ideal SWR.

demodulators, and the like. In the ideal software radio [144], analog-to-digital converter (ADC) and digital-to-analog converter (DAC) convert digital signals to and from radio frequencies (RFs) directly, and all RF channel modulation, demodulation, frequency translation, and filtering are accomplished in software.

Since the ideal software radio is not readily implemented, the SDR has comprised a sequence of practical steps from the baseband DSP of the 1990s toward the ideal. As the economics of Moore's Law and of increasingly wideband RF and IF devices allow, implementations move upward and to the right in the SDR design space (Figure 5-9).

This space consists of the combination of digital access bandwidth and programmability. Access bandwidth consists of ADC/DAC sampling rates converted by the Nyquist criterion and practice into effective bandwidth. Programmability of the digital subsystems is defined by the ease with which logic and interconnect may be changed after deployment. Application-specific integrated circuits (ASICs) cannot be changed at all, so the functions are "dedicated" in silicon. Field programmable gate arrays (FPGAs) can be changed in the field, but if the new function exceeds some parameter of chip capacity, which is not uncommon, then one must upgrade the hardware to change the function, just like ASICs. Digital signal processors (DSPs) are typically easier or less expensive to program but are less efficient in power use than FPGAs. Memory limits and instruction set architecture (ISA)

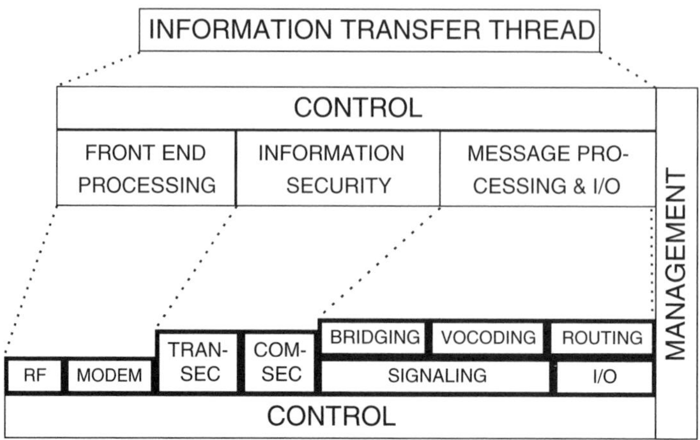

FIGURE 5-10 SDR Forum (MMITS) information transfer thread architecture.

complexity can drive up costs of reprogramming the DSP. Finally, general purpose processor software, particularly with reduced instruction set architectures (RISC), are most cost-effective to change in the field.

5.5.2 Radio Architecture

For SDR, the critical hardware components are the linear wideband analog RF transmitter, IF receiver, ADC, DAC, and processor suite. The critical software components are the user interface, the networking software, the information security (INFOSEC) capability (hardware and/or software), the RF media access software, including the physical layer modulator and demodulator (modem) and media access control (MAC), and any antenna-related software such as antenna selection, MIMO beamforming, pointing, and the like. INFOSEC consists of transmission security (TRANSEC), such as frequency hopping, plus communications security (COMSEC), typically encryption.

The SDR Forum defined a very simple, helpful model of radio in 1997, shown in Figure 5-10. This model highlights the relationships among radio functions. The CR has to "know" about these functions, so every CR must have at least an internal model of radio like this.

This model and the techniques for implementing various degrees of SDR are addressed in depth in the various texts on SDR [114, 151].

The self-referential model of a wireless device used by the CRA, RKRL 0.4 and the RXML <Self/>, is illustrated in Figure 5-11. This radio knows about sources, source coding, networks, INFOSEC, and the collection of front-end services needed to access RF channels. This model includes multiple channels and their characteristics (the channel set), so that the radio may

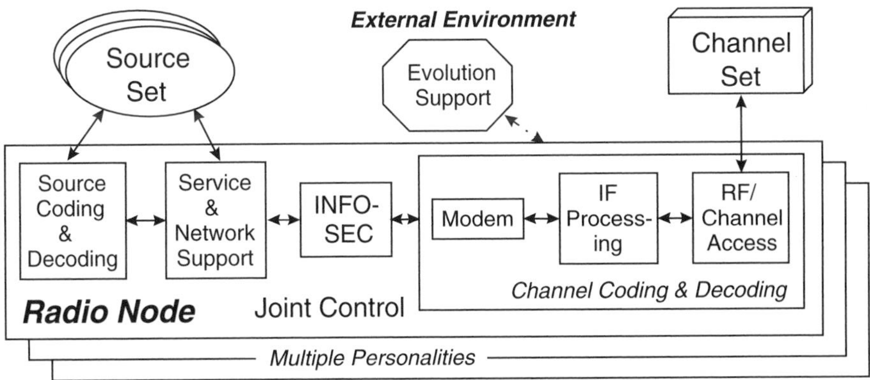

FIGURE 5-11 Functions–transforms model of a wireless node.

have many alternative personalities at a given point in time. Through evolution support those personalities may change over time.

Since CR reasons about all of its internal resources, the CRA requires a computational model of analog and digital performance parameters and how they are related to features the <Self/> can measure or control. MIPS, for example, may be controlled by setting the clock speed. A high clock speed generally uses more total power than a lower clock speed, and this tends to reduce battery life. The same is true for the brightness of a display. The CR only "knows" this to the degree that a data structure captures this information and algorithms, preprogrammed or learned, deal with these relationships to the benefit of QoI. Constraint languages may express interdependencies, such as how many channels of a given personality are supported by a given hardware suite, particularly in failure modes.

The ontological primitives of the above may be formalized as follows:

Expression 5-3 SDR Subsystem Components

```
<SDR>
  <Sources/> <Channels/> <Personality>
    <Source-coding-decoding/> <Networking/> <INFOSEC/>
    <Channel-codec> <Modem/> <IF-processing/> <MIMO/>
      <RF-access/> </Channel-codec>
    </Personality>
  <SDR-platform/> <Evolution-support/>
</SDR>
```

This text leaves the formal ontology of SDR to industry groups like the SDR Forum and OMG, focusing instead on ontological constructs that enhance QoI.

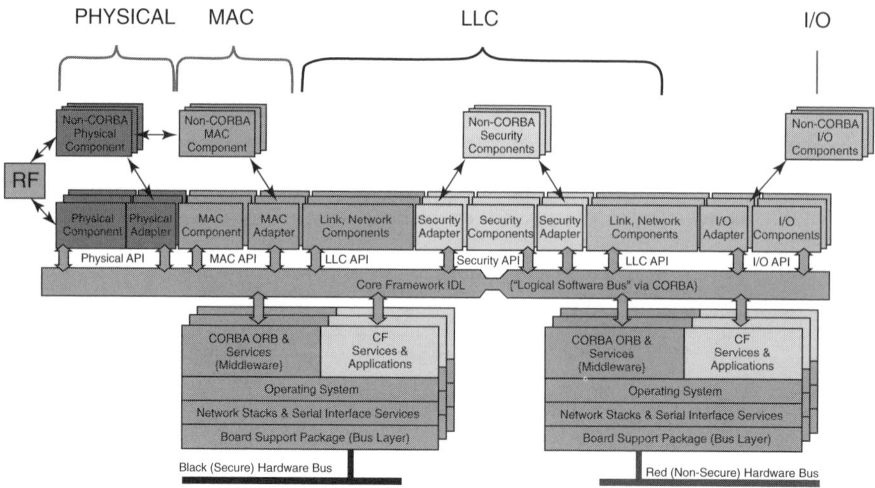

FIGURE 5-12 SCA Version 1.0. (© SDR Forum, used with permission.)

5.5.3 The SCA

The U.S. DoD developed the SCA for its Joint Tactical Radio System (JTRS) family of radios [79]. The SCA identifies the components and interfaces shown in Figure 5-12. The APIs define access to the physical layer, to the media access control (MAC) layer, to the logical link control (LLC) layer, to security features, and to the input/output of the physical radio device. The physical components consist of antennas and RF conversion hardware that are mostly analog and that therefore typically lack the ability to declare or describe themselves to the system. Most other SCA-compliant components are capable of describing themselves to the system to facilitate plug-and-play among hardware and software components. In addition, the SCA embraces POSIX and CORBA, although the SCA has also been implemented in Java as well.

The SCA evolved through several stages of work in the SDR Forum and Object Management Group (OMG) into a UML-based object-oriented model of SDR (Figure 5-13). Waveforms are collections of load modules that provide wireless services, so from a radio designer's perspective, the waveform is the key application in a radio. From a user's perspective of a wireless PDA, the radio waveform is just a means to an end, and the user doesn't want to know or have to care about waveforms. Today, the cellular service providers hide this detail to some degree, but consumers sometimes know the difference between CDMA and GSM, for example. With the deployment of the third generation of cellular technology (3G), the amount of techie jargon consumers need to know is increasing. So the CR increases access to ad hoc networks and the wireless Web but insulates the user from those details, unless the user really wants to know.

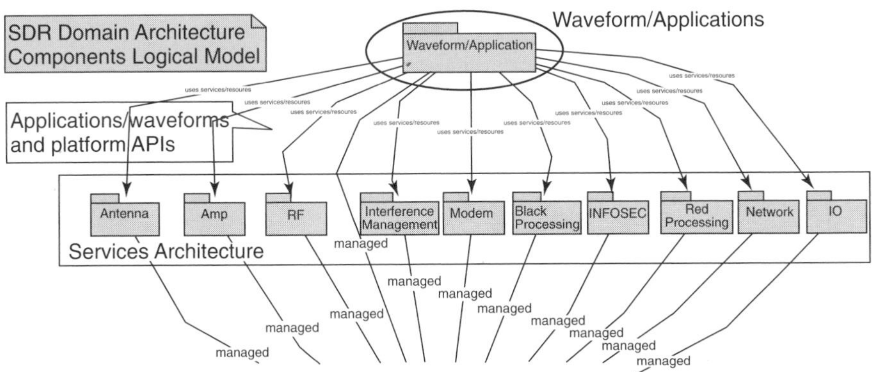

FIGURE 5-13 SDR Forum UML model of radio services. (© SDR Forum, used with permission.)

In the UML model, Amp refers to amplification services, RF refers to RF conversion, and Interference Management refers to both avoiding the generation of interference and filtering it out of one's band of operation. In addition, the jargon for U.S. military radios is that the "red" side contains the user's private information, but when it is encrypted it becomes "black" or protected, so it can be transmitted. Black processing occurs between the antenna and the decryption process. In this figure there is no user interface. The UML model contains a sophisticated set of management facilities, illustrated further in Figure 5-14, to which human–machine interface (HMI) or user interface is closely related.

Systems control is based on a framework that includes generic functions like event logging, organized into a computational architecture. The management features are needed to control radios of the complexity of 3G and of the corresponding generation of military radios. Fault management features deal with loss of a radio's processors, memory, or antenna channels. CR therefore interacts with fault management to determine what facilities may be available to the radio given recovery from hardware and/or software faults (e.g., error in a download). Security management protects the user's data, balancing convenience and security, which can be very tedious and time consuming. The CR will direct virtual channel management (VCM) and will learn from the VCM function what radio resources are available, such as what bands the radio can listen to and transmit on and how many it can do at once. Finally, SDR performance depends on the parameters of analog and digital resources, such as linearity in the antenna, millions of instructions per second (MIPS) in a processor, and the like.

5.5.4 Architecture Migration: From SDR to AACR

Given the CRA and contemporary SDR architecture, one must address the transition of SDR, possibly through a phase of AACRs toward the iCR. As

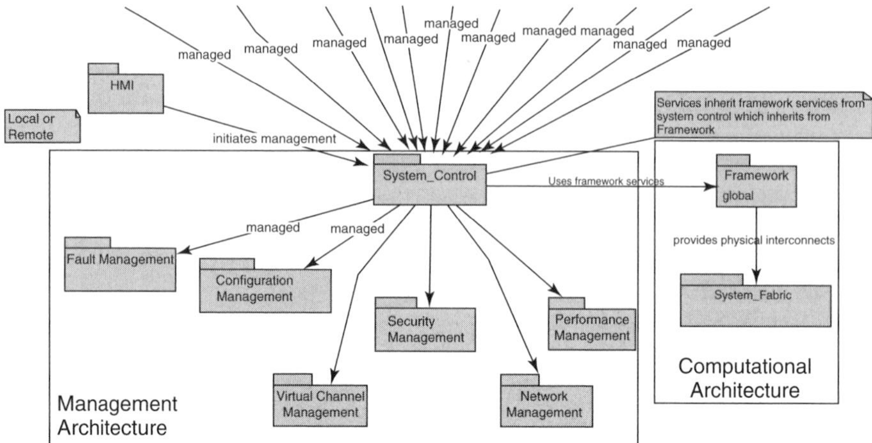

FIGURE 5-14 SDR Forum UML management and computational architectures. (© SDR Forum, used with permission.)

the complexity of hand-held, wearable, and vehicular wireless systems increases, the likelihood that the user will have the skill necessary to do the optimal thing in any given circumstance decreases. Today's cellular networks manage the complexity of individual wireless protocols for the user, but the emergence of multiband multimode AACR moves the burden for complexity management toward the PDA. The optimization of the choice of wireless service between the "free" home WLAN and the for-sale cellular equivalent moves the burden of radio resource management from the network to the WPDA.

In the migration process the CRA could increase the computational intelligence of a wireless laptop. It could know about the user by observing keystrokes and mouse action as well as by interpreting voice and the images on its camera, for example, to verify that the Owner is still the user since that is important to building user-specific models. It might build a space–time behavior model of any user or it might be a trustable single-user laptop.

In 1999, Mitsubishi and AT&T announced the first "four-mode handset." The T250 operated in TDMA mode on 850 or 1900 MHz, in first generation Analog Mobile Phone System (AMPS) mode on 850 MHz, and in Cellular Digital Packet Data (CDPD) mode, a multiband, multimode, multimedia wireless handset. These radios enhanced the service provider's ability to offer national roaming, but the complexity was not apparent to the user since the network managed the radio resources in the handset.

As the number of bands and modes increases, the SDR becomes a better candidate for the insertion of cognition technology. But it is not until the radio or the wireless part of the PDA has the capacity to access multiple ad hoc RF bands such as nodes of a ubiquitous wireless Web that cognition technology begins to pay off. With the liberalization of RF spectrum-use rules, the early

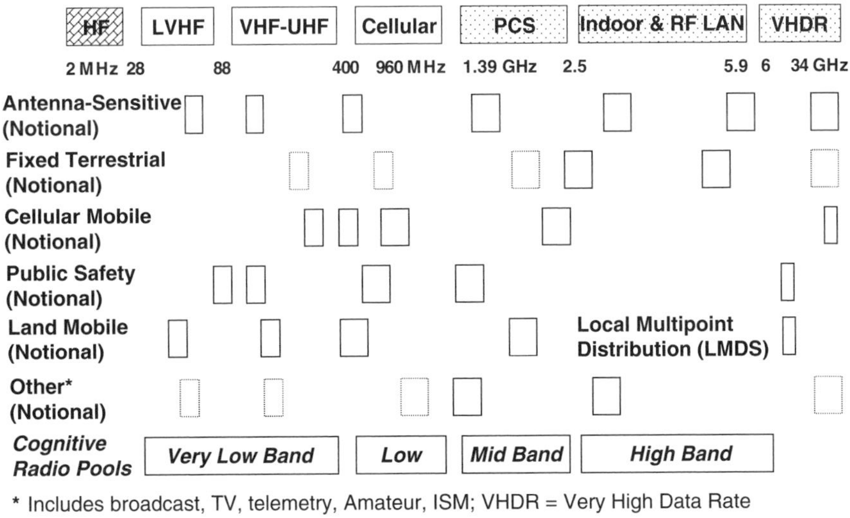

FIGURE 5-15 Fixed spectrum allocations versus pooling with cognitive radio.

evolution of AACR may be driven by RF spectrum-use etiquette for ad hoc bands like the FCC use case. In the not-too-distant future, SDR PDAs could access satellite mobile services, cordless telephone, WLAN, GSM, and 3G bands. An ideal SDR device with these capabilities might affordably access octave bands from 0.4 to 0.96 GHz (skip the air navigation and GPS band from 0.96 to 1.2 GHz), from 1.3 to 2.5 GHz, and from 2.5 to 5.9 GHz (Figure 5-15). Not counting satellite mobile and radio navigation bands, such radios would access over 30 mobile subbands in 1463 MHz of potentially sharable outdoor mobile spectrum. The upper band provides another 1.07 GHz of sharable short-range, indoor and RF LAN spectrum. This wideband radio technology will be affordable for military applications, for base station home and business infrastructure, for mobile vehicular radios, and later for handsets and PDAs. When a radio device accesses more RF bands than the host network controls, CR technology can mediate the dynamic sharing of spectrum. It is the well-heeled conformance to the radio etiquettes afforded by iCR that makes such sharing practical.

Various protocols have been proposed by which radio devices may share the radio spectrum. The U.S. FCC Part 15 rules permit low power devices to operate in some bands. In 2003, a Rule and Order (R&O) made unused television (TV) spectrum available for low power RF LAN applications, making the manufacturer responsible for ensuring that the radios obey constraints [64]. DARPA's neXt Generation (XG) program developed a language for expressing spectrum-use policy [152]. Other more general protocols based on peek-through to legacy users have also been proposed [157].

Does this mean that a radio must transition instantaneously from the SCA to the CRA? Not at all. The six-component CRA may be implemented with minimal sensory perception, minimal learning, and no autonomous ability to modify itself. Regulators hold manufacturers responsible for the behaviors of such radios. The simpler the architecture, the simpler the problem of explaining it to regulators and of getting concurrence among manufacturers for open architecture interfaces. Manufacturers who fully understand the level to which a highly autonomous CR might unintentionally reprogram itself to violate regulatory constraints may decide to field aware–adaptive (AA) radios, but may not want the risks of a self-modifying CR just yet.

Thus, one can envision a gradual evolution toward the CRA beginning initially with FCC CR for XG′ and a minimal set of functions mutually agreeable among the growing community of AACR stakeholders. Subsequently, access to new radio access points on the wireless Web will reduce costs for new services to drive the evolution toward iCR with additional APIs, perhaps informed by the CRA.

5.6 COGNITION ARCHITECTURE RESEARCH TOPICS

The cognition cycle and related inference hierarchy imply a large scope of hard research problems for cognitive radio. Parsing incoming messages requires next-generation natural language text processing to overcome jargon, anaphora, and elipsis in semi-structured message exchanges. Scanning the user's acoustic environment for voice content that further defines the communications context requires next-generation speech processing. Planning technology offers a wide range of alternatives in temporal calculus [153], constraint-based scheduling [163], task planning [162], and causality modeling [146], that have yet to be leveraged for iCR. Resource allocation includes algebraic methods for wait-free scheduling protocols [142], open distributed processing (ODP), and parallel virtual machines (PVMs). Finally, machine learning remains one of the core challenges in artificial intelligence research [45]. The focus of this cognitive radio research, then, is not on the development of any one of these technologies per se. Rather, it is on the organization of cognition tasks and on the development of cognition data structures needed to integrate contributions from these diverse disciplines as applicability to AACR becomes clear.

5.7 EXERCISES

5.1. What is the FCC's (or your local regulatory authority's) current policy regarding spectrum sharing?

(a) Regarding the use of cognitive radios?

(b) Do regulatory authorities differentiate adequately among adaptive, aware, and cognitive radio?

(c) How could additional clarity enhance sharing?

(d) Assess the degree to which FCC policy on cognitive radios has stimulated and/or stifled SDR innovation.

5.2. At what point does it make sense to identify a cognition component for radio architecture?

5.3. Is sensory perception in visual or auditory domains or both a prerequisite for cognition?

5.4. Can machine learning be integrated into a radio that lacks auditory or visual perception or both? How?

PART II

RADIO-DOMIAN COMPETENCE

CHAPTER 6

RADIO-DOMAIN USE CASES

This chapter takes the next step in AACR by quantifying the value proposition. It also develops additional use cases along with the next level of detail in the technical issues of implementing the use cases.

6.1 RADIO USE-CASE METRICS

Today's radios function primarily as bit pipes, wireless paths for voice and data. Although data now includes short messages, pictures, video clips, and the Web, the radio moves these bits but does not quantify their QoI. And mobile radios typically are not as good at bit-pipe data delivery as their hard-wired infrastructure counterparts. Since connectivity drives QoI, the value of mobility is high. These use cases therefore quantify the value of mobility and multiband multimode radio flexibility, where the radios themselves perceive the scene and process the bits to quantify <RF/> contribution to <User/> QoI.

6.1.1 QoS As a RF Use-Case Metric

Quality of service (QoS) is often emphasized as a limitation of wireless technologies. Of course, the bit error rate (BER), data rate (R_b), delay (dT), and delay jitter (σT) typically fall short of wireline performance by orders of magnitude as illustrated in Table 6-1. Mobility has its price.

TABLE 6-1 Wireless and Wireline QoS Parameters

QoS[a]	Dial Up	WLAN	Core Networks ISP and PSTN	Cellular	3G
R_b (bps)	56k	0.1–1.5M	1–100G	8k	100k
BER (10^)	–5	–6	–9	–3	–4
dT (ms)	100	10	1	30	10
σT (ms)	10	1	0.1	300[b]	100[b]
GoS (%)	99	99.9	99–99.999	70–90	80–95[b]

[a] Values are illustrative and not intended to be pejorative.
[b] Periods of moderate fading in a heavily loaded cellular system.

Different digital connectivity technologies have better or worse QoS, as quantified in Table 6-1 with illustrative values. Expression 6-1 internalizes via RXML the QoS parameters of alternatives for the <Self/> to obtain <Connectivity/> per Table 6-1, associating each with a <Path/> that consists of one or more communications <Modes>.

Expression 6-1 Connectivity Abstractions Differentiate <Paths/>

```
<Abstractions> <Connectivity/>
<Connectivity> <Path> <Mode/> </Path> </Connectivity>
<Schema> <Connectivity> <Path> <Mode>
  <QoS> <Rb (bps)/> <BER (10∧–)/> <dT(–ms)/> <σT(ms)/>
    <GoS%/> </QoS>
  </Mode> </Path> </Connectivity> </Schema>
<Connectivity>
  <Mode> <Dial Up/> <DSL/> <Core Network/> <Cellular/> <3G/>
    <BT/> </Mode> </Connectivity>
<Connectivity> <Path> <Mode> <Dial Up>
  <QoS> <Rb> 56k </RB> <BER> 5 </BER> <dT> 100 </DT> <σT>
    10 </σT> <GoS> 99</GoS> </QoS> </Dial Up> </Mode> </Path>
      </Connectivity>
<Connectivity> <Path> <Mode> <Cellular> . . . (see CD-ROM) . . .
</Connectivity/> </Abstractions>
```

The XML expression <BER> 5 </BER> means BER = 10∧(–5) = 10^{-5}. The other QoS metrics are similarly formatted per <Schema/>. In addition, the QoS of a <Path/> may be estimated as the worst QoS metrics of the constituent modes when connected.

6.1.2 Context Sensitivity of QoS

For many use cases there is high value of mobile connectivity. From the user perspective, the cell phone probability of connection is often much bigger

TABLE 6-2 Probability of Proximity (P_p)

Scene	Office	Home	Car	Restaurant
ISP	0.99	0.1	0.0	0.01
PSTN	0.99999	0.3	0.0	0.01
Cellular (urban)	0.9	0.99	0.99	0.99
Cellular (rural)	0.9	0.5	0.7	0.5
Cellular (remote)	0.1	0.01	0.05	0.01
Iridium (remote)	0.1	0.1	0.9	0.1

than zero, which is the probability of wireline access from the car, the boat, and the beach. In other words, while the quantitative measures of QoS of Table 6-1 make wireline bit pipes appear much better than wireless bit pipes, these metrics don't account for the user's experience.

Proximity probabilities in Table 6-2 are illustrative, based on the assumption that the ISP and PSTN are BlueTooth enabled so the CWPDA can access wireline and wireless modes for the user. In this model the CWPDA must be within 2 meters of an access point and personal offices are less than 4 meters across; homes have at least two rooms wired with PSTN out of a six-room house, but only one has an ISP port (e.g., DSL). Cellular and satellite telephone probability of proximity is estimated as probability of network connectivity. If the values of the QoS in Table 6-1 were weighted for the probability of proximity, the wireline values become nearly zero much of the time in a business day. This metric accurately reflects the dramatic success of cellular services in the 1990s with the right combination of price and performance for the marketplace and cultural inclinations. A CWPDA takes into account the situation dependence of the probability of connection. This strategy is expressed computationally in the QoI metric augmented with Tables 6-1 and 6-2 so CWPDA's can use <RF/> accordingly, seeking out WLAN-ISP access points, for example.

Knowledge of the likelihood of RF connectivity of various types underlies nearly all the RF use cases. Although all the knowledge could be learned, AACR evolution may be accelerated via a priori models updated by experience. The a priori knowledge includes abstract models of the scenes with QoS and QoI estimates in stereotypical scenes. These models may be updated by the CWPDA's ability to connect experience in a given scene, such as in an office or home.

Expression 6-2 <Scenes/> Represent Stereotypical QoS

<Abstractions> <Scene/>
<Scene> <Generic-setting/> <Office/> <Home/> <Car/> <Restaurant/>
<Pedestrian/> <Traffic-accident/> <Earthquake/>
<Severe-weather/> </Scene>

<Scene> <Generic-setting> <Mode/> </Generic-setting> </Scene>
<Abstractions/>

The CRA <Self/> scene abstractions define <Scene/> as one or more places like the office, home, or car to be parameterized by QoS and probability & proximity through experience. A <Pedestrian/> scene may characterize the mobility of the AACR but does not correspond to a specific place. Other generic scenes may be characterized by physical and social imperatives such as <Traffic-accident/>, <Earthquake/>, and <Severe-weather/>. In each of these scenes, human actors with AACRs are characterized by roles, such as <Injured/> or <Care-giver/>. The scene <Generic-setting/> is a place-holder to which <Scene/> attributes may be assigned through a discovery process. Each <Scene/> has one or more communications <Modes> and a probability of mode presence in a scene. The CRA doesn't require probability per se representing instead relative frequency of occurrence K/N, integer data that is easier to learn and on which probability-like inferences can be based. These may be integrated in reinforced hierarchical sequences.

6.1.3 Reinforced Hierarchical Sequence Describes RF Context

In the CRA whenever a stimulus is encountered, both the number of times that stimulus has been observed and the number of cognition cycles since last observation are updated in the CRA inference hierarchy to form reinforced hierarchical sequences (RHSs). These observations are referenced in RHSs to time and space.

Expression 6-3 Scene Model Showing RHSs

<Scene> <Office> <Observations> 1000 </Observations>
<ISP-WLAN> <OBS> 990 </OBS> </ISP-WLAN> <PSTN-BT>
<OBS>
999 </OBS> </PSTN-BT> . . .
</Office> </Scene>

The probabilities of Table 6-2 are readily expressed as RHS ratios of the presence of a communications mode in a scene (<OBS/>), reinforced independently of the experience of the scene itself (<Observations/>). Specifically, in Expression 6-3, the probability that a BlueTooth PSTN connection is available in the office is approximated by the relative frequency of occurrence of the observation of PSTN-BT in <Office/> versus the experience of the <Office/> scene. As the number of encounters with <Office/> increases, the precision of the reinforcements interpreted as probabilities increases. With, say, two visits to an office and two observations of <PSTN-BT>, one could write $P(\text{Office} \cap \text{PSTN-Terminal}) \cong 1.0$, interpreting relative frequency as an estimate of the probability of the joint events Office $\cap$ PSTN-BT. Keeping the raw counts of observations enables alternate hypothesis analysis (e.g., during a sleep cycle), for better QoI. Thus, the CRA stipulates RHSs from which applications may infer probability.

6.1.4 Autonomous Reasoning About RF in Use Cases

The iCR perception of the RF domain should estimate GoS and QoS in a way that most accurately reflects QoI so that the AACR may reason from the user viewpoint. This implies multidimensional matrices of probability densities of <Scene/>, <Location/>, <Mode/>, < GoS/>, and <QoS/> for QoI. Those probabilities are reflected in the RHS ratios of reinforcement of <Scene/> versus <Mode/> and in combinations of <Location/> and <Mode/> versus <QoS/>. Thus, a priori RHS models readily integrate experiential updates.

AACR <RF/> use cases enhance the probability of staying connected. Wireline telephony has optimized grade of service (GoS) to 99% or better probability of connecting on the first attempt, now expected of wireline systems in developed nations. Wireless service providers tend to not quote GoS in part because, unlike the wireline system, GoS is only partially under the control of the service provider. The rest is up to "Mother Nature." Even the best built-out cell phone network drops calls because of network loading and multipath fading. In urban settings, the many moving metal objects exacerbate multipath fading. Along with per-cell traffic imbalances, dead zones, co-channel interference, and other less common channel impairments conspire to cause dropped calls, which by now are expected by the cell phone user. AACR's multiband multimode radio (MBMMR) technology can mitigate network shortfalls by autonomously seeking other <Paths/>. GSM hops among its 200 kHz channels because when a given channel is faded, others are not as deeply faded at a given location and time. Thus, frequency hopping mitigates fade depth. AACR's MBMMR mode hopping can choose a WLAN-ISP access point in a cellular dead zone, for reduced cost or QoI optimization. When <Connectivity/> is based on choice, GoS and thus QoI increases. Specifically for independent ergodic paths, the probability that either Path1 or Path2 is available is the contrapositive of the probability that both are not available at the same time, or

$$\begin{aligned} P(\text{Path1 or Path2}) &= P(\text{Not }\{P(\text{Not(Path1) and Not(Path2)})\}) \\ &= 1 - (1 - P(\text{Path1}))\,(1 - P(\text{Path2})) \end{aligned} \quad (6\text{-}1)$$

These engineering principles are expressed to the <Self/> in RXML to quantify criteria for autonomous path selection in Expression 6-4.

Expression 6-4 Path Choice Yields Improved GoS

```
<Connectivity-choice>
  <Path> Path1 <GoS> X </GoS> </Path>
  <Path> Path2 <GoS> Y </GoS> </Path>
```

<OR> Path1 Path2 <GoS> 1-(1-X)(1-Y) </GoS> </OR>
</Connectivity-choice>

From the equation, it is clear that, like HF automatic link establishment (ALE), band agile SDR can transform a larger number of reasonably reliable (90%) RF channels in different bands into an aggregate more reliable meta-channel, theoretically approaching the "five-nines" level of the wireline as WLAN-ISP access points build out. Connectivity with flexibility enables new kinds of information services as developed further in the use cases.

6.1.5 Focusing RF Use Cases

The use cases of this chapter therefore generate ideas for AACR that go beyond the bit pipe to the MBMMR-platform with computational intelligence for situation-dependent improved connectivity, QoS, and QoI. An AAR detects the absence of a TV channel to form an ad hoc wireless LAN. An AACR knows that TV is just one of several MBMMR capabilities to form parallel paths for increased GoS. AACRs that perceive location infer the presence of hidden nodes, not transmitting and in need of interference-clear spectrum. AARs learn the space–time distribution of radio energy for control of high spatial-density wireless technologies like multiple-input multiple-output (MIMO), using independent multipath reflections as independent paths for multiplexed bitstreams [109–113, 154]. These technologies also enable new use cases with not just better data rates and connectivity, but with autonomous use of spatial knowledge, to improve spectrum use in ways not feasible with MIMO alone. Enthusiasm for such use cases can engender unreasonable expectations.

One of the key lessons learned from the artificial intelligence hype-bust cycle of the 1980s is that razor sharp focus on the value proposition of the use case with realistic expectations of results was the key to successful AI projects. Healthy skepticism would treat AACR as another AI technology with the risks of another boom–bust cycle. By sequencing use cases starting with the smallest subset of capabilities that are of the greatest value, and plotting a path of affordable and opportunistic evolution, one may deploy selected AACR technologies with reasonable expectations in a customer-driven way. The enabling user perception technologies are so limited that substantial effort (analysis, experimentation, and field trials) is required to insert awareness, adaptation, and learning into real-world settings. The use cases therefore identify (1) the significant entities (AACR, legacy, and human), (2) the significant information exchanges among those entities, (3) the behaviors required of the entities, and (4) the specific QoI enhancement in each of the use cases.

The industrial-strength method for analyzing use cases includes computer-aided software engineering (CASE) tools like the Unified Modeling Language (UML). High end UML tools such as ROSE from IBM Rational™

map UML to executable code. UML facilitates the formalization of use cases, but one need not employ UML for every use case. SDL, the Specification and Description Language of ITU-T (Standard Z.100), might be used instead. Analytica™, Matlab™, and Mathcad™ also offer powerful capabilities for quantifying the benefits and challenges of use cases.

The CASE tools can accelerate the implementation of a use case in software, but CRs must learn from experience. At present no UML tools address AML needs like training scripts that differentiate skill acquired through experience from preprogrammed behaviors. The source code of CRI in the companion CD-ROM/web site isn't based on UML, but on a CR learning-machine design environment for self-modifying CRs. This CR environment includes a simulation in which they can be trained and diagnosed. Since the diagnosis of skill gained from AML is complex, this text introduces ML training and diagnostics with suggestions for research addressing the deeper issues.

6.2 FCC UNUSED TV SPECTRUM USE CASE

Much of the interest in cognitive radio today entails the secondary use of radio spectrum that happens to be unoccupied at a given place and time. Therefore, this is a good use case to develop further. The FCC conducted a series of meetings and issued findings regarding secondary spectrum markets between 1999 and 2003. At present, the FCC supports cognitive radio for secondary use of radio spectrum. Plots like those presented to the FCC (see Figure 6-1) helped make the case [155].

The author first presented the idea of CR for spectrum management to the FCC on 6 April 1999 (see the companion CD-ROM/web site for the text of this statement). Later, the ideas were refined in a public forum on secondary spectrum markets [156] in a layperson's version of the core research ideas. The FCC's technical advisory committee recommended that the Commission pursue CR as a method for enhancing secondary spectrum markets. The FCC conducted an inquiry and posted a ruling and order encouraging CR as an enabler for secondary markets [155].

6.2.1 Single-Channel Spectrum Rental Vignette

This idea of CR for spectrum rental was first presented publicly at the IEEE Workshop on Mobile Multimedia Communications (MoMUC) [157]. In this paper, a "polite" spectrum rental protocol was defined and termed an etiquette. In addition to the basic idea for posting the availability of spectrum in a given place and time, the protocol defined a way of listening for the occurrence of legacy primary users of the spectrum that lack the ability to know that their spectrum is being used by others. The legacy users push to talk and the etiquette detects the legacy user and makes the channel available

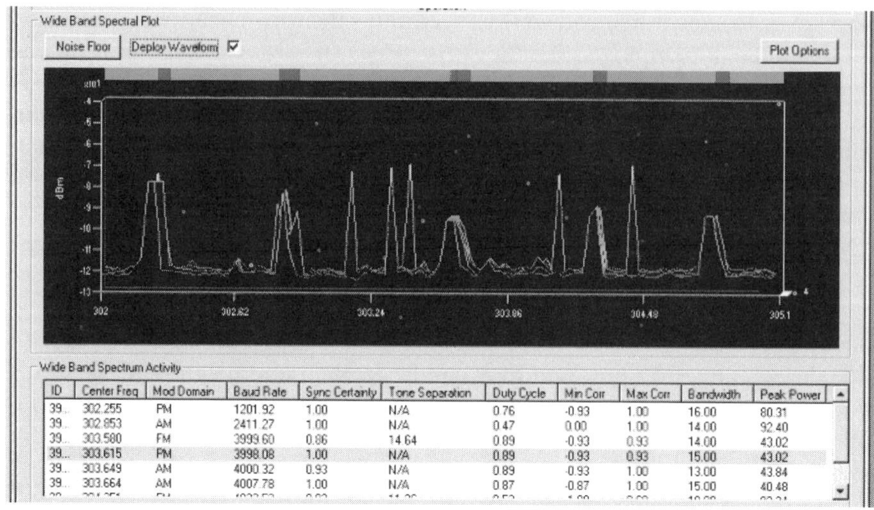

FIGURE 6-1 Illustrative spectrum occupancy. (Courtesy of Dr. Bruce Fette, "SDR Technology Implementation for the Cognitive Radio," FCC 19 May 2003, General Dynamics, Falls Church, VA.)

within 30–60 milliseconds. This self-contained etiquette use case appears in [158] and the companion CD-ROM.

To formalize this etiquette use case, one must describe the entities to the <Self/> as in Expression 6-5. The significant entities are the regulatory authority (<RA/>), multiple RA agents (<RAAgents/>), legacy radios that are mobile (<LRM/>), legacy radios that are fixed (<LRF/>), cognitive radios that are mobile (<CRM/>), and cognitive radios that are fixed (<CRF/>), along with the <Users/> and network <Operators/>, the people who take responsibility for the actions of the radio networks.

Expression 6-5 Spectrum Rental Ontology (Simplified)

```
<Abstractions> <RF> <Spectrum-rental> <Scene/>
   <Scene> <Entity/> <Services/> </Scene>
      <!– Entities provide services –>
   <Scene> <Entity> <Behavior/> <Message/> <!– via behaviors and
      messages –>
       </Entity> <Services/> </Scene>
   <Scene> <Entity> <RA/> <RAAgents/> <LRM/> <LRF/> <CRM/>
      <CRF/>
       <Behavior/> <Message/> </Entity> <Services/> </Scene>
</Spectrum-rental> </RF> </Abstractions>
```

Each of the entities of Expression 6-5 collaborates with the others by generating and receiving messages, behaving in accordance with the local knowledge obtained from the knowledge exchange (Expression 6-6). The RAs electronically <Publish/> a <Policy/> that enables the secondary spectrum user to <Rent/> spectrum in a given place and time. The place is an area, not a point, so the policy implies power and spatial controls on RF propagation. Those wishing to rent their spectrum <Publish/> a more detailed offering typically from fixed infrastructure. CRMs <Rent/> the spectrum for a specified <Time-interval/> using <Spectrum-cash/>, confirming <Satisfaction/> if the spectrum is available as advertised. Otherwise, the renter may register a <Complaint/> to the RA.

Expression 6-6 Spectrum Rental Instance

```
<Abstractions> <RF> <Spectrum-rental/> </Abstractions>
<Physical-universe> . . . <RF> <Spectrum-rental/> <Scene> <Location>
  "Fairfax, VA" </Location> </Scene>
  <Entity> <RA> FCC <Publish> <RF> 400MHz </RF>
    <Policy> "Enable spectrum rental"</Policy> </Publish> </RA>
    </Entity>
  <Entity> <RAA> Agent1 <Behavior>
    "Observes and enforces policy compliance" </Behavior> </RAA>
      </Entity>
  <Entity> <LRF> <Behavior>
    <Publish> <Rent> <RF> 405MHz </RF> <Time-interval>
      5minutes </Time-interval>
    <Spectrum-cash> $0.07 </Spectrum-cash> </Rent> </Publish>
      </Behavior> </LRF>
  <Entity> <CRM> AACR <Message> <Rent/> <Satisfied/>
    <Complain/> </Message>
    <Behavior> <Rent> <RF> 405MHz <Time> 1105 </Time> </RF>
      </Behavior>
    <Behavior> <Satisfied> <RF> 405MHz <Time> 1110 </Time> </RF>
      </Behavior>
    <Behavior> <Rent> <RF> 405MHz <Time> 1115 </Time> </RF>
      </Behavior>
    <Behavior> <Rent> <RF> 405MHz <Time> 1118 </Time> <Defer/>
      </RF> </Behavior>
    <Behavior>
      <Complain> <RF> = 405 MHz <Time> 1118 <Time/> <Reason>
        <Legacy/> </Behavior>
    </CRM> <CRF/> <Entity/> </Scene>
</Spectrum-rental> </RF> . . . </Physical-universe>
```

In the experience remembered in Expression 6-6, the AACR rented spectrum at 1105 AM for 5 minutes, paying $0.07 at 1110. The AACR was interrupted by a legacy user at 1118, so it had to <Defer/> the spectrum to the legacy user and therefore decided to <Complain/> to the offeror as shown. The signal structure for such spectrum sharing is outlined in the companion CD-ROM.

In this vignette the spectrum is rented for a small fee collected in spectrum-cash, funds that can be traded among parties without necessarily being converted to actual cash. Police, fire, rescue, and others that now radiate 10 W to reach their own sparse towers could employ spectrum with greater efficiency radiating 100 mW to nearby PCS towers on their legacy frequencies with legacy radios connected via ISP or cellular backhaul in exchange for secondary commercial use of public use spectrum. Such public–commercial spectrum pooling has great value at peak intensity hours of the day when cellular providers are turning away paying customers (e.g., 10 am). Such cognitively rented secondary spectrum could be a low-cost source of long-term revenue. Many social and regulatory issues must be resolved for this promising technical case to become deployed.

6.2.2 OFDM Spectrum Management Vignette

This vignette shows how a renter could occupy multiple empty channels at once using orthogonal frequency division multiplexing (OFDM). OFDM senses vacant space–time epochs of spectrum and formulates broadband energy to fill in the gaps. Friedrich Jondral and Timo Weiss, for example, have investigated opportunities for more agile use of spectrum via their spectrum measurement campaigns conducted specifically for OFDM spectrum pooling [159]. (See Figure 6-2.)

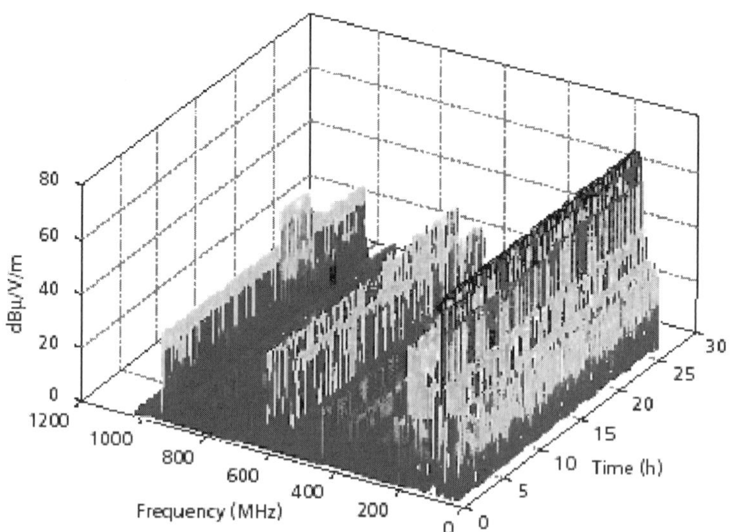

FIGURE 6-2 Spectrum use in Lichteneau, Germany in 2003. (Courtesy of Dr. Friedrich K. Jondral, Universität Karlsruhe, Germany; used with permission.)

Although subbands are allocated to licensed users, secondary OFDM users may employ unused spectrum either by regulatory permission or by renting from the primary user. The waveforms that optimize sharing fit in the vacancies between instantaneously used spectrum. OFDM features [160] enable the tailoring of multiple orthogonal frequency-domain carriers to fit in the instantaneous cracks between legacy users.

To complete this use case one must abstract the significant entities and formalize their behaviors for the <Self/>. In an interesting companion to the Jondral–Weiss ideas, the Communications Research Centre (CRC) of Canada examines the 5250 to 5320 MHz license-exempt band [161, 162].

The CRC shows empirically the relationship between wind and signal strength for small RF cells on an optical backbone.

6.2.3 Regulatory Authority Vignettes

The perspective of the RAs loom large in the various spectrum management use cases. The many regulatory issues to be addressed in flexible spectrum management include the certification of hardware–software configurations to transmit; licensing entities and collecting tariffs; monitoring for abuses; and enforcing penalties [163].

6.2.3.1 European Regulatory Perspective

In September 2003, the U.K. Regulatory Authority's Regulatory Round Table on SDR addressed European concerns about the management of SDR and cognitive radio. The workshop was organized by Walter Tuttlebee, Director, Mobile Virtual Communications Enterprise (VCE) for the U.K. RA. At that time, regulators stated a need for clearer definitions to facilitate regulation.

European regulators planned to keep the burden of type certification on the original equipment manufacturers (OEMs) as defined in the EU telecommunications regulatory framework. This framework wiped out "type certification" by the RAs in favor of rules ("R&TTE") that consolidated the markets of the member states into a single integrated market, third largest in the world after the United States and Japan, with China growing fast. This means that OEMs certify that they comply with published constraints like spectrum masks and co-use limits. This regulatory framework is one of the most permissive regulatory styles, intended to promote market consolidation and economies of scale.

Expression 6-7 Liability Is Assigned by the RA to the OEM Who Certifies a Configuration

<Abstractions> <Liability/>
<Liability> "Legal responsibility for failure to comply with policy"
</Liability>

<Liability> <Entity> <RA/> <OEM/> <Operator/> <User/> </Entity>
</Liability>
<Liability> <Behavior> <Assign> <Type> <Spectrum/> </Type> <RA/>
<OEM/> </Assign>
</Behavior> </Liability> <! – RA assigns liability to OEM –>
<Liability> <Behavior> <OEM> <Certify> <Type> <Spectrum/> </Type>
<Configuration> <Software/> <Hardware/> </Configuration> </Certify>
</OEM>
</Behavior> </Liability> <! – OEM certifies a HW-SW configuration –>
</Liability> </Abstractions>

This framework hinges on the assignment of liability for spectrum conformance to the OEM, who implements this responsibility by type certification. The framework is expressed in the CRA in RXML (Expression 6-7). As hardware–software <Configurations/> become more complex, one must define a standard way of keeping track of certified configurations. The RA of Japan offers an approach to this chronic problem with SDR.

6.2.3.2 Japanese Perspective

Harada-San-se, IAI (Independent Administrative Institution of Japan), Communications Research Laboratories (CRL), Yokosuka Radio Communications Research Center, Leader in Wireless Research, presented the Japanese thinking in 2003 and 2004 at various RA convocations throughout the world. The Telcom Engineering Center (TELEC) organized a study group on technical regulation conformity on SDR October 2000 to March 2003. The Ministry of Public Management, Home affairs, Posts, and Telecommunications (MPHPT) was considering regulation on SDR. At that time, Japanese radio law did not permit any change to equipment once licensed; to make a change, OEMs or users must get a new license. Japanese IEICE, sister organization to the IEEE, had a companion study group. Physical layer "security" has regulatory issues, monitored by the government. Security layer 2 includes a right to protection for service quality, which is provided by the OEM or service provider, while layer 3 brings data and privacy protection.

In the Japanese future technical certification framework, they would allow independent sales and purchases of hardware that have been jointly certified by the certification agency. Harada-San-se listed information to be provided by the OEM at the time of application for type certification. This included whether the device is a hardware-defined radio (HDR) or SDR; various administrative data; and a statement of the confirmation method (typically testing, the same as current methods). The data also requires a definition of the larger communications system, modulation method, data rate (R_b), center frequency (f_c), and such technical parameters of the radio part supporting the hardware.

The Japanese framework then includes a "tally," a registry of successes and failures in hardware and software type certification. This includes a bitmap of success (1) or failure (0), for software certification but no hardware certification, or the converse, both being certified, or both failing. The tally is made for various hardware and software configurations in a matrix with hardware configurations on one axis and software on the other, defining $N \times M$ conformance test cases. (See the componion CD-ROM for details.) There are other proposals, but the Japanese tally matrix method could be a very important idea for download management, checking unlicensed, tampered, and unsuitable software. MPHPT has also studied SDR via questionnaires; trends in other countries and standards bodies and forums; as well as TELEC, IEICE, and other technical bodes. This major undertaking is covered in the *IEICE Special Issue on Software Radio and Its Applications* [164].

6.2.4 XG Spectrum Policy Vignette

DARPA's neXt Generation (XG) radio program envisions broadcast of spectrum policy so that radios can know what spectrum-use methods are permitted at a given time in a given place. The XG model of current spectrum use observes that there are policy, physical access, and air interface technology limits on agile use of radio spectrum.

DARPA's approach to relaxing these constraints may be summarized as follows:

1. Enhance policy flexibility.
2. Accept and manage more risk.
3. Increase capability to dynamically sense and adapt.
4. Develop faster spectrum analyzers with more instantaneous bandwidth.
5. Develop radio and waveform standards that can adapt to meet sharing requirements.
6. Develop wider coverage, better antennas.
7. Develop adaptive waveforms.

To develop this use case further, one may access DARPA's public XG web page. Future XG notices regarding spectrum-use policy language and broadcasts could be on public Web sites. One could extract from these descriptions the identities and features of those entities (e.g., the FCC) and classes of objects (TV stations and TV receivers) that would interact in XG spectrum-use policy language. Some features could be built into the next AACR product cost-effectively. One also needs to identify the modifications that these features undergo as spectrum policies change and as the XG language evolves. Machine learning and adaptation in the field could facilitate the evolution of such radios after they are deployed.

To fully develop the XG use case, one might analyze links among personal/wearable AACRs, vehicular AACRs, and an AACR fixed infrastructure.

The vehicular AACRs may relay content between mobile and fixed infrastructure using steerable millimeter wave gigabit links. Entities and roles change over time, changing GoS, QoS, and QoI parameters among the policy community, OEMs, service providers, and commercial users. Quantitative models characterize the trade-off between relaying lower power messages shorter distances on the one hand or increasing the operational radius of a relay point on the other.

6.2.5 Scaling Up Spectrum Sharing

The cost of operations and maintenance support of a large number of cognitive radios must be considered. This is sometimes called "scaling up" of AACR from a few isolated radios to a large population. The algorithm by which ants forage for food scales up from small colonies consisting of a queen plus a few ants to megacolonies consisting of literally billions of ants. Will XG scale up like that as increasingly more AACRs forage for available radio spectrum? One might consider the relevance of the artificial life research community in addressing this question [165]. A radio that follows spectrum-use policy broadcast in a specific format will be simple, like a worker ant, and not very expensive to develop and deploy. But as the framework for spectrum use continues to expand, these radios may need new adaptation software. The ants will need to learn new tricks. The ideal cognitive radio is supposed to learn new spectrum-use policies from experience, so if this ideal can be achieved, there would be no need for reprogramming. The radios would need to learn from individual experience, validate learned behaviors by autonomous peer review, and collaborate by the autonomous downloading of group experience. Although the AML foundations showed how to enable one radio to discover <Interesting/> features of a scene and to use that to focus AML, many technology challenges remain about how to realize autonomous extensibility of spectrum-use policy. Positive and negative reinforcement by RAAs could guide the evolution of a collection of iCRs if that ant colony were enabled to evolve within strict bounds. One approach to the CR ant colony suggested by genetic algorithm technology expresses spectrum-use policy in an algorithmic genome, evolving policy in a simulation incrementally validated by the experience of AACRs in the field.

This book addresses the potential technology leap and architecture challenges of such ideas powered by AML. Others also will pursue the complementary incremental approaches, deploying the required software staffs as the regulatory framework, user needs, technology, and products for evolution emerge.

6.3 DEMAND SHAPING USE CASE

The use case of this section examines the economic impact of shaping demand between "free" wireless LAN spectrum and "monthly billed" 3G or 4G wireless networks. It postulates the wide availability of low cost dual-mode PDAs and examines the potentially disruptive nature of cognition technology in those nodes. The key difference between today's networks and those posed in this use case is that the AACR nodes work *on behalf of the consumer*, not on behalf of the network. In the past, this was impractical for many reasons. CR technology along with MBMMR cell phones and ad hoc network technology overcome many of the technical challenges to such a progression.

6.3.1 Use-Case Setting

This use case examines the behavior of AACRs in a realistic scenario. Figure 6-3 shows the spatial structure of a notional small to moderate sized urban area. Its daily pattern of use of the mobile terrestrial radio spectrum includes demand offered by government entities; police, fire, and other public entities; and consumers. Although commercial services like taxis and delivery vehicles constitute a potentially important distinct niche, for simplicity they are not modeled in this use case.

The daily pattern of activity includes commuting from the suburb and residential areas to the city center and industrial areas. The airport has a high concentration of business travelers on weekdays and of vacationers on days before and after holidays. The stadium area offers relatively low demand except during sporting events. Each of the eight places shown in the figure corresponds to the coverage of one RF macrocell.

6.3.2 Analytical Model

The top-level structure of the analytical model is shown in Figure 6-4. The space–time–context distribution allocates classes of user to parts of the city

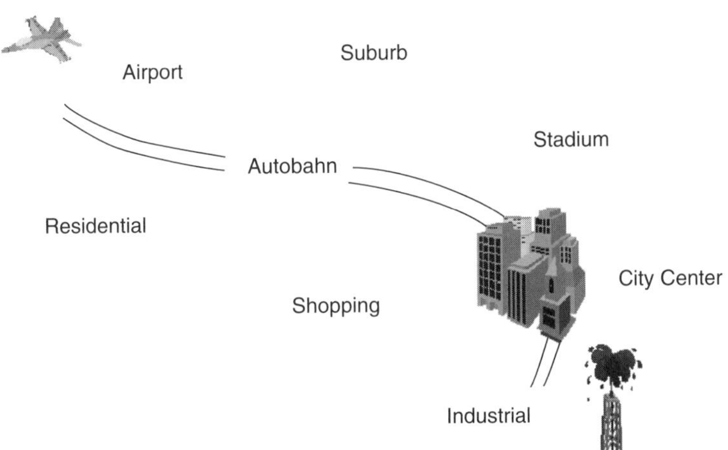

FIGURE 6-3 Spatial structure of demand shaping use case.

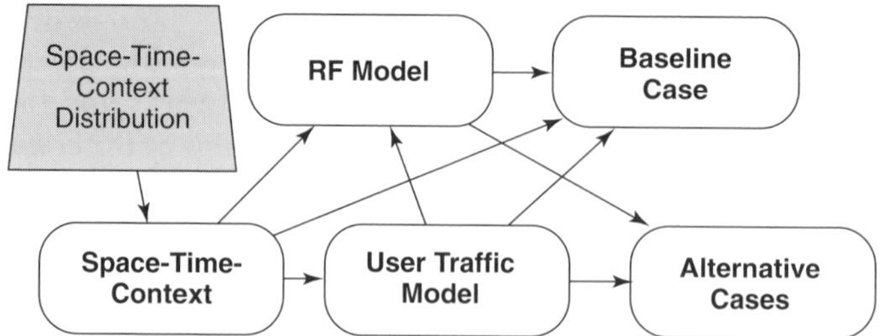

FIGURE 6-4 Analytic model of RF, space, time, context, traffic, and economics.

as a function of time of day. The Space–Time–Context model characterizes space and time in terms of densities of users during diurnal epochs. It also translates the distributions into fractional allocations of demand to space–time epochs. The RF Model defines the radio bands, cell sites, channels available to class of user, and tariff structure. The tariff structure is notional but estimates the relative economic value of the potential AACR behaviors. The User Traffic Model includes email and files submodels that model multimedia demand in parameters like the number of emails a user receives and sends per day and the size of attachments. The Baseline Case model allocates demand to channels and computes total revenue generated and revenue lost due to lack of capacity.

Users are classified as <Infrequent/>, <Commuter/> <Power-commuter/>, <Police/>, <Fire & Rescue/>, <Government-users/>, <Emailers/>, <Browser/>, and <Telecommuters/>. The spatial distribution of users covers all eight spatial regions with statistical apportionment according to the proportion matrix of Figure 6-5. The normal day is partitioned into nine epochs of 2 or 3 hours each (Morning Rush, etc).

Figure 6-6—from the Analytica model that accompanies the text—shows how fractions of population redistribute as commuters move from the suburbs to the city center and back to the residential areas late at night, modulating the offered demand. Offered demand is defined by class of user and type of content with separate Beta distributions for each period of time. Police, for example, can offer substantial voice traffic during some periods of the night as illustrated in the Beta distribution of Figure 6-7.

The corresponding cumulative probability densities show that demand is likely to peak at about 0.15 erlang in the "Late Night" epoch (from 1 until 3 am). These temporal variations of demand are multiplied by the spatial fractions to yield probability distributions of demand as a function of space and time. The notional wireless infrastructure to which this demand is directed consists of cell sites in each of the eight urban regions. The commercial sector has 100 traffic channels per site in this notional model. The police, fire, and rescue have four sites serving the city center and industrial region, the shopping and stadium, the autobahn and suburbs, and the residential and airport

Edit Table - Space-Time-Context Distribution

Edit Table of Space-Time-Context Distribution

User Classes: Commuter

Context: Normal

Daily Epochs / Spatial Regions

	Airport	Autobahn	CityCenter	Industrial	Shopping	Residential	Suburb	Stadium
Morning Rush	0	4	0	0	0	1	1	0
Morning	1	0	5	4	0	0	0	0
Lunch	0	0	5	4	2	0	0	1
Afternoon	0	0	5	5	0	0	0	0
PM Rush	0	5	1	1	0	0	0	0
Evening	0	1	1	1	0	5	4	0
Night	0	2	0	0	1	5	4	0
Late Night	0	0	0	0	0	5	4	0
Wee Hours	0	0	0	0	0	5	4	0

FIGURE 6-5 Space–time–context distribution for normal commuters.

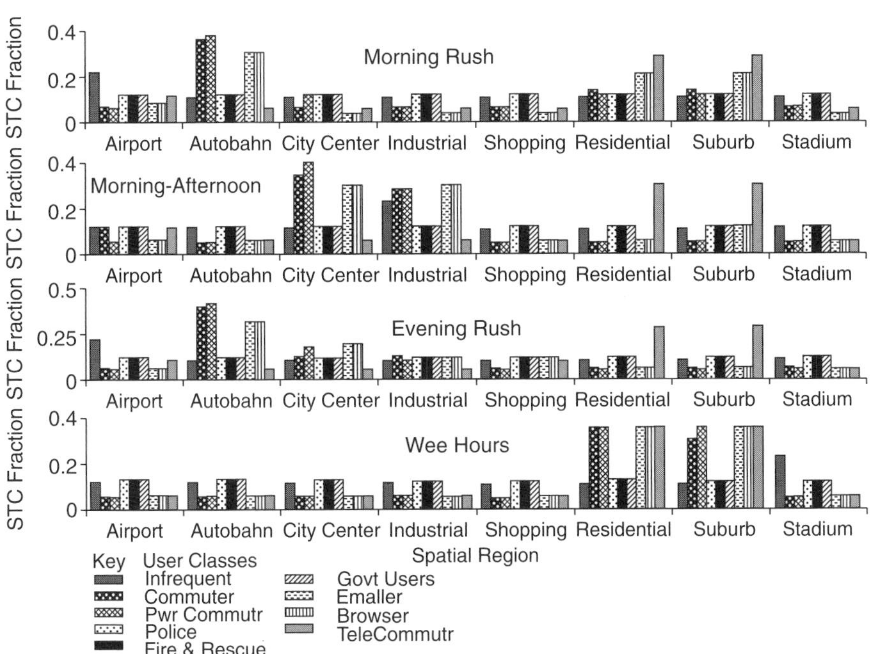

FIGURE 6-6 Diurnal pattern of commuter locations (four of nine epochs).

regions, respectively. They allocate N erlang capacity (N full-time traffic channels) among the regions. The model sets $N = 100$ for both cellular and civil infrastructure. The government has one site for the entire urban area, allocating its channels uniformly to demand from any region (presumably a

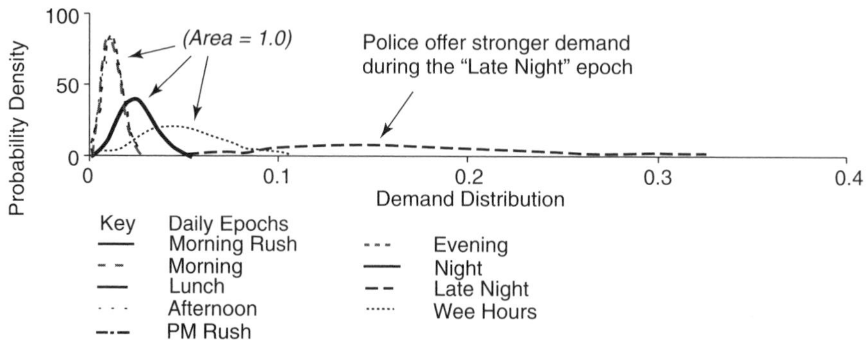

FIGURE 6-7 Probability density of demand offered by police.

TABLE 6-3 Analysis Cases for Pooling

	Allocated Case		Pooled Case	
	Revenue	Lost Revenue	Revenue	Lost Revenue
Baseline (mostly 1G)	\$136k	\$132k	\$242k	\$47k
Growth with 2G	\$145k	\$213k	\$305k	\$87.9k
Explosion with 3G	\$147k	\$243k	\$335k	\$109k

site on a tall hill or building overlooking the entire area). The analytical model has a notional tariff structure, with tariff tables for each of three RF bands (cellular, public, and government). Since there are 167 modules in this Analytica model, a detailed description is beyond the scope of this treatment. The model includes cross-checking to assure internal consistency. Illustrative results are presented in the following sections.

6.3.3 Value of Pooling

The cases shown in Table 6-3 characterize the contributions of spatial and temporal demand on pooling. This illustrates the benefits of etiquette and is the baseline for demand shaping below.

In the "Baseline" case, the population of 37,200 subscribers offers 24-hour traffic of 11,600 erlang hours per day. About half the potential revenue is lost because of the statistical overload conditions at peak hours. When this demand is pooled, all offered traffic is spread across cellular, government, and public use bands (and all infrastructure towers), a total of 84.6% more channels. The lost revenue shrinks to only 16% of the total while the total revenue-generating capability of the system increases with the revenue-bearing role of the pooled government and public bands.

The second and third cases of Table 6-3 show revenue as email and multimedia traffic increase over time. Although spectrum pooling helps, almost

25% of the potential revenue is lost even with 3G technology (third case). All of this revenue growth is from wireless email, electronic maps, stock broker services on the move, and other services that appear attractive to consumers.

6.3.4 Cognitive Shaping of Demand

The postulated AACRs learn accurate user models in communications contexts to autonomously shape demand. They balance value to the user (e.g., rapid delivery of email, rapid retrieval of items from the Web) against cost (e.g., offering the demand during off-peak periods). The third vignette in Table 6-3 exhibits a pattern of digital traffic where 68% of the afternoon demand is email or file transfers. If only 30% of this demand is delayed by from 1 to 4 hours, total traffic increases by about 5%. This is attractive to network operators but not necessarily to consumers.

Instead of being merely shifted in time AACRs divert the digital traffic that is not time sensitive to the corporate RF LAN where airtime is (regarded as) free. If only 25% of subscribers find the high-band RF LAN convenient, the cellular service providers lose 4.8% of their revenue-bearing traffic, not an insignificant loss.

These value systems can be expressed in RXML, learned by AACRs and used for specific users for situation-dependent load shaping.

Expression 6-8 Knowing that the User's Value System Is Situation Dependent

```
<Abstractions> <Radios>
  <iCR> <Software> . . .
    <Learns> <User> <Value-system/> </User> </Learns>
      </Software> </iCR> </Radios>
  <User> <Value-system> <Cost/> <Delay/> <Speed/> </Value-system>
    </User>
. . . </Abstractions>
<Physical-universe> . . . <Instances>
  <User> Charlie <Value-system>
    <NOT> <CONTAINS> <Situation/> <Unusual/> </CONTAINS>
      </NOT>
    <Cost> <Lowest/> </Cost> </Value-system> </User>
  <User> Charlie <Value-system>
    <CONTAINS> <Situation> <Urgent/> </Situation> </CONTAINS>
    <Delay> <Lowest/> </Delay> </Value-system> </User>
</Instances> </Physical-universe>
```

As illustrated in Expression 6-8, specific iCRs learn the specific value systems of their users through a combination of a priori models and an ability to recognize and tailor user value systems to specific features of a situation. Ontological primitives like <Urgent/> must be defined in terms of observables in a scene. In addition, temporal reasoning must translate <Lowest/> cost or delay into the acts of buffering the traffic to the RF LAN.

Introducing AACRs in large numbers diverts traffic to RF LANs, increasing total traffic by 8.9% with cognitive delay shaping of 25% resulting in a loss of 4.8% of the revenue-bearing cellular/3G traffic to WLANs. As more traffic is shaped to corporate RF LANs, more revenue is shifted from the service provider to the corporate clients in the form of savings of the costs of mobile traffic. The simple economic model does not include price breaks for corporate clients or the costs of acquiring and operating the corporate RF LANs, but such simplifications may not be unrealistic if the RF LANs are installed and operated to reduce costs of wiring (and rewiring) office spaces or manufacturing floors; this represents a small but beneficial use of an existing corporate infrastructure. This use case does not reflect any particular business case and its simplifications may render the results inapplicable in many cases. Still, millions of autonomous decisions can shift revenue measurably if autonomy moves millions of bits off the cellular networks and onto the corporate RF LAN, or the wireless Web.

6.4 MILITARY MARKET SEGMENT USE CASES

Globally, there have been many military radio programs that envision the fielding of military SDRs. These include the U.S. Joint Tactical Radio System (JTRS) [79], the U.K. Bowman Program, and the European Future Multiband Multimode Modular Tactical Radio (FM3TR). Without necessarily addressing any one of these in detail, one may define a generic Coalition Tactical Radio System (CTRS).

In this use case, CTRS would be the basis for a global open architecture standard for intelligent radios to facilitate the formation of military teams, for example, to assist in humanitarian relief operations like the tsunami disaster of 2004. CTRS addresses military radio modes between 2 MHz and 6 GHz, LF/HF through SHF. These include HF Morse, HF ALE, frequency hop (FH), direct sequence spread spectrum (DSSS), analog AM/ FM push to talk, GPSGlonass navigation, digital data links, FM FDM, and digital PCM. Suppose there are four antenna sets per military vehicle: (1) LF/HF dual whip antennas (2–30 MHz), (2) dual midband whip antennas (30–500 MHz), (3) dual high-band whip antennas (0.5–2 GHz), and (4) quadrant directional antennas (0.9–6 GHz). CTRS enables communications among coalition vehicular radios such as ships, aircraft, and land vehicles. The functions of CTRS include agile radio spectrum management, mode management (the association of modes to specific vehicles), priority (by vehicle, by user, and by content), preemption of a lower priority by a higher priority, power management, communications security (COMSEC), transmission security

(TRANSEC), and ad hoc networking. CTRS meets coalition needs for gateways to interconnect legacy users. This use case and the next complement each other in humanitarian relief operations.

6.5 RF KNOWLEDGE THAT SAVES LIVES

In October 2003, UCSD announced WISARD, the Wireless Internet Information System for Medical Response in Disasters [166]. WISARD replaces the felt pen and whiteboard of triage (e.g., in a humanitarian relief operation) with digital triage using RF identification (RFID) tags. Some RFID tags measure vital signs when attached to a finger. Among the motivations cited by project director Leslie Lenert was the recent scenario in which the Russian government used a gaseous agent to disable the terrorists in a Moscow theater. More than 100 of their hostages died. According to Lenert, medical personnel later reported that most deaths were due to lack of vital signs monitoring at the scene and an inability to organize care to determine who was breathing and who wasn't. Immediate application of the RFID tags would enable digital triage, hopefully saving lives.

Lenert characterizes problems with current field care where there are mass casualties as in Figure 6-8. Victims flee, some get dramatically better while others become dramatically worse; responders have trouble coordinating; tags are lost; clinical data is not available for the aid giver, triage, or surgical team; and specialized equipment is hard to locate, power, or operate when

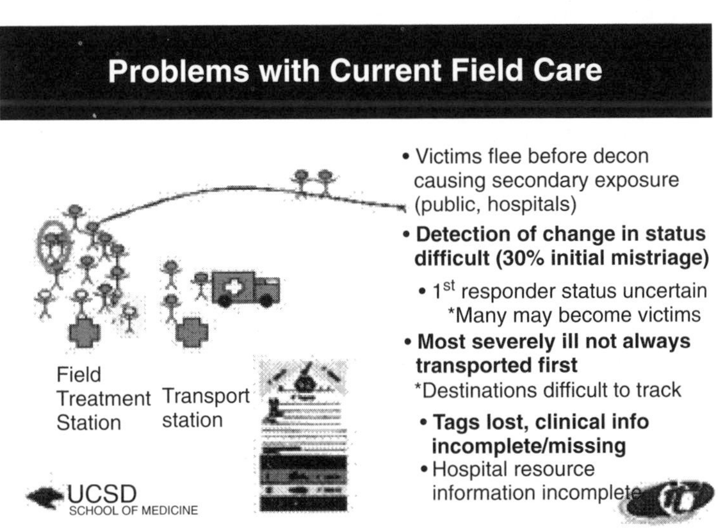

FIGURE 6-8 CR can enable digital triage, overcoming RF problems with field care. © UCSD. Used with permission.

needed. How could CR enable such a vision of digital triage? The entities of the use case may be expressed in RXML as in Expression 6-9.

Expression 6-9 Disaster Relief Entities

```
<Abstractions>
  <Scene> <Disaster>
    <Earthquake/> <Tsunami/> <Severe-weather/> </Disaster> </Scene>
  <Scene> <Disaster> <People> <Casualties> <Dead/>
    <Injured> <Minor-injury/> <Serious-injuries/> <Critically-injured/>
      </Injured> <Seriously-ill/> </Casualties>
    <Relief-personnel> <NGO/> <Medical/> <Sustenance/>
      <Infrastructure/> <Military/> </Relief-personnel>
    <Indigents/> </People> </Disaster> </Scene> </Abstractions>
```

Sustenance relief personnel provide food, water, and other supplies. Infrastructure relief personnel reconstruct roads, bridges, housing, and the like. Military relief personnel sent from other countries assist with disaster recovery. <NGO/> are nongovernmental organizations like the International Red Cross and Red Crescent. In this taxonomy, there are just three kinds of people: <Casualties/>, <Relief-personnel/>, and <Indigents/>. These are the actors or <Entities/>. The behaviors of medical personnel, casualties, and indigents with respect to medical attention are the focus of the use case. RFID tags provide the raw data; iCRs locate the tags and exchange information about tag location and patient status via their ad hoc CWN.

Expression 6-10 RF Entities in the Spectrum Scene

```
<Scene> <Abstractions>
  <RF> <RA> <Spectrum-plan> </RA>
  <RF-devices> <Transmitters/> <Receivers/> <Transceivers/>
    <CWPDAs/>
      <Transponders> <RFID/> <Transponders/> </RF-devices>
    <RF-environment> <Signals/> <Noise/> </RF-environment>
  </RF> </Abstractions> </Scene>
```

RFID tags come in many flavors. The ontological primitives of Expression 6-10 describe them as <Transponder/> entities in a RF scene that includes regulatory authority (<RA/>), RF devices, and a radio propagation environment of signals and noise. RFID tags for digital triage from different suppliers may have different air interfaces, sensors, and data storage and transfer capabilities. RFID tags of different national origin may be brought to a given humanitarian relief operation. Since radio spectrum management differs by region of the world, country, or political subdivision, RFID tags will be incompatible with each other's home RF networks, creating RFID bedlam.

Envisioned iCR behaviors that mitigate these challenges, for more efficient medical assistance, are as follows:

1. The iCR RFID base stations probe newly arriving RFID tags to discover their air interfaces, registering them and their capabilities into the <Scene/>.
2. The iCRs form an ad hoc network to track locations of the unpowered transponder RFID tags.
3. AACRs monitor emissions and advise RAs to reallocate interfering radio emissions away from RFID tag frequencies.
4. AACRs learn the priority of different content (e.g., nominal vital signs versus values that are out of safe medical norms), prioritizing the data flow from unsafe RFID tags to the medical teams.
5. AACRs allocate the more reliable physical layer RF <Paths/> to higher priority data to reduce the GoS and QoS of less essential data, enhancing net QoI by buffering the less critical data for queued delivery or for retrieval by a medical team when and where needed (versus delivering the data to a display that is not being monitored by medical staff).
6. When new types of RFID tags appear on the scene, the AACRs determine their capabilities and limitations sharing this with medical teams using a medical XML.

Technologies like XML enable future RFID tags to express their abilities using open-standard medical ontologies. Privacy concerns may dictate the adoption of a privacy system similarly based on an open architecture standard such as PKI. The CWN could segregate RFID tags of a given type to a geographical subdivision of the disaster area to limit co-channel interference. The CWN members could mutually agree on a specific AACR to serve as the hub for the automatic coordination of wireless spectrum among AACRs and legacy radios. Criteria for the hub iCR would include the number, types, and flexibility of its RF access, the processing capability to handle multiple RFID channels in real time, enhanced antennas, longer endurance power supplies, and better physical access of the lead medical teams.

The image-capable cell phone is already finding quasimedical uses. For example, researchers at Tokyo University [167] have developed a sales-support system for door-to-door cosmetics sales based on skin-image grading. Skin-CRM (Skin Customer Relationship Management) analyzes the customer's skin from a picture taken by the salesperson's cell phone. Skin grade is assigned by rules generated by data mining a baseline of grades given by human skin-care experts. Skin-CRM uses a cellular phone with a camera, email software, and a web browser. The skin picture is sent to the analysis system by email. The picture is analyzed by the Skin-CRM server, resulting in a web page plus an email advising of the results "within minutes." Salespeople browse the results on their cell phone, including skin grade and recommendations for care and cosmetics suitable for the customer's skin. Skin-CRM suggests AACR humanitarian applications in indigent personal hygiene, diagnosing skin conditions, remotely analyzing wounds, and otherwise enabling people who cannot physically arrive on a scene to apply specialized expertise via wireless information services. Derivative commercial and military AACR use cases could combine patient tags, equipment tags, digital cameras, and wireless connectivity for point-of-action information services tailored to communications and network server capabilities.

6.6 PROGNOSTICATION

Chapters of this text develop the technologies needed to learn from experience so that future CPDAs based on AACR technology do not fall victim to the expense of a huge logistics tail of hundreds or thousands of knowledge engineers needed to keep up with a customer base. This was called the knowledge engineering bottleneck in the 1980s. Hopefully the use cases above and the related exercises demonstrate the relative futility of anticipating all the permutations and combinations of knowledge needed to customize each AACR for the specifics of the radio environment. On the other hand, even this introductory set of radio-domain oriented use cases suggests a trade-off between a priori knowledge and ML, autonomous or not. Well engineered products will tend toward fewer, simpler, and easier to understand functions. The kids (the computer scientists) may "get it" before the parents (the RF engineers). But even the parents will see through a product that really isn't very smart, even if it says Cognitive on the label. Thus, the next chapter introduces the store of a priori knowledge that is foundational to bootstrapping radio knowledge by AML, while the subsequent chapters relate contemporary AML technologies to this prior knowledge, for lower cost and more flexibility of AACR applications as the technologies mature.

The radio knowledge chapter that follows consists of dozens of knowledge vignettes that focus mostly on the radio, especially on the potential connectivity, data rates, and networking opportunities offered by successful employment of the physical layer of the protocol stack. Subsequently, both radio knowledge and user-driven knowledge are formalized for AACR applications, which conclude the text. This chapter therefore builds a progression that, taken as a whole, introduces CR technology towards an evolution of AACR applications.

6.7 EXERCISES

6.1. Complete Expression 6-1 using the data from Table 6-1.

6.2. Describe in RXML the experience of reinforcement of path connectivity for 1000 attempts to connect if each attempt yields exactly one reinforcement of attempt and if connected yields only reinforcement success given that 80 connections occur out of 100 attempts on Path1 and only 70 connections occur out of 100 on Path2.

6.3. Further develop a particular use case of this chapter by building your own analytical model of the use case. Extract those entities that will interact in order to achieve the demand-shaping aspect of the use case and describe them in RXML, inventing ontological primitives in your own taxonomy.

6.4. Develop the XG UML use case under the assumption that spectrum management policy undergoes a period of rapid change. At first, the FCC advocates the use of unused TV channels for ISM networks. There are heart monitors already that use those bands. Suppose the FCC's policy enables any reasonable use, but industry decides to defer to legacy medical users in order to avoid civil liability. Model the impact of these changes on policy language and computational intelligence of deployed nodes (personal, vehicular, and infrastructure), with particular emphasis on the support costs. How much could it cost to maintain a staff of computer scientists, ontologists, and radio engineers to create the new personality downloads for the deployed FCC CR radios? What is the impact on this logistics tail of machine learning?

(a) Include techniques from Chapter 4 on AML to create UML for this use case. What aspects of AML are not readily captured in UML? What other software tools can capture such situations? How can you work around the limitations of UML to represent learning?

(b) After reviewing the CD on CR1, teach CR1 a relevant subset of the XG policy language. How much can it learn by itself using its CBR? What is the minimum set of additional CBR nodes needed to learn the heart-monitor example? If you accomplished this by simply training CR1 about heart monitors without programming any new PDANode classes, then give yourself an A. If not, can your new PDANode class extend to other classes of radio?

6.5. What CASE tools are available for developing the use cases of this chapter? What is their value? How does that compare to the costs? How do you measure the cost of the time it will take you to learn to use the tool?

6.6. Develop a UML model of your favorite use case from this chapter.

(a) What are the entities of the UML model? The user and the "self" of the CWPDA are clearly entities. What other entities are critical? Should there be an entity for each member of a family? If so, how will the CWPDA come to know those family members? Same for a friend from work?

(b) What other entities should be in the UML model? Is there a model for entities at work such as the boss or co-workers? With conventional knowledge engineering technology, there is no right answer because almost no matter what you preprogram, you are wrong. Can your marketing department give you a good enough profile to sell to early adopters?

(c) Write an algorithm by which your CWPDA could learn from experience who is present and which entities need to be represented with separate identities. This ability to learn from experience separates the cognitive from the merely aware–adaptive radios.

6.7. Answer the following question for the narrowband spectrum rental etiquette of the FCC spectrum rental use case. What is the relationship between the RF physical layer and the amount of computational intelligence in the nodes?

Include spatial modeling of radio propagation, sharing of observations of spectrum occupancy and multipath signatures (delay tap settings for time-domain tapped delay line equalizers), and detection of user states such as commuting, shopping, going to a football game, and getting mugged. What is the order of implementation of computational intelligence features? How would you distribute computational intelligence to the users' individual nodes or home network? Would there necessarily be a centrally controlled network, or could all of the required networks be ad hoc? How about a local or regional repository for cognitive knowledge that itself is rented or paid for on a per-use basis? Could some kind of nonmanaged network architecture emerge to meet the need at lower cost than today's centrally controlled and relatively expensive cellular networks?

6.8. Consider the XG policy use case. Review the companion Excel spreadsheet or XML files of Radio Knowledge Representation Languages (RKRL), Version 0.4. What ideas does RKRL suggest as alternatives to XG? Think about policy modeling in space–time–RF as an alternative or supplement to a static policy such as enabling the Part 15 use of unused TV channels. In this use case trade off the XG approach with other approaches suggested by RKRL.

6.9. For CTRS, extract from the text description of the use case those entities that should be abstracted to a RXML ontology, modeled in the ontological treatment, and modeled in detail (e.g., physical models over space–time such as aircraft flight paths) in order to establish and maintain the gateway information exchanges among coalition partners.

- **(a)** Suppose each CTRS in a humanitarian relief <Situation/> registers with the situation management authority.
- **(b)** Define in UML a CTRS network coordinated by a management agent.
- **(c)** What is the difference between a <Situation/> and a <Scene/>?

6.10. Suppose you have made your radio both location aware with GPS and speech aware with a speech-to-text package. Go to Dr. Lenert's UCSD web page to explore his ideas further. What additional technical capabilities must be available in the CR beyond location and language awareness in order to enable digital triage? Is the speech-to-text capability needed? Explore the Web to learn more about RFID tags. How many RFID standards and products can you find?

6.11. SDR technology to enable RFID tags in medical emergencies includes the mobile multiband base station technology [144, Chapter 15]. Describe the modifications to the mobile base station of that reference needed for the medical RFID tags use case of this chapter.

CHAPTER 7

RADIO KNOWLEDGE

Radio knowledge is the body of knowledge of the world's radio engineers. This body of knowledge continues to expand and to be revised, fueled by $1 trillion per year in related global economic activity. This chapter organizes that knowledge into pairs of <Knowledge/> chunks with a corresponding <Use/> for that chunk. For the full collection of <Knowledge/>–<Use/> pairs see the companion CD-ROM.

7.1 RADIO-DOMAIN OVERVIEW

Radio science embraces the many broad classes of radio propagation and signal modalities suggested in Figure 7-1. A few radio bands such as ELF are not listed because there are few classes of transmitter and receiver and the antennas are miles long. In addition, the near-terahertz (THz) bands used in medical research are not listed. Those specialized bands do not benefit much from the technologies of AACR. Commercial, civil, aeronautical, scientific, medical, and military radios employ the bands and modes of the figure.

The types of radio implied in Figure 7-1 suggest the breadth of radio knowledge that must be formalized to enable algorithms of AACRs to be autonomously competent in radio domains. In SDR jargon, a waveform corresponds to an air interface and protocol stack. A radio mode is a set of parameters of the waveform, including the waveform class, allocated RF band, multiple access scheme, channel symbols, timing, framing, and control signals, traffic

Cognitive Radio Architecture: The Engineering Foundations of Radio XML

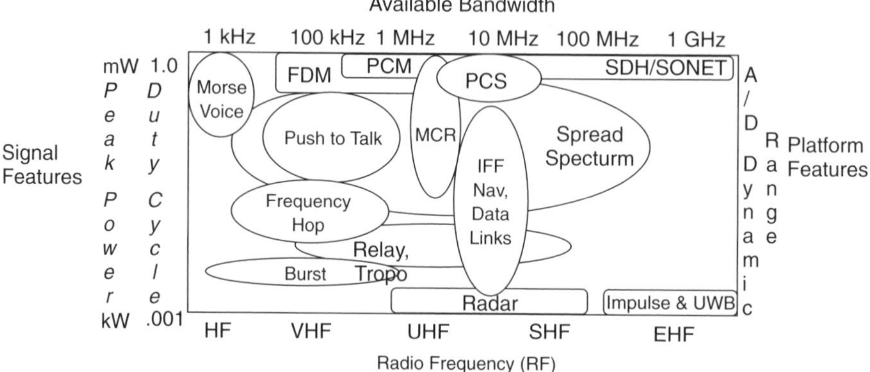

FIGURE 7-1 Radio spectrum overview. (See ontology of Expression 7-1.)

channels, protocol stack, and the services provided by the mode such as voice, data, fax, video, and location.

7.1.1 Structuring the Radio Domain

The first step in formalizing the radio domain is to define the scope of the domain. Minimum-competence radio-domain knowledge has both breadth and depth. The breadth of that knowledge embraces:

1. *Bands*: RF bands determine radio propagation characteristics and thus the available connectivity, operating range, bandwidth, the need for relay points, and the types of channel impairments.
2. *Modes*: The air interfaces ("waveforms") define node and network power, organization, and services; standard operating parameters ("modes") constrain quality of service (QoS).
3. *Networks*: The network architectures from satellite communications (SATCOM) to terrestrial cellular to ad hoc peer networks.
4. *Protocols*: The protocol stack and its interactions with the radio physical layer.
5. *Services*: The information services supported by a given band and mode, such as location finding, navigation, time reference, instrumentation, or broadcast media (voice, TV, closed caption, etc.), with associated specialized users (taxi, police, fire, rescue, military, civil, sports, etc.)

The open taxonomy of Expression 7-1 defines the initial scope of the AACR radio domain. Open taxonomies are extensible at every level.

Expression 7-1 Radio Bands, Modes, Services, and Broad Category of User

<Abstractions> <RF>
<Band> <ELF/> <LF/>
<HF/> <LVHF/> <VHF/> <UHF/> <SHF/> <EHF/> <TH/> </Bands>
<Mode> <Parameters> <Waveform-class/> <Allocated-band/>
<Multiple-access/> <Channel-symbol/> <Timing/><Framing/>
<Control/><Traffic-channel/>
<Protocol-stack/> <Services/> </Parameters>
<Analog-voice/> <Morse-code/>
<MCR> <Definition> Mobile Cellular Radio </Definition> </MCR>
<FDM> <Definition> Frequency Division Multiplexing </Definition>
</FDM>
<PCM> <Definition> Pulse Code Modulation </Definition> </PCM>
<PCS> <Definition> Personal Communications Systems </Definition>
</PCS>
<IFF> <Definition> Identification Friend or Foe </Definition></IFF>
<UWB> <Definition> Ultra Wideband </Mode></Definition> </UWB>
<Troposcatter> <Definition> Refractive tropospheric propagation
</Definition> </Troposcatter>
<Radar> <Definition> RAdio Detection And Ranging </Definition>
</Radar>
<Product> <AACR/> <CWPDA/> <RF-device/> <Node/>
<Infrastructure/> </Product>
<Network> <! – A coordinated set of communications nodes –>
</Network>
<Service> <Duplex> <Definition> Two way communications
</Definition> <Voice/> <Data/> <Short-message/> <FTP/> </Duplex>
<Conference/> <Location-finding/> <Navigation/> <Time-reference/>
<Broadcast> <News/> <Music/> <TV/> </Broadcast> </Service>
<User-category> <Commercial/> <Civil/> <Aeronautical/> <Scientific/>
<Medical/> <Military/> </User-category>
</RF> </Abstractions>

Since radar is a complete domain in itself entailing specialized propagation models, detection criteria, tracking algorithms, special-purpose antennas, and the like that dominate the radio system architecture, radar is developed only to the degree necessary for AACR to recognize it and avoid interference.

7.1.2 Components of Radio Knowledge

Each radio band exhibits characteristic channels driven by the physics of the radio carrier frequencies of the band. Therefore, each of the radio knowledge sections that follow is organized by RF band. The four knowledge components per band are:

1. *Physical Parameters*
   (a) The ITU frequency limits of the band.
   (b) Wavelengths—the converse of frequency with respect to the speed of light.
   (c) Propagation modes—engineering parameters of the way energy is propagated such as multipath delay spread, the range of variation of multipath delays, and Doppler spread. <Propagation-modes/> are statistical aggregates distinct from <RF> <Modes/> </RF>.
2. *Spatial Distribution of Energy*: The details of energy reflections from objects in the scene, including terrain features, buildings, vehicles, and people.
3. *Available Communications Modes*: Channel symbols and network organization and protocols representative of communications in the band. These are the <RF> <Modes/> </RF>.
4. *Services and Systems*: Services available in the band and how they are obtained. Historically, services were intimately associated with communications systems. With particular hardware, one could get specific services. For location information, one needed a LORAN or, subsequently, a GPS receiver. To find the airport required a TACAN or microwave landing receiver. SDR enables a broader scope of available services on general purpose communications hardware–software platforms. AACR that bridges across legacy systems, modern SDR, and iCR must know about the legacy hardware. Antennas, legacy communications products, and evolving SDR products therefore are part of this knowledge base of this chapter.

General uses of these four radio knowledge components are outlined in the next few sections.

7.1.3 Physics-Related Knowledge

The ITU frequency limits define a radio band. AACRs must communicate with spectrum management authorities with precision and accuracy. A database of radio spectrum allocations as a function of relevant regulatory authority (RA) would be a comprehensive way of embedding the basic knowledge of band allocations. This knowledge can be used by a CR in employing radio bands for which it had not been previously programmed, such as in creating a new ad hoc network in a band previously off limits. AACRs with limited memory may acquire new physics-related knowledge from trusted sources like RAs using physics-related ontological primitives.

7.1.3.1 Antenna Knowledge

Wavelengths are the converse of carrier frequency, f, with respect to the speed of light:

$$\lambda = f/c \tag{7-1}$$

Given f, an iCR can estimate the wavelength from this equation. Since antennas resonate at full, half, and quarter wavelengths, the wavelength implies the size of an *efficient* antenna. An iCR can use this knowledge of wavelengths and antennas to reason about the transmission and reception efficiency of its own antennas, for example, in determining whether to attempt to use a download that is slightly out of band of its antenna. This is reflected in RXML via a <Physics/> <Abstraction/> as follows.

Expression 7-2 AACRs Know About Frequency–Wavelength Relationships

<Physics> <Propagation> <Speed-of-light *c*>
<Frequency *f*> <Wavelength λ> $\lambda = f/c$ </Wavelength> </Frequency>
</Speed-of-light> </Propagation>
<Radiation> <Antenna> <Half-wave> $\lambda/2$ </Half-Wave> . . . </Antenna>
</Radiation>
</Physics>

The level of detail of Expression 7-2 is sufficient for the recognition and shallow reasoning about antenna type, but not for deeper reasoning about <Efficiency/>. With a more complete ontology and related naïve physics skills, an iCR could offer antenna advice to a nonexpert. Police, fire, and rescue spectrum managers might use such knowledge in emergency situations to rapidly configure available antennas onto available radios in a way that enhances communications, improves battery life, or otherwise tailors the available radios to the emergency. Configuration advice should be of the breadth and depth typical of a competent radio operator for the iCR to be perceived by nonexpert users as helpful. The cost of sustaining such iCR expertise may be unworkable unless the antenna knowledge is relatively stable over time. A self-modifying iCR could use its knowledge of wavelength and antenna resonances to propose, synthesize, and test new waveforms on its conventional antennas to improve radiation metrics.

7.1.3.2 Propagation Mode Knowledge

RF propagation is fundamental to the design of air interfaces. Delay spread defines time delay of tapped delay-line equalizers. AACRs with propagation knowledge may tailor the operating parameters of a basic waveform to the propagation parameters of a RF <Scene/>. AACR that autonomously creates and deploys an improved time-domain equalizer capable of equalizing unanticipated multipath delay spreads obtains higher data rates than equalizers in legacy SDR personalities. Deploying an equalizer with more taps requires reasoning about computational resources and the complexity of the compiled equalizer.

Expression 7-3 Delay Spread and Distance

<Physics> <Propagation> <Speed-of-light c> <Scene> <Distance x>
<Time-delay dT(x, c)> dT = x/c </Time-delay>
<Reflection> <MAXIMUM Min-x> <Distances xi/> </MAXIMUM>
<Delay-spread> Min-x/c </Delay-spread>
<Delay-spread> <Time-delay (Min-x) </Delay-spread> </Reflection>
</Distance> </Scene>
</Speed-of-light> </Propagation>
</Physics>

Operation on higher speed vehicles—like aircraft—may warrant compensation for unexpectedly large Doppler spread entailing autonomous modification of loop gain and bandwidth. Historically, the radio mode parameters related to delay and Doppler spread have been set at waveform design time. If the time-domain equalizer isn't up to the multipath encountered, today's radios can't do anything to help the user. An iCR could reallocate unused processing capacity from an unused channel to improve equalization, rendering unusable propagation conditions usable. In addition, an AACR could precompensate the transmitted waveform to overcome other channel impairments about which it can reason. Knowledge of delay and Doppler spread relationships in a given band therefore enables autonomous adaptation to those parameters of propagation experienced in the field.

7.1.3.3 Spatial Distribution of Energy

Delay and Doppler spread are statistical aggregates sufficient for planning equalizing, but not sufficient for MIMO, where individual paths carry independent parallel bitstreams. With multiple transmit and receive antenna elements between two subscribers, MIMO algorithms project multipath onto statistically independent subspaces, each of which carries an independent bitstream. Planning the siting of MIMO endpoints can benefit from 3D scene models with high fidelity ray-tracing propagation models such as that of Figure 7-2.

As AACRs transition from passive objects sited by people to active participants in the planning of siting decisions, AACR location planning subsystems will increasingly benefit from the integration of 3D <Scene/> models. Thus, throughout the chapter, AACRs reason about spatial distribution of RF energy, fixed reflectors, and moving objects in scenes. Geospatial computing includes mapping, geodesy, cartography, 3D modeling, and image registra-

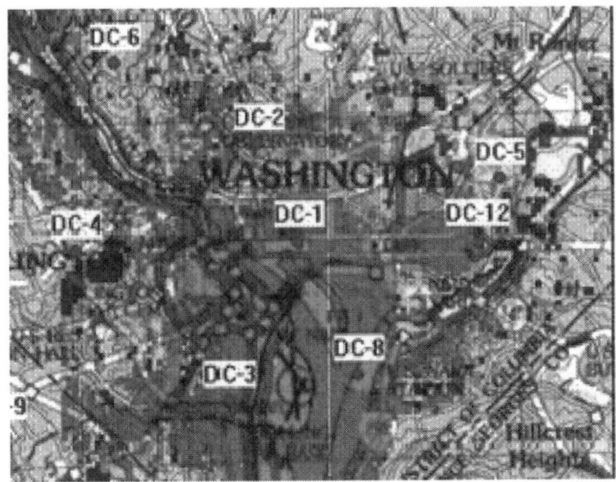

FIGURE 7-2 Urban spatial distribution of energy (Courtesy of WrAP; reprinted with permission.)

tion among others. Geospatial tools enable AACR to reason about maps and spatial distribution of radio energy, but that has just the minimum needed for a given use case.

Expression 7-4 Spatial Reasoning Requires Knowledge of Maps

<Abstraction>
<Space> <Map/>
<Map> <Definition> Computational analog to earth </Definition>
<Example> <Map> <Dimensions> 2 </Dimensions> washingtonDC.gif
<Legend> <Color> <Red> <RSSI> <Strong/> </RSSI> </Red>
</Color> </Legend> </Map>
</Example> </Space> </Abstraction>

Expression 7-4 is very shallow, sufficient for a radio to reason about RSSI in the Washington, DC area using the map of Figure 7-2. To do this the AACR extracts pixels from the <Map> based on x–y pairs and uses the gif2LL program to <Project/> the map to a standard coordinate system like (Latitude, Longitude). The AACR knows the color value of "Red" and the RSSI values associated with the tag <Strong/>. The sample map washingtonDC.gif is calibrated spatially and in the power dimension by WrAP™.

Commercial geospatial tools like Arc Explorer [168] enable one to view and query ESRI® shapefiles, ArcInfo® coverage, map layers, CAD drawings, and street addresses, measure distances, and find features (e.g., points of interest like parks). These tools may be embedded into AACR knowledge space by binding the API to semantic primitives of the <Space/> abstraction.

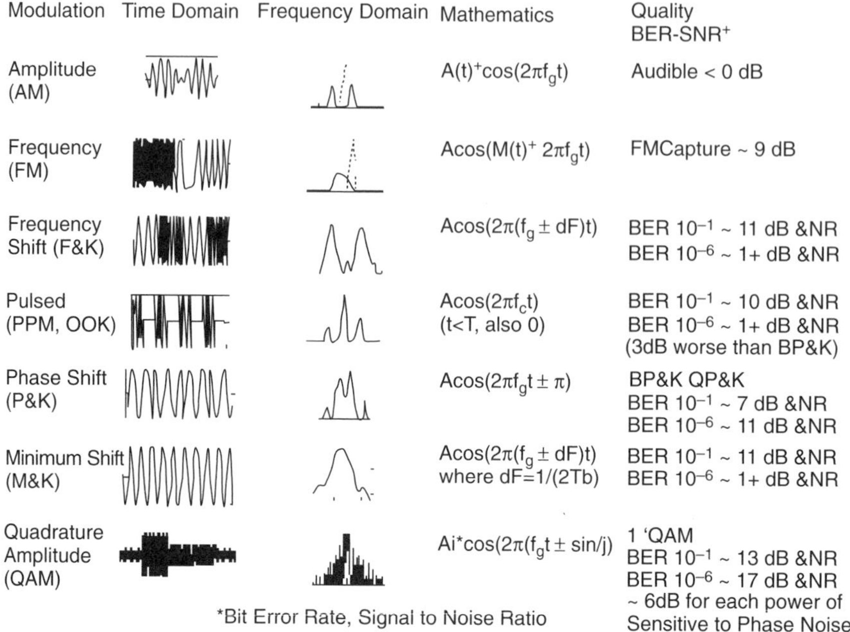

Modulation
Time Domain
Frequency Domain
Mathematics
Quality
BER-SNR+
Amplitude (AM)
A(t)+cos(2πfgt)
Audible < 0 dB
Frequency (FM)
Acos(M(t)+ 2πfgt)
FMCapture ~ 9 dB
Frequency Shift (F&K)
Acos(2π(fg ± dF)t)
BER 10−1 ~ 11 dB &NR
BER 10−6 ~ 1+ dB &NR
Pulsed (PPM, OOK)
Acos(2πfct)
(t<T, also 0)
BER 10−1 ~ 10 dB &NR
BER 10−6 ~ 1+ dB &NR
(3dB worse than BP&K)
Phase Shift (P&K)
Acos(2πfgt ± π)
BP&K QP&K
BER 10−1 ~ 7 dB &NR
BER 10−6 ~ 11 dB &NR
Minimum Shift (M&K)
Acos(2π(fg ± dF)t)
where dF=1/(2Tb)
BER 10−1 ~ 11 dB &NR
BER 10−6 ~ 1+ dB &NR
Quadrature Amplitude (QAM)
Ai*cos(2π(fgt ± sin/j)
1 'QAM
BER 10−1 ~ 13 dB &NR
BER 10−6 ~ 17 dB &NR
~ 6dB for each power of
Sensitive to Phase Noise
*Bit Error Rate, Signal to Noise Ratio

FIGURE 7-3 Basic channel symbols. (From *Software Radio Architecture* by J. Mitola III, Wiley-Interscience, 2000.)

7.1.4 Mode Knowledge Chunks

Autonomous reasoning about air interfaces and protocol stacks requires knowledge of channel symbols and network organization.

7.1.4.1 Basic Channel Symbols

Except for UWB and radar, the channel symbols of wireless communications are based on the modulation of sine waves. Digital modulation techniques impart a state to the carrier for the duration of the channel symbol. The carrier is continuously modulated for analog modes. A digital channel symbol also is called a baud. The fundamental methods of modulating a sine wave modify frequency, amplitude, and/or phase. Combinations of these features yield the basic channel symbols of Figure 7-3.

As shown in Figure 7-3, these channel symbols exhibit characteristic time- and frequency-domain structure. In addition, each of the basic channel symbols has many variations. So there are not just 2-PSK and 4-PSK but also 8-PSK, 16-PSK, and so on. Each variation has associated Bit Error Rate (BER) signal-to-noise ratio (SNR) or signal-to-noise and interference ratio (SINR). In the ultra-wideband (UWB) channel symbol, the subnanosecond pulse with very high instantaneous power spreads energy over multiple gigahertz [169].

Expression 7-5

<Channel-symbol> <FM> <Definition> Frequency-modulation
</Definition>
<Analog-FM> <FM-voice> </Analog-FM>
<FSK> <Definition>Frequency-shift-keying </Definition> </FSK>
</FM>
<AM> <Definition>Amplitude-modulation </Definition>
<Analog-AM> <AM-voice> </Analog-AM>
<ASK> <Definition> Amplitude-shift-keying </Definition>
<OOK> <Definition> On–off keying </Definition>
<Morse-code/> </OOK> </ASK> <QAM/> </AM>
<PM> <Definition> Phase-modulation </Definition>
<PSK> <Definition> Phase-shift-keying </Definition> . . .
<QPSK> <Definition>Quaternary PSK </Definition></QPSK>
<QPSK> <PSK States = 4> </QPSK> . . .
<QAM> <Definition> Quadrature amplitude modulation
</Definition>
<16-QAM> <QAM States = 16> <16-QAM> . . .
<1024-QAM> <QAM States = 1024> <1024-QAM>
</PSK> </PM> </Channel-symbol> <–! See CD-ROM –>

To communicate with a CWN regarding the presence of unexpected channel symbols in its band, an iCR may use the equations of Figure 7-3. To autonomously synthesize waveforms, iCR must know also how filtering shapes the time and frequency responses of the synthesized channel symbol. In order to manage its SDR resources, iCR must be able to relate the basic channel symbols to computational resources of its own SDR algorithms for generating and filtering each of these symbols in real time.

7.1.4.2 Sharing the Radio Spectrum

AACR needs basic knowledge of multiple access as in Expression 7-6.

Expression 7-6 TTD in RXML (Simplified and Readable Format)

<Multiple-access> <Analog>
<TDD> <Definition> Time-domain duplexing </Definition>
<PTT> <Definition> Push-to-talk <Definition>
<Principle> Analog channel symbols are modulated for the duration of the information signal <Diplexing> Users listen to the channel and transmit when no other users are present </Diplexing>
</Principle></PTT> </TDD> </Analog> </Multiple-access>

For FDMA, AACRs also need to know that FDMA assigns each subscriber voice channel to a dedicated carrier offset from the others by an increment slightly greater than the (analog) bandwidth of the channel. Since the intelligibility bandwidth of a 20kHz auditory acoustics channel is only 3.6kHz for voice, FM FDM air interfaces separate FM subcarriers by 4kHz to allow guard-bands between subscribers, while FM broadcast separates channel centers by 200kHz.

Expression 7-7 Analog FDD Includes FM FDM

<Multiple-access> <Analog>
<FDD> <Definition> Frequency-domain Duplexing </Definition>
<FDM> <Definition> Frequency division multiplexing </Definition>
<Principle> Analog FM FDM channel symbols are continuously modulated with each subscriber channel offset from the group carrier, a FM FDM carrier hierarchy <Duplexing> Two groups of FM FDM carriers are required for full duplex communications, one uplink group and one downlink group. </Duplexing>
</Principle> </FDM> </FDD>
</Analog> </Multiple-access>

The CRA <Self/> includes basic knowledge of TDD/PTT and FDM of Expression 7-7.

Digital multiple access formats may employ a digital control channel in the radio access protocol that manages the traffic channels bearing subscriber streams. The CRA <Self/> includes the following methods of sharing the radio channel.

Expression 7-8 Multiple Access Schemes

<Multiple-access> <Digital>
<TDMA> <Definition> Time division multiple access </Definition>
<Framed> Time slots defined by a frame </Framed>
<Synchronous> Bit periods synchronous to the frame </Synchronous>
<Asynchronous> Bit periods not isochronous to the frame </Asynchronous>
<Example> <GSM/> </Example>
</TDMA>
<FDMA> Frequency division multiple access <FDMA>
<AMPS> <Control-channel/> <Traffic-channel/> </AMPS>
<CDMA/> <Noise-temperature/> </Digital> </Multiple-access>

The mix of RXML and text may range from almost no RXML to nearly all RXML. The balance of this section tags the content with RXML where paragraphs are related specifically to ontological primitives and are structured for use in AACR explanation databases, such as the next paragraph. Try to read Expression 7-9 without being distracted by the RXML tags.

Expression 7-9 Readable RXML Definition of DSSS and CDMA

<CDMA> <Definition> Code division multiple access (CDMA) applies direct sequence spread spectrum (<DSSS>) <Coding/> and <Processing-Gain/> to enable multiple <Narrowband/> subscribers to share a single <Wideband/> <DSSS> <Channel/> </DSSS> by <Coding/> the narrowband channels orthogonally. </Definition> <Example> <3G/> <Air-interface> based on <CDMA/> <Example> <WCDMA/> <TD-SCDMA/> <CDMA2000/>
</Example> </Example> </Coding> </CDMA>.

One may use <DSSS <Channel/>> to abbreviate <DSSS> <Channel/> </DSSS>. While the content of these statements is the same, the alternate syntax is instructive (Expression 7-10).

Expression 7-10 CRA Knowledge of Frequency Hopping (Illustrative)

<Frequency hopping> (<FH>) <Definition> <TDMA> <FDMA> with signal dwells for some number of <Bits> <Typically> one packet </Typically> on <Carrier-frequency/> then changes <Carrier-frequency/> enabling others to use that <Carrier-frequency/> at other points in time </Definition>.

7.1.4.3 Co-channel Interference (CCI)

AACRs need a vocabulary of interference.

Expression 7-11 Defining and Illustrating Co-channel Interference

<Interference> <CCI> <Definition> Cochannel interference (<CCI>) is interference on <Self <Carrier-frequency/>/> from a distant <Transmitter/> of the same <Network/> </Definition><Example> <Counter> TDD works well in lightly used spectrum bands such as marine mobile radio </Counter>. As the number of simultaneous TDD users increases, the likelihood of collision increases rendering <Reception/> <Unreliable/> </Example> </CCI> </Interference>

7.1.4.4 Specialized Channel Symbols

Some channel symbols are tailored to specific services. WWV broadcasts timing with one second pulses. In order for a CR to advise a user which bands and modes could provide needed services, the CR must be able to recognize the modes and specialized channel symbols of a given band. Location and Ranging for Air Navigation (LORAN) was the predecessor to the GPS DSSS signal. Military settings are replete with such specialized waveforms. In short, the military AACR should accomplish the radio connectivity tasks historically assigned to a skilled radio operator.

Expression 7-12 Specialized Channel Symbols

```
<Channel-symbol>
  <WWV> <Function> <Timing/> </Function> </WWV>
  <LORAN> <Function> <Location-finding/> </Function> </LORAN>
  <GPS> <Function> <Navigation/> </Function> </GPS>
    </Channel-symbol>
```

7.1.5 Services and Systems

Services available in given bands historically have been the function of specialized hardware. The industry's transition to SDR mitigates that hardware dependence but will exacerbate the learning curves associated with a deep enough understanding of the capabilities and limitations of a given SDR to fully employ its bands and modes. Characteristics of radio systems (*as systems*) about which a CR should be aware include antennas, integrated communications products, and evolving SDR products.

7.1.6 Knowledge Chunks

Fundamental radio knowledge for AACR may be expressed as a sequence of alternating "*Knowledge Chunks*" followed by advice on "*Using That Knowledge*." Each chunk of knowledge is significant enough that its use enables an AACR to either *be aware* of radio in a way that would inform a non-radio-expert user; or *be adaptive* to a feature of radio by detecting the feature and changing behavior with respect to it (with respect to the user's context, which may entail timing the adaptation appropriately); or bootstrap knowledge by *learning* more and autonomously extending its own knowledge and capabilities accordingly.

7.1.7 Exercises

7.1.1. Define Doppler spread in RXML, introducing ontological primitives without referring to the version in CRA <Self/>. How does your treatment differ from CRA <Self/>? Why did you do it that way? If your way is better, then submit a recommendation to www.sdrforum.org, www.w3c.org, and www.omg.org. Which of these bodies responded to your suggestion quickest, and what did each have to say about your suggestion?

7.1.2. If AACRs reason about space, they could ask people to move communications devices or multipath reflectors to improve communications. Some use cases envision an AACR coaxing the user into moving things around until the communications path is workable. Define such a use case, emphasizing the RF aspects of the scene including the physical setting, the propagation differences of different RF bands and modes, multipath, commercial products (legacy and visionary) in the scene, and the role of AML. Enhance Expression 7-4 to include the related three-dimensional reasoning.

7.1.3. Reasoning about multipath could also enable AACR to time its transmissions and to set parameters to capitalize on predictable path conditions. List scenes in which AACR reasoning about fine-grain radio propagation features of a scene enables improved communications, such as near an airport. How is an airport like meteor-burst communications?

7.1.4. AACRs that monitor the weather may reason about absorption, picking those bands and modes that support the user's current situation. Suppose a vacationer is using an outdoors EHF hot-spot access point from a distance of 20 meters and has moved under a tent because it might rain. Build a spreadsheet model of the significant RF engineering parameters of the use case. What engineering parameters should be predicted and measured for an AACR to reason about the EHF link versus other wireless modes? One might like bad weather to result in advice from the AACR to the user that the link to the EHF hot-spot WLAN is likely to be impaired with the oncoming thunderstorm, recommending that the user turn on the (older, slower but more reliable) 802.11 WLAN that at that particular time happened to be turned off or that the user send the large email attachment right away because the user is likely to lose high data rate EHF connectivity for the duration of the storm. Define in RXML the knowledge needed for such reasoning.

7.1.5. Modify the spatial knowledge ontology to enable an AACR to embed a commercial tool like Arc Explorer by establishing links between the spatial ontology and the tool's API. Specify the semantic bindings between the RXML ontology and the API extending RXML.

7.1.6. Write RXML enabling the iCR to reason about the limits over which BER–SNR values are representative of realizable signals in space for HF and SHF. What additional ontological primitives are needed to constrain BER–SNR in the cellular radio bands?

7.1.7. Consider an emergency relief setting. How can staff get from a network of iCRs the kind of spectrum management support typically needed from radio experts? Since the radio experts' knowledge may be relatively limited, how will AACR broaden and strengthen the contributions of those experts to basic radio connectivity, or potentially replace them at establishing connectivity among disparate legacy nodes?

7.2 KNOWLEDGE OF THE HF RADIO BAND

Highlights: HF is still the least costly way to connect two people, who are 3000 miles apart, by voice or low speed data. Its near vertically incident

(NVI) skywave propagation mode is one of the few ways of communicating wirelessly between users in two mountain valleys in a tropical rain forest. Although narrowband modes are the most common, wideband operation is possible under special conditions. HF ALE enables Internet connectivity at modest data rates (e.g., 9.6kbps). The knowledge of HF needed by AACR includes the following knowledge chunks.

7.2.1 HF Physics-Related Knowledge

ITU frequency band: 3–30MHz [170]

Wavelengths: 100m to 10m

Propagation modes:

- Skywave: single "mode" (single layer reflection), multiple "mode" single hop, multiple hop, within lowest usable frequency (LUF) and maximum usable frequency (MUF), potentially near vertically incident (NVI).
- Ground wave (short ranges, <50km).

Multipath delay spread: 1–10ms

Doppler spread: 5Hz

7.2.1.1 RF Knowledge Chunk

HF extends from 3 to 30MHz according to international agreement. The length of a full cycle radio wave in these bands is 100 meters at 3MHz and 10 meters at 30MHz. HF antennas resonate well across bandwidths that are less than 10% of the carrier frequency. To cover a full HF band using efficient resonant elements requires multiple elements. These may be packaged into a single log-periodic antenna to access the entire HF band efficiently and with 6 to 10dBi of gain. Loaded whip antennas between 10 and 30ft long also access the entire HF band with less efficiency and near 0dBi of gain.

Using That Knowledge: An AACR needs to know the efficiency and directional gain that can be expected of an antenna of a given nomenclature to predict the <Self/> radio's distribution of energy in space. These predictions are essential in following spectrum-use policy for mobile HF radios. Military and amateur radio operators manually employ radio propagation prediction software tools that project likely HF propagation. The propagation is sufficiently complex that continuing calibration is required for reliable high performance connectivity.

7.2.1.2 AM Broadcast Knowledge Chunk

In the United States, there is a commercial AM broadcast band between 800 and 1600kHz, just below the lower edge of the ITU HF band. Typical HF antennas receive this broadcast. The locations of AM broadcast stations are generally very well known since they employ antenna arrays hundreds of feet tall. In addition, the typical multi kilowatt radiated power may be well known as well.

Using That Knowledge for Location Estimation: Knowledge of location and radiated power of fixed site transmitters could be used by an AACR to estimate its own position. This can help overcome GPS or other location-finding impairments. For example, in the humanitarian assistance use case, if a victim of a disaster is huddling under rubble, GPS may be unavailable or unreliable, but the RF distribution of AM reception may indicate the location of the victim.

Using That Knowledge for Self-Calibration: In addition, an AACR could calibrate the low band edge of its antenna by measuring the received signal strength from a known AM broadcast at a known location with a known radiated power. Calibration informs choice of modes. For example, antennas that have deteriorated with age suffer signal loss selectively across the bands. Calibration guides the adjustment of mode parameters to avoid using parts of the band in which reception will be less effective. By collaborating in a CWN, iCRs can calibrate radiated power to identify inefficient transmission. To do this, the iCRs need detailed knowledge of transmission chains and of the RF environment so that they calibrate with few multipath reflections and interferers, while not violating RF-quiet bands like radio astronomy.

7.2.1.3 Antenna Knowledge Chunk

A multiband radio may employ a distinct antenna for HF along with other antennas for the higher frequency bands, or it may employ multiband antennas.

Using That Knowledge: The AACR may explain to the user that it needs a better antenna if it is to be effective in an anticipated HF use case. It may ask other CRs what antennas they are using and how they perform. If an associate CR has a spare antenna, the CR could ask the user to borrow that antenna.

7.2.2 HF Spatial Distribution of Energy

7.2.2.1 Ionosphere Knowledge Chunk

As Figure 7-4 suggests, HF radio waves are usually reflected from the ionosphere, resulting in communications beyond line-of-sight (BLoS). An AACR should know that the HF band may radiate in ground wave mode across short distances and that it uses ionospheric reflection for skywave transmission. The AACR should know that skywave mode provides a user with BLoS communications at relatively low data rates. In addition, the AACR should learn from experience exactly what data rates it can achieve and should remember

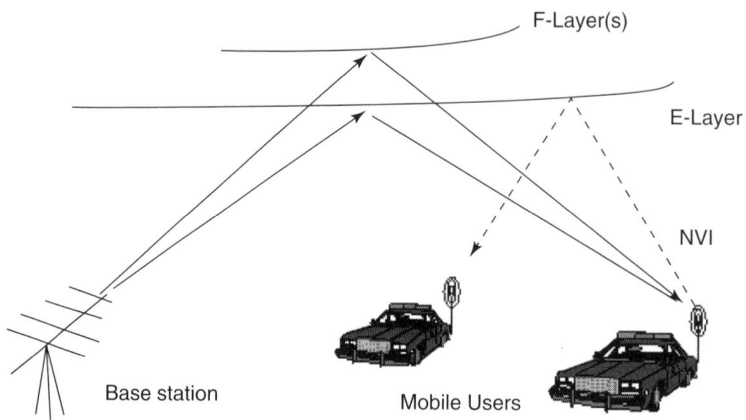

FIGURE 7-4 HF skywave propagation from layers of the ionosphere.

where/who it can connect the user to when located at a given point in space at a given time. Since reflection is a function of solar energy and this varies with the sunspot cycle, the radio should know time with respect to the sunspot cycle.

Using That Knowledge: The AACR's temporal-reasoning capability should relate the current propagation conditions to the sunspot cycle and time of day. It should update a priori ionosphere calibration data from CWNs that are HF aware. The AACR should advise the user on alternate time or place to overcome HF propagation impairments if connection is not achieved changing waveform parameters, HF waveforms, or switching to an alternative band/mode for relay via a trusted network.

7.2.2.2 Multimode Knowledge Chunk

The ionosphere has several layers from which HF waves may reflect. These are identified as the D, E, and F layers in order of increasing altitude. Two or more such skywaves may be received in "multimode" propagation illustrated in Figure 7-4.

Using That Knowledge: The AACR should know that HF multimode reception concerns the reflection of radio waves from multiple layers of the ionosphere. The AACR should at least detect multimode propagation and explain to its user why the link is impaired.

7.2.2.3 MBAA Knowledge Chunk

The AACR should further know that a HF multibeam antenna array (MBAA) and associated algorithms can overcome some mode-interaction impairments.

Using That Knowledge: Therefore, if the user intends to employ HF for an extended time or for an important task, the AACR should request permission from the user to obtain a HF MBAA.

7.2.3 Signal in Space Reasoning

7.2.3.1 Delay Spread Knowledge Chunk

The better informed CR will know that reflected skywaves add as complex vectors at the receiver, resulting in phase and amplitude variability. The time differences between two reflected waves ("HF propagation modes") will be about 1 ns per foot of altitude separation. Since the reflecting layers may be from less than 1 to over 10,000 miles apart, this corresponds to 1–10 ms of delay spread. The CR should know that typical delay spread reflects norms, but that the parameters of any given transmission may be significantly different.

Using That Knowledge: This better informed HF AACR will estimate delay spread from calibration signals. It should know that a HF ALE probe is such a signal, and it should know how to interpret the resulting HF propagation mode parameters with respect to diurnal and sunspot cycles and with respect to available mode compensation algorithms.

7.2.3.2 Doppler Knowledge Chunk

In addition, the ionosphere is typically approaching or receding from fixed transmitters, imparting Doppler shift onto the RF carrier. Since the layers of the ionosphere may be moving in different directions, the Doppler spread at HF can be plus or minus 5 Hz.

Using That Knowledge: The HF AACR should be able to use Doppler shift to estimate D-, E-, and F-layer trends. It should be able to adapt the carrier tracking loop algorithms of its HF SDR to typical and abnormal Doppler conditions.

7.2.3.3 LUF MUF Knowledge Chunk

If the RF carrier is too low or too high, it will not reflect, but will pass through the ionosphere. Beyond LoS, reflections from the ionosphere are only possible on radio frequencies between the lowest usable frequency (LUF) and the maximum usable frequency (MUF).

Using That Knowledge: AACR should be able to explain the LUF and MUF. It should know how to use LUF–MUF tables, and it should be able to obtain updates from a CWN. It should be able to advise the user that the HF signal will be ground wave or skywave or will simply radiate into space if the frequency of transmission is not between the LUF and MUF. As conditions change through the diurnal cycle, it should be able to employ alternate frequencies, narrowband modes, or other subbands for HF ALE to probe.

7.2.3.4 NVIS Knowledge Chunk

Specific combinations of RF and antenna configuration can result in near vertically incident skywave (NVIS or NVI) propagation in which the waves reflecting from the ionosphere propagate only a few tens of miles. NVI is useful in mountains for communications between subscribers in adjacent valleys, for example.

Using That Knowledge: The AACR should be able to explain NVIS for BLoS communications over relatively short distances. A pair of iCRs should be able to experiment with BLoS HF in atypical terrain such as urban canyons and sporting arenas, should the opportunity present itself. These CRs should learn from these experiments following a hypothesize-and-test protocol so that they can share that knowledge with others, including CWNs.

7.2.3.5 Multihop Knowledge Chunk

In addition, HF will reflect from water and from some landmasses, enabling multihop communications (ionosphere–water–ionosphere–land).

Using That Knowledge: AACR should be able to explain models that simulate the gross characteristics of multihop HF propagation, including sources of error. The iCR should be able to recommend experiments by which to discover such multihop HF propagation modes for connectivity needs.

7.2.4 Available HF Communications Modes

Typical HF Band Communications Modes include:

- Manual Morse code (amateur radio logistics, shipping)
- Automatically generated Morse code (e.g., historically data broadcasts to ships at sea)
- Amplitude modulated (AM) including voice—double sideband (DSB), upper sideband (USB), lower sideband (LSB), vestigial sideband (VSB), 4 kHz IBW
- Narrowband data (frequency shift keying (FSK), ~100–40,000 Hz IBW and on–off keying (OOK), 5 Hz IBW)
- Spread spectrum (wideband BPSK direct sequence (DS), and 1–10 MHz IBW, 1–100 kbps 2000 km)
- Automatic link establishment (ALE) link-quality agility.

7.2.5 HF Narrowband Morse Code

7.2.5.1 Modes Knowledge Chunk

Historically, HF communications consisted primarily of narrowband voice and Morse code, generated automatically or manually. Within the past few years, ALE introduced a higher grade of availability and reliability of HF for digital networking. Thus, HF today is replete with the chirped probing signals of HF ALE.

Using That Knowledge: An AACR should be able to list the typical modes of HF communication, which it could do with a small database. The AACR should be able to determine which class of HF modes would be most appropriate for a particular GoS need. It should know which of the modes it can generate with its own waveform and RF capabilities. A CR could observe the user, manually controlling the radio to connect to a given counterpart. The CR could observe the communications between its user and the counterpart(s) available from this connection. An introspective CR could analyze the speech or data during a sleep cycle. It could verify with its user its association of QoI with the HF mode. Finally, when the user again needs to get connected to that same counterpart, the AACR should be able to connect the two by reconstructing and adapting the RF and mode parameters of the prior connection. If the parameters of the prior and current connection were informed by the sunspot cycle or time of day, then it should analyze that context and project the times and places at which those parameters are likely to be successful again.

7.2.5.2 Morse Code Knowledge Chunk

Morse code has been used since the 1800s for ship to shore and transoceanic communications. Automatically generated Morse code became popular with the emergence of microprocessors. PC-based software readily translates text into Morse code automatically.

Using That Knowledge: The HF expert AACR should know the history of HF along a time line so that it can explain it to the interested user. It should be able to generate Morse code and to find a trustable download if needed. It should advise the user regarding the data rate by explaining how long it would take to send a message using Morse code from one CR to another.

7.2.5.3 OOK Knowledge Chunk

Morse code might be formulated as an instance of the more general on–off keying (OOK) signal class with the channel code information carried both in the duration ("pulse width") of the channel symbol, a simple sinusoid, and in the order of the channel symbols to form a source symbol. Because of the relatively low rate at which people can compose and send Morse code, it occupies a bandwidth of approximately 5 Hz. This yields a plethora of such narrowband signals packed into the cluttered HF bands.

Using That Knowledge: The Morse-aware AACR should be able to explain that the code translates letters to "dits" that are three to four times shorter than "dahs." It should be able to generate and receive low speed Morse code from a counterpart AACR to conserve radio bandwidth.

7.2.5.4 HF Narrowband FSK Knowledge Chunk

Other common HF data modes include frequency shift keying (FSK). The FSK channel code consists of Mark and Space, corresponding to a negative or positive frequency shift. The frequency shift may be as small as a few hertz, on the order of the inverse of the data rate. Data rates ranging up to 1200 bits per second require FSK shifts of several hundred hertz. A FSK channel symbol also called a "baud" encodes one bit of information.

Using That Knowledge: The AACR should be able to explain FSK to a user, including the terminology and data rates. It should be able to select FSK as an appropriate mode, for example, for a text message.

7.2.5.5 HDR Knowledge Chunk

During very short time intervals (from a few milliseconds to a few tenths of a second), the ionospheric transfer function is approximately constant. Higher data rates (HDRs, e.g., 10–40 kbps) may be used for such short intervals to "burst" small amounts of data over long distances using FSK modems.

Using That Knowledge: The AACR should be able to advise the user of the possibility of higher data rate communications and get permission to opportunistically employ a higher data rate if the spectrum allocation and propagation conditions permit. It should be able to interpret the spectrum management authority and user constraints on use of the higher speed FSK modes. The sophisticated AACR may experiment with other channel symbols and compare results with other AACRs regarding channel symbols like PSK or QAM, potentially evolving new HF-ALE-class waveforms.

7.2.5.6 HDR Hardware Resources Knowledge Chunk

Morse, low speed, and HDR FSK waveforms can be implemented using less than 25 MIPS and low speed/high dynamic-range ADCs.

Using That Knowledge: The AACR should be able to analyze the MIPS and ADC of the <Self/> to acquire, test, and employ FSK waveforms for HF when the distances, propagation conditions, timeliness, and data quantities warrant.

7.2.5.7 Voice Modes Knowledge Chunk

Voice transmission at HF uses amplitude modulation (AM) to accommodate the limited bandwidth of the HF channel. Double sideband (DSB) AM creates two mirror image replicas of the voice waveform—one above and one below the carrier—using twice the bandwidth required for the information content. Upper sideband (USB) filters the lower of these two voice bands, suppressing any residual carrier. Lower sideband (LSB) is the converse of USB. Vestigial sideband (VSB) allows a small component of carrier to be transmitted, simplifying carrier recovery in the receiver. Voice intelligibility requires only 3–4 kHz for the principal formants (sinusoidal information-bearing components of the speech waveform). Speech processing at HF was one of the first commercial applications of ADCs and DSPs.

Using That Knowledge: Consequently, the AACR should know that each of these voice modes may be implemented digitally with an ADC rate of typically 10–25 kHz using modest processing power (10–25 million instructions per second (MIPS)). An AACR should be able to select the mode appropriate to the counterpart AACR's capabilities and to the ionospheric conditions. A flexible AACR need not have a set of predefined HF voice waveforms but could synthesize a voice waveform tailored to the operating environment and user's needs, such as VSB to match the counterpart radio. An iCR could experiment with other variations of voice and Morse coding, for example, forwarding digital voice mail via Morse code.

7.2.5.8 HF ALE Knowledge Chunk

HF automatic link establishment (ALE) equipment probes the propagation path in a prearranged sequence to identify good frequencies on which to communicate. The ALE signals include "chirp" waveforms that linearly sweep the RF channel so that the receiver can estimate the channel transfer function. The two ends of the link choose RF based on reception quality.

Using That Knowledge: The AACR should be able to explain and employ HF ALE for the user. The user-aware/adaptive AACR should be able to recommend HF ALE as the preferred method of communications based on distances, type of message, quality of service, and other QoI parameters.

7.2.5.9 Wideband Research Knowledge Chunk

The literature also presents successful research in the use of wideband spread spectrum at HF, including thousands of chips per bit and millions of chips per second (MHz) bandwidths [171]. In addition, HF radar may use direct sequence spread spectrum in a frequency-hopped pulsed signal structure. Neither of these relatively exotic waveforms is typically available in SDR.

Using That Knowledge: The AACR should know that high data rates can be achieved at HF over skywave paths with low grazing angles and wide instantaneous bandwidths. It should also know that to spatially limit the transmitted signal to the grazing layer requires a large antenna such as a HF log-periodic antenna. It should know that there are waveforms that can achieve 100 kbps at the right time of day. A more imaginative CR could experiment with the synthesis of such waveforms between appropriate CR endpoints at appropriate times of the day. A less imaginative CR could request the download of such waveforms if it infers that its antenna could control the transmitted waveform and that it has sufficient computational resources for such a waveform, based on self-inspection.

7.2.6 HF Services and Systems

HF services and systems include:

Service band allocations [120]—broadcast, maritime mobile (12 MHz), aeronautical mobile, fixed, amateur satellite (7–7.1), frequency/time (20);

Antennas (log periodic ~20 m × 25 m; "elephant cage" ~3 km diameter; whip 8–15 m; loops 2–10 m [172])

Illustrative systems include the TRC331 HF SSB, 280,000 channels (microprocessor cont), 2, 10, 20 W, 5.9 kg, 200 ms antenna tune time, Thomson CSF, France; J3E USB LSB telephony; J2A, A1A, F1B telegraphy; SEFT 001A; MIL-STD-810C; DEF-133 L3, United Kingdom

Emerging SDR products include the Rhode & Schwarz M3TR with 1.5–512 MHz coverage, ALE, STANAG

7.2.6.1 HF Radio Services Knowledge Chunk

Amateur radio ("Ham"), commercial broadcast, aeronautical mobile, and timing/frequency standards are provided at HF. Amateur satellites are non-HF satellites that interoperate with Ham HF links.

Using That Knowledge: An AACR must know the definitions of each HF service. In addition, its knowledge must include technical parameters to access these services for enhanced QoI.

7.2.6.2 Timing and Frequency Standards Knowledge Chunk

Timing and frequency standards include WWV broadcast.

Using That Knowledge: If an AACR does not have an ability to decode timing and frequency standards, it must know the function of these broadcast signals and must know that it does not have these personalities. A network-supported AACR should also know how to obtain the timing signal decoding download.

7.2.6.3 HF Antennas Knowledge Chunk

HF antennas and power amplifiers often dominate the size, weight, and power of HF radio systems. Antennas matched to HF wavelengths are large—some research antennas extend for over a kilometer.

Using That Knowledge: An AACR must know that, unlike many radios such as cell phones, the HF antennas can be large, if not huge. It must be able to contrast the HF antenna with other antennas in terms of size, weight, and power since these are major technical determinants limiting HF employment in a given <Scene/>. The CR must be able to associate these parameters with a user's situation, and to reason about these features of HF for user planning of logistics support.

7.2.6.4 Military Antennas Knowledge Chunk

Military applications employ circularly disposed array antennas (CDAAs) for long-haul communications and location finding using triangulation. Reliable long-haul communications are also possible using small log-periodic antennas (e.g., 20 × 25 ft horizontally mounted on a 50 to 100 ft mast).

Using That Knowledge: A military AACR should be able to list, describe, and supply a digital image of military CDAAs of current and historical interest. It should be able to estimate the ability to connect via conventional antennas to CDAAs in known locations. It should also be able to describe and provide a digital image of a high performance log-periodic HF antenna.

7.2.6.5 HF Propagation Modeling Knowledge Chunk

The U.S. Navy has software tools for the modeling and simulation of HF radio propagation.

Using That Knowledge: A military AACR should be able to estimate connectivity among CDAA, log-periodic, whips, and other types of antennas using authorized radio propagation modeling and simulation tools. If the AACR lacks sufficient accuracy, it should be able to email the appropriate center for modeling and simulation support. It should know conditions under which propagation modeling is typically useful to enhance QoI.

7.2.6.6 HF Whip Antennas Knowledge Chunk

Whip antennas 8–15 ft long may also be inductively loaded to match HF wavelengths.

Using That Knowledge: An AACR should know HF whip antenna characteristics. A Ham radio operator may be interested in antennas available from eBay® or Radio Shack® while a military user may be interested in antennas normally used by the unit or coalition partners. When a CR meets a CR of an associate (e.g., QSL partner or coalition partner), the CRs should offer to exchange technical data on their antennas to facilitate accurate modeling of connectivity.

7.2.6.7 HF Direction of Arrival (DoA) Measurement Knowledge Chunk

Loop antennas in the 2–10 meter bands can measure direction of arrival.

Using That Knowledge: An AACR should know whether it can measure DoA. If not, but if it is given a location aware or location adaptive task, then it should seek knowledge of such an antenna for location knowledge. It should know the accuracy of DoA as a function of multiple estimates from spatially diverse antennas for location estimates of a required accuracy. It should know whether or not it can use DoA by itself or in conjunction with other location estimation methods, such as GPS. It should be able to aggregate DoA data to estimate location if it has a DoA capability.

7.2.6.8 HF Signal Enhancement Knowledge Chunk

Although software radios cannot change the laws of physics that cause HF antennas to be large, they can enhance signals received using smaller, less optimally tuned antennas to achieve quality approaching that of the larger antennas.

Using That Knowledge: An AACR should know what digital filters and enhanced signal processing improve HF transmission and reception with additional processing resources. It should know what kinds of communications are enhanced by what classes of algorithm and should be able to employ those methods to enhance QoI.

7.2.6.9 Mercury Talk System Knowledge Chunk

Mercury Talk [172] exemplifies the short range, low power HF radios. With 2 watts of output power, this radio can close a voice link on a 10 km path. With its 3.5 watt output, it can close a Morse code link over a 160 km path.

Using That Knowledge: An HF AACR should know how to interoperate with a legacy Mercury Talk radio. It should acquire radio parameters and recognize its legacy protocol.

7.2.6.10 TRC331 Knowledge Chunk

The TRC331, by Thomson CSF of France, weighs less than 10 kg and accesses 280,000 HF channels.

Using That Knowledge: An HF AACR should know that Thomson CSF is now part of Thales Corporation. It should know whether its user is likely to need to communicate with a user of one of these radios. If so, then it should know the details of frequency plan and operation of the legacy TRC331 and new Thales radios.

7.2.6.11 Narrowband Standards Knowledge Chunk
Published standards for HF communications include J3E, USB, LSB, telephony, J2A, A1A, F1B telegraphy, SEFT 001A, MIL-STD-810C, and DEF-133 L3 from the United Kingdom. They meet additional narrowband communications standards for military interoperability [172].

Using That Knowledge: An HF AACR should know the standards for HF as well as the relationships between these standards and its own waveform templates on board and within reach via CWNs. It should know that J3E is a voice telephony standard, relating that to a specific user's need for voice QoI. The HF AACR should know the standards authority and those authorized to employ the standards (e.g., NATO for STANAGs) and should monitor its own use of these standards to assure that it complies with the standards' owner intent as published. Limiting use of NATO standards to NATO radios exemplifies this principle.

7.2.7 HF SDR Products

With the rapid emergence of HF SDR products and systems, this section merely hints at the knowledge possibilities for AACR and emerging HF SDR.

7.2.7.1 Rhode & Schwarz M3TR
Rhode & Schwarz M3TR MBMMR operates from 0.5 to 512 MHz with ALE, STANAG, and other HF modes.

Using That Knowledge: An HF AACR should know that a SDR can be extended in the field by a downloaded waveform personality and that a download may be received via storage media or over the air (OTA). It should recognize the M3TR as a SDR with the parameters of the typical M3TR. If the <Self/> is an M3TR, then it should know this and the ways in which it differs from other M3TRs.

7.2.7.2 Harris PRC-117
The Harris PRC-117 is a family of HF radios. The A and B models were hardware-defined radios, while the F model has SDR personalities conforming to the SDR Forum's SCA. Certain personalities of the F model are not available for export.

Using That Knowledge: The HF AACR should be able to exchange HF waveforms with a trusted PRC-117F with limited user tedium but trust appropriate to the military situation.

7.2.8 Exercises

7.2.1. List the unique features of HF in terms of (a) unique service offered by the physical properties of HF propagation; (b) unique challenges in establishing reliable links and networks in this band; and (c) unique size, weight, and power of HF RF devices.

7.2.2. Create and embed RXML tags for the following knowledge chunk using the ontological primitives of CRA<Self/>:

(a) "The length of a full cycle radio wave in these bands is 100 meters at 3 MHz and 10 meters at 30 MHz. HF antennas resonate well across bandwidths that are less than 10% of the carrier frequency. To cover a full HF band using such resonant elements requires ten or more such antennas. These may be packaged into a single log-periodic antenna to access the entire HF band efficiently and with about 6 dBi of gain. Loaded whip antennas between 10 and 30 ft long also access the entire HF band with less efficiency and near 0 dBi of gain."

(b) Using the tags from (a), convert the following to Radio XML: "An AACR needs to know the efficiency and directional gain that can be expected of an antenna of a given nomenclature in order to predict the radio's distribution of energy in space."

(c) Write a computer program to associate power efficiency and directional gain with names of antennas. Evaluate MS Excel, Analytica, MS Access, Java, and C (C++ or C#) for this task, identifying the advantages and disadvantages of each.

7.2.3. Consider Section 7.2.2.1, Ionosphere Knowledge Chunk. Write RXML so that an AACR that accesses HF, VHF, and UHF has the formal radio knowledge needed to achieve the performance of this knowledge chunk. Suppose the user interface is written in Java. What software tool would you use to fully integrate this knowledge into such a user interface? Do you need a logic programming subsystem like PROLOG? Why or why not?

7.2.4. Consider Section 7.2.6.1, HF Radio Services Knowledge Chunk. "Amateur radio ('Ham'), commercial broadcast, aeronautical mobile, and timing/frequency standards are provided at HF. Amateur satellites are non-HF satellites that interoperate with amateur radio operators HF links." Search the Internet to find the frequencies and other operating parameters of at least one commercial broadcast, aeronautical mobile, amateur satellite, and timing/frequency standard. Write RXML to make this knowledge available to AACR. How much of this knowledge is an <Abstraction/> and how much of it is part of the <Physical-universe>? See the CD-ROM for related questions.

7.3 KNOWLEDGE OF THE LVHF RADIO BAND

Highlights: The lower VHF (LVHF) band from 28 to 88 MHz has traditionally been the band of ground armies because of the robust propagation offered among ground-based subscribers in rugged terrain. Amateur radio and the U.S. Citizens Band also use LVHF. LVHF knowledge is cumulative with HF knowledge. Ideas for CR use of radio knowledge introduced in HF will usually not be repeated in LVHF even if applicable. Instead, LVHF and subsequent bands develop new ideas enabled by the physics, systems, and services of the other bands.

7.3.1 LVHF Physics-Related Knowledge

Frequency band: 28–88 MHz
Wavelengths: 10.71 m to 3.4 m
Propagation modes:

- Skywave ($W_c < 50$ kHz): direct and multipath (e.g., air–ground), ducting beyond line of sight (LoS), refracting beyond LoS ($0.5 < K < 5$; $R = (3Kh/2)^{1/2}$), and diffracting near obstacles (Fresnel zones)
- Ionospheric scatter [170, p. 33–12] is possible between 30 and 60 MHz; with −85 dB D path loss; and $W_c < 10$ kHz
- Meteor burst [170, p. 33–12] modes typically occur in 50–80 MHz; over 600–1300 km; with ~kW power; and $W_c < 100$ kHz.
- Ground wave (short ranges, <10 km)

7.3.1.1 Band Edge Knowledge Chunk

The lower band edge at 28–30 MHz may propagate in skywave mode. The upper edge of this band is defined by the commercial broadcast band from 88 to 108 MHz. The upper band edge can have a much shorter radio horizon than the lower band edge.

Using That Knowledge: An AACR should know that the upper band edge of LVHF is defined in practice by the lowest frequency commercial FM broadcast station in most parts of the world. A typical multiband multimode SDR will cover FM broadcast and LVHF. The AACR should be able to find the lowest FM channel broadcasting in a given place and time. It should find other broadcast stations in its database, such as time or frequency standards. Two CRs could experiment with propagation near the lower and upper band edges to determine whether the CRs are within geometric line of sight (LoS) or refracted. They should know to operate closer to the lower band edges for non-LoS unless interference precludes this.

7.3.1.2 LVHF Antenna Knowledge Chunk

Wavelengths from 10.7 to 3.4 meters are efficiently transmitted in smaller antennas than HF, with a quarter-wave dipole having a length of 3–10 feet.

Using That Knowledge: The AACR should know whether it is a wearable, portable, vehicular, or fixed site radio. It should know which antennas the <Self/> can use for transmission or reception. The helpful AACR should assist the user in selecting a suitable LVHF antenna or with improvising if necessary. It should know that efficient LVHF transmission is practical with whip antennas from a vehicle, but that a fixed site can use a larger and hence more efficient and higher gain antenna, while an individual soldier's antenna is probably a loaded whip or fractal mesh that is the least efficient. When the user is in or near a large vehicle, it should try to find out whether the vehicle's CR has a larger, more efficient LVHF antenna, and if so, should ask the vehicular CR to share one of the vehicle's LVHF transmission paths via WLAN or wired (e.g., Ethernet) connectivity.

7.3.1.3 LVHF Military User Knowledge Chunk

Historically, LVHF military users have employed single-channel half-duplex PTT AM and FM modes. The commercial success of Racal's® Jaguar frequency-hopped radio with its digital vocoding and digital air interface resulted in a proliferation of FH modes for military users during the late 1980s.

Using That Knowledge to Interoperate With Coalition Partners: A LVHF AACR should know if coalition forces own Racal Jaguars by being told by a military user or by looking it up, for example, in *Jane's Military Communications* [172]. The AACR should be able to compare observed air interfaces with published specifications for the signal structure and vocoder packets. It should propose interoperating in Jaguar mode via SDR download with Jaguar coalition partners. The CR should share this interoperability knowledge with ad hoc CWNs and RAs. The CR could advise its CWN to download a trusted Jaguar personality. The Jaguar is offered as a common radio among global military organizations that might cooperate in a humanitarian operation. Other radios and signal structures could be recognized and treated in the same way as the Jaguar. Since military radios are constantly evolving, a CR could collaborate to resolve unknown interference.

7.3.2 LVHF Spatial Distribution of Energy

Spatial distributions of energy at LVHF are illustrated in Figure 7-5.

7.3.2.1 LVHF BLoS Knowledge Chunk

Although LVHF frequencies do not reflect from the ionosphere with the reliability of HF, these waves scatter from the lower D layer. D-layer scatter at 30–60 MHz with a bandwidth of less than 10 kHz often has only about 8.5 dB greater loss than LoS propagation. In addition, LVHF propagation beyond geometric LoS is common via tropospheric refraction.

Using That Knowledge: An AACR could reason about D-layer scattering to connect a CR to a legacy LVHF radio. This may entail downloading a terrain map from a trusted web site, estimating the radio propagation over the terrain for LoS frequencies, comparing these to D-layer scattering, and beamforming to enhance BLoS QoI.

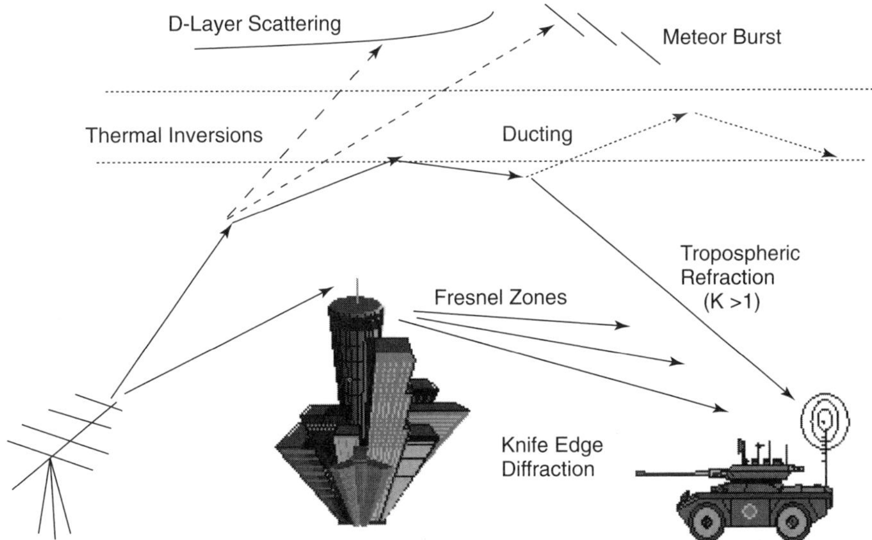

FIGURE 7-5 LVHF scattering, ducting, refraction, and diffraction.

7.3.2.2 LVHF Radio Horizon Knowledge Chunk

Since the atmosphere is denser at lower altitudes, the speed of light is less near the ground than at higher altitudes. Since typical LVHF whip antennas provide an omnidirectional radiation pattern with relatively large vertical extent, the waves propagate across significant differences in index of refraction. Therefore, the waves emitted just above the geometric grazing angle propagate beyond the geometric LoS, having been bent down as they traverse the path. This effect can be modeled as an increase in the effective radius of the earth. The approximation of radio horizon is given by

$$R = \sqrt{4Kh/2} \tag{7-2}$$

Range, R, is in miles, K is the effective radius of the earth, and h is the altitude of the transmitter in feet. K, the effective earth radius, is defined experimentally. $K = 1$ defines geometric LoS propagation, while $K = \frac{4}{3}$ in temperate climates, but K may range from $\frac{1}{3}$ to 3 with climate and weather.

Using That Knowledge: An AACR should be able to apply the radio horizon formula for a range of K. It should know that if its GPS indicates temperate latitudes, $K \sim \frac{4}{3}$. It should also know that if its weather <Scene/> includes high humidity or precipitation that the radio horizon will shift accordingly. It should know how the radio horizon affects received signal strength, relating the radio horizon to the link budget aquation. It should suggest mobile user or vehicle movement to enhance the probability of communications. In other words, the AACR should have basic skill in deployment management. In exercising this skill, the CR must be able to explain its reasoning in simple clear terms that address user questions about the trade-offs of radio deployment versus other objectives (such as finding birds if one is on a bird watching trip but would also like to see babycam video of the kids and babysitter at the lodge).

7.3.2.3 LVHF Ducting Knowledge Chunk

At night, particularly in subtropical climates, LVHF waves may propagate by ducting, where the refractive index of the atmosphere inverts (air density increases with increasing altitude instead of decreasing as usual). Ducting can extend the range of LVHF 200 miles or more beyond LoS.

Using That Knowledge: An AACR should know the positive and negative aspects of ducting and when it is likely to occur. It should be able to verify BLoS ducting by listening to distant broadcast stations or to transmissions from known locations. It should know that ducting increases interference, and it should know how to mitigate that interference by adjusting the parameters of the air interface. Specifically, it should be able to establish analog notch or digital filters to notch out the strongest ducting interference, and it should track the filter parameters against parameters of the interference.

7.3.2.4 LVHF Ground-to-Air Communications Knowledge Chunk

Ground-to-air radios also experience skywave multipath scattered from the D layer or refracted through tropospheric ducts.

Using That Knowledge: An AACR should know whether it is on the ground or in the air, and whether it is associated with a person, vehicle, or ground facility. It should know whether its user needs to be in contact with airborne entities, and negotiate with the AACR on the aircraft to employ anomalous propagation to enhance QoI temporarily given diurnal variations and changing position of transmitter, receiver, and ducting. It should learn how ducting behaves in an operations area, and it should share that experience on CWNs so that all the CRs can employ air-to-ground ducting better than conventional radios.

7.3.2.5 Diffraction Knowledge Chunk

In knife edge diffraction, waves bend around sharp obstructions as if the entire wavefront above the obstacle were a point source forming Fresnel zones.

Using That Knowledge: An AACR should know that large obstacles such as hills, tress, and buildings can interfere with radio propagation. It should model the location and size of such obstructions in its visual scene. It should register its LoS propagation model to the buildings and warn the user when the mobile is likely to be shadowed by a building. This requires knowledge of both the user's location and the communicant's location, so the AACR should strive to keep track of both locations in spite of lack of connectivity, GPS impairments, and the like. An iCR should have a kinetic model of each communicant by which to estimate planned movement to predict and correct communications impairments. For example, it may be possible to correct shadowing of a 2.4 GHz LoS link by moving the transmission temporarily to LVHF, albeit at a lower data rate or with higher power consumption. If the data rate will be lower, the CR should so advise the users and systems applications to compensate or at least understand the connectivity and data rate situation.

7.3.2.6 Fresnel Zones Knowledge Chunk

Point sources induce an interference pattern of reinforcement (waves on the average in phase) and cancellation (waves on the average 180° out of phase) called the Fresnel zones. A receiver in the Fresnel zones experiences alternating strong and weak signals as the receiver moves through multiples of a wavelength. LVHF radios may maintain reception continuity across Fresnel zones using diversity in space (e.g., multiple antennas) and frequency (e.g., slow frequency hopping) with error control coding.

Using That Knowledge: An AACR with sufficient processing capacity could use computer-aided design (CAD) models of an urban scene to predict Fresnel zones. It could use this knowledge to advise a pedestrian user where to walk to obtain best connectivity to a communicant whose location is known by the CR. The CR might have a top-level goal of knowing the locations of all entities within its radio access, tempered by the far reach of HF, and thus focusing on those entities enjoying LoS or nearly LoS propagation.

7.3.2.7 LVHF Meteor Burst Communications Knowledge Chunk

Each minute a dozen meteors penetrate the earth's atmosphere, where they burn up. This creates trails of ionized gas from which radio waves may be reflected. Meteor burst communications (MBC) use trails that endure for periods of 10 milliseconds to over a second.

Using That Knowledge: An AACR should know about MBC and about SDR personalities for MBC. It should be able to explain the operation of MBC to an inexperienced user, including the low data rate and unusual protocol by which the receiver transmits first. For additional MBC knowledge chunks, see the companion CD-ROM.

7.3.2.8 LVHF Ground Wave Knowledge Chunk

LVHF, like HF, may also be propagated via ground wave over short ranges (e.g., 10 km). Ground wave mode generally suffers large attenuation, with a path exponent of 2.5–4. That is, instead of path loss proportional to $1/R^2$, the path loss will be proportional to $1/R^{2.5}$ to $1/R^4$.

Using That Knowledge: An AACR should be able to estimate the path loss and link budget given the approximate distances between two potential communicants. Using calibrated delays in electronics, two AACRs should be able to estimate distance between each other to within hundreds of meters or less based on 100 ns calibration errors. An AACR should be able to use that knowledge to manage transmitted power proactively to minimize interference generated by the ground wave network. Peer protocols that lack explicit power management could be managed at the macrolevel as the CR estimates and then measures BER or E_b/N_o in a given situation, tracking statistics and reducing power when practicable. The more effective AACR could insert redundant bits or forward error control (FEC) "outside the protocol" by repeating bits or packets or by precoding data offered to an ad hoc ground wave network.

7.3.3 Available LVHF Communications Modes

Typical LVHF communications modes include AM (LSB, USB, VSB), FM (voice, fax), narrowband data (FSK, PSK, 1.2–10 kbps typical), spread spectrum, frequency hop (FH) (slow hop (<100 Hps); medium (100–1000 Hps), fast (>1000 Hps); narrow hop (6 MHz), wide hop (60 MHz)), and wideband DS/frequency hop hybrids.

7.3.3.1 Single Channel Per Carrier LVHF Knowledge Chunk

AM (DSB, USB, LSB, and VSB) and analog modulated FM voice are common at LVHF. The analog modes arose in the 1960s, when signal processing was limited to analog frequency translation, filtering, automatic gain control, and simple RF selection. In single-channel-per-carrier (SCPC) modes each subscriber has a unique RF carrier. Ground-based military forces prefer SCPC for squad-level manpack and vehicular radios. LVHF fills in low-lying terrain where higher frequency waves do not penetrate.

Using That Knowledge: An AACR should be able to list and explain each AM and FM SCPC mode to an inexperienced user. It should analyze channel content to know which modes are available to the user's likely communicants so that it knows which mode to use to find a given communicant by name or function, particularly if that communicant has a legacy radio. It should be able to cooperate with users and with other CRs to manage legacy modes. It should be able to participate in an ad hoc CWN. Each CR should be able to measure and to forward interference information and recommendations for changes of mode (particularly frequency or power setting) of legacy radios to manage the communications environment. Each CR should be able to employ or download personalities that enable terrain filling when needed to support the user's QoI, but should avoid LVHF unless needed for disadvantaged terrain. Multichannel CRs should volunteer to relay signals from a disadvantaged user to the intended recipient, taking advice and direction from the authorized users, AACRs and CWNs.

7.3.3.2 PSK Data Mode Knowledge Chunk

FSK and phase shift keying (PSK) are common data modes. Simple PSK formats such as binary (BPSK) and quaternary PSK (QPSK) offer reliable data service at LVHF from 1.2kbps to about 10kbps within the coherence bandwidth of LVHF.

Using That Knowledge: An AACR should know about FSK, BPSK, and QPSK as LVHF modes. It should be able to match available data rates to applications and quality of service (QoS). The CR should map data rate and connectivity to requirements of applications like location, mapping, navigation, trip planning, and web browsing. Applications parameters need to be tuned to these low data rates.

7.3.3.3 TETRA Knowledge Chunk

Digital vocoding and private networks (e.g., TETRA [173]) are increasing in LVHF.

Using That Knowledge: A LVHF AACR should know about TETRA and should be able to explain it to the nonexpert. It should be able to learn how TETRA supports taxi and public safety users. It should be able to communicate with the TETRA regulatory authority for download and passwords. The CR should know that the TETRA users in some locations are police and should be able to identify police networks for emergency assistance. It should note the language of the local law enforcement and obtain an emergency signaling download in the native language from the local CWN so that emergencies may be declared, such as how to say "chest pains" in Swahili.

7.3.3.4 LVHF Frequency Hop (FH) Knowledge Chunk

Contemporary LVHF military radios may employ FH. LVHF propagates well in rugged terrain since the waves penetrate vegetation and reflect, refract, and diffract over and around obstacles. FH provides additional fading compensation and also is more difficult for an adversary to jam than pure SCPC waveforms. Jane's and other on-line and spectrum management resources have databases of radios, which employ FH.

Using That Knowledge: An AACR should be able to explain FH to a nonexpert user as the systematic variation of physical layer RF to resist multipath fading (e.g., in GSM) and to be more difficult to jam (e.g., in military radios). The AACR should be able to detect and characterize known coalition FH modes for interoperability and counterfratricide services. A CWN should be able to configure a low-resource FH SDR download for a coalition partner for CR-to-CR cooperation.

7.3.3.5 LVHF Spread Spectrum Knowledge Chunk

Spread spectrum modes include FH, DSSS, and hopped-spread hybrids such as JTIDS [174]. Some FH radios hop over subbands of LVHF, employing 1–6 MHz hopping bands. Others provide the full 60 MHz hopping agility from 28 to 88 MHz.

Using That Knowledge: A FH AACR should know that spread spectrum uses more bandwidth over a transmission epoch than the instantaneous bandwidth needed by the source. It should know that FH and DSSS are examples of spread spectrum. It should be able to explain the difference between the two to a nonexpert. It should be able to explain JTIDS, and why JTIDS operates in the UHF band versus the LVHF band (insufficient hop bandwidth for its 250 MHz hop range). It should be able to synthesize a JTIDS-like LVHF waveform and to request that transmission be authorized by the local RA.

7.3.3.6 LVHF FH SDR Knowledge Chunk

The narrower hop bandwidths may be implemented digitally via SDR (e.g., using fixed tuned medium bandwidth RF conversion and a 6 MHz ADC/DAC).

Using That Knowledge to Use the Appropriate Waveform: A FH AACR should know how to synthesize LVHF FH waveforms.

Using That Knowledge to Synthesize a More Appropriate Waveform: A more sophisticated CR might synthesize a FH physical layer tailored to the user's needs. It would have to cooperatively test the waveform with another CR before using it for communications. Design drivers include specific interference, multipath fading parameters, and the capabilities and limitations of the partner CR. For details see the companion CD-ROM.

7.3.3.7 FH Vocoding Knowledge Chunk

The FH radios are typically vocoded. The speech waveform is represented digitally using a vocal tract model such as Linear Predictive Coding (LPC). Waveforms based on subband coding [175] and adaptive LPC were implemented in DSP chips in the mid- to late-1980s. Other voice codecs like Vector Excited Linear Prediction (VELP) and Codebook Excited Linear Prediction (CELP) have better perceptual properties.

Using That Knowledge: A vocoder-aware FH AACR should be able to explain the idea behind a voice coder to a nonexpert user, as a mapping from the analog microphone to a packetized digital bitstream. The more expert CRs may be able to explain the principles of vocoding to a maintenance technician or radio engineer. The AACR should be able to mix and match vocoders to optimize QoI. For example, if the user's native language is causing vocoder errors, the AACR should be able to identify a more compatible vocoder for better intelligibility and enjoyment. See the companion CD-ROM for additional FH vocoder knowledge.

7.3.3.8 LVHF Multichannel Air Interface Knowledge Chunk

FM frequency division multiplexing (FM FDM) for military LVHF applications includes modes with four channels per RF carrier. Some of these modes are analog while others are digital. Digital PCM at 64 kbps enables 4×16 kbps continuously variable slope delta modulation (CVSDM) digital voice channels in a FM FDM spectrum allocation. These modes meet the needs of relatively low-echelon military forces [172].

Using That Knowledge for Extended Ad Hoc Networking: An AACR should know whether its waveform library includes multichannel modes. It should know if it has more than one channel and multichannel waveforms, and how to act as a radio relay. Its CR should be able to estimate the impact on battery life or power drain on the vehicle if it enables a multichannel mode.

7.3.3.9 LVHF FM FDM and TDM Knowledge Chunk

Due to the relatively narrow coherence bandwidths of LVHF, conventional FM FDM is limited to about 60 channels. LVHF digital backbone radio formats often observe T- or E-carrier, or NATO STANAG digital formats. Thus, a T1 provides 24 channels in a 1.544 Mbps data stream that might occupy between 1 and 3 MHz of bandwidth, while an E1 provides 30 channels in a 2 MHz bandwidth.

Using That Knowledge: An AACR should know how many subscribers can be accommodated with a given analog FM FDM waveform and for a given T- or E-carrier digital format. It should be able to explain its capabilities and limitations to a nonexpert user attempting to configure a network. It should know if there is any dependence on or preference for directional antennas for such multichannel waveforms. Military radios should be able to explain the security issues, approaches, and components to the user and to a designated security authority. The CR should be able to reason about its own signal processing bandwidth and internal connectivity needed for multichannel modes. See the companion CD-ROM for additional TDM-PCM knowledge.

7.3.4 LVHF Services and Systems

LVHF service band allocations [170] include broadcast, fixed/mobile, radio astronomy, aeronautical radio navigation (74.8 MHz), and commercial FM broadcast (87.5–108 MHz). LVHF antennas typically are log-periodic

[172, p. 597], whip (ground), blade (aircraft), passive network arrays, or biconical horns [172, p. 613]. Illustrative systems include the Rockwell Collins AN/ARC-210 (10–22 W, ECCM, 4.5 kg), and the Racal Radio Ltd. (UK) Jaguar-V (10 mW, 5 W, 50 W; ECCM, 6.6–7.5 kg) [172, p. 69]. Emerging SDR products include Harris AN/PRC-117F (30–512 MHz), and the Motorola WITS 6000 Series.

7.3.4.1 LVHF Services Knowledge Chunk

LVHF supports broadcast, fixed and mobile applications, radio astronomy, aeronautical radio navigation (74.8 MHz), and commercial FM broadcast (87.5–108 MHz), among others.

Using That Knowledge: An AACR should have a database of authorized LVHF transmissions along with the locations of known fixed broadcast stations. It should know the radio astronomy bands where transmission is prohibited and should avoid those bands. On the other hand, if a prohibited band must be used in an emergency, it should describe the user situation that justifies the exception and advise the RA that it employed the band in an unauthorized way, posting this event to a web site to coordinate the transmission with radio astronomers.

All AACRs should model aeronautical radio to avoiding generating interference in those bands for flight safety.

7.3.4.2 LVHF Antennas Knowledge Chunk

Antenna products include log-periodic arrays with broad bandwidth and high gain (e.g., the Allgon Antenn 601 [172, p. 597]).

Using That Knowledge: An antenna-aware AACR should have a knowledge base of LVHF antennas with computational models of antenna technology that enable it to explain the capabilities and limitations of antennas to users. It should know about connector types and VSWR and should be able to measure the efficiency of wave propagation to and through the antennas it is using. It should cooperate with other CRs to diagnose antenna behaviors. It should be able to recommend antenna choices to the user, and to requisition needed antenna(s) from appropriate sources. For knowledge chunks and use for LVHF aircraft and high gain antennas, see the companion CD-ROM.

7.3.4.3 Aeronautical Radio Knowledge Chunk

The AN/ARC-210 from Rockwell Collins illustrates airborne LVHF products. It radiates 10–22 W of power, weighs 4.5 kg, and supports a variety of electronic counter-countermeasures (ECCM) including FH.

Using That Knowledge for Federated System Management: An AACR should know which nomenclatured radios like the ARC-210 are in its federated suite. In the transition from discrete radios like the ARC-210 to SDRs like the JTRS or the M3TR, larger aircraft may employ a mix of discrete radios, SDRs, and CRs in a federated radio communications suite for a relatively long period of time. The CRA cognition component managing such complicated suites must know the operating parameters of the discrete radios, ECCM mode settings, manufacturer, and the authorized maintenance download suppliers. It should know how modes of this discrete radio complement SDR so that the CR can configure the suite for connectivity and protection of information. The aeronautical CR should also be expert in the rules of countries in which it could transit, such as European narrowband requirements and flight safety channels to autonomously generate alerts when a channel becomes active.

7.3.4.4 Jaguar Knowledge Chunk

The Jaguar-V from Racal Radio Ltd., UK [172, p. 69] popularized LVHF FH. This affordable manpack configuration produces power of 10 mW, 5 W, and 50 W with the Jaguar's own advanced FH ECCM in a compact 6.6–7.5 kg package.

Using That Knowledge for Power Management: A CRA cognition component managing a federated suite of legacy radios on a military vehicle may control a Jaguar legacy radio. In addition to the tasks the CR performs to manage an airborne suite, the ground vehicular CR should be adept at power management. Today's tactical radios typically are not aggressively power managed. Power settings often are set to maximum power output in part because military crews know maximum power is needed "most of the time," particularly if everybody else has set their power to the maximum screaming in their own ears, when in close proximity. The cognition component should control the final power stage of all the radios in the suite. Such power-aware and power-adaptive CRs would then enhance the available radio spectrum. Through location awareness, the CRs would reduce power when the communicants are adjacent to each other or would route traffic through a high band HDR channel, such as an SHF WLAN offloading LVHF.

Using That Knowledge for Commercial Applications: The commercial equivalent would be the use of the 5.4 GHz ISM band in lieu of Citizens Band LVHF at maximum power when two sportsmen are within a hundred meters of each other.

7.3.4.5 LVHF Software Radio Knowledge Chunk

A LVHF SDR draws more power from a battery than a customized microprocessor-controlled analog/digital hybrid product.

Using That Knowledge for Battery Planning Via Topic-Spotting Traffic Shaping: An AACR should manage power cooperatively with the user. Algorithms estimate battery life from initial charge, history of use, and typical performance. AACRs learn the user's typical patterns and plan, to predict battery endurance. If the battery is predicted to not meet that time line, then the CR could advise the user and develop a mitigation plan. If the organization has authorized the AACR to override talkative users, the AACR could indicate "Channel unavailable—use text messaging" to force the user to limit nonessential conversation. Some organizations may authorize the CR to monitor conversations for content. If, for example, the AACRs have reliable location data on all personnel yet the user expends exorbitant talk time regarding location, "Are you sure that is where we are? I don't think so. I can't see that water tower. . . . Oh, there it is," and so forth, then the CR could detect this communications state using speech topic-spotting algorithms. There is a trade-off between expending processing resources monitoring the content of the user's conversation and reserving that processing resource for communications. Suppose it is only one-fifth as expensive to topic-spot as to transmit the content. The CR could advise the user that it is switching to "rapid voice mail mode," a power management mode in which the radio listens to a 2 second chunk of speech before sending it. If the chat is about location, the CR could advise the user that the location shown on the map is correct and it could highlight the water tower in the user's heads up display when he tries to transmit "I can't see the water tower." In other words, the CR could assist the military user locally in understanding where everybody (self included) is located, shaping the traffic by answering questions locally that can be answered or assisted by the CR. Of course, no military user would want his own radio to behave like this, preferring to talk to whoever, whenever. But in training, the difference could be clear if it were to turn out that those with the assistance from the CR are more efficient and capable and more likely to accomplish the mission because of the conservation of resources (one battery for a mission with the CR, but you have to lug two along for the legacy radio).

Using That Knowledge for Personal Safety: If a sportsman is in the woods from dawn till dusk, then the battery of the integrated GPS–LVHF–cell phone should last until dusk as well. Some sportsmen will use the radio more than others, so the commercial AACR could proactively advise the user when it is likely that there will be no battery left at dusk. Speech recognition can generate text for low power transmission. For example, the common phrase, "Where are you?" can be reliably converted from speech to text and Huffman coded for very low power transmission. The companion CR could convert that text to speech, yielding a voice request to the other sportsman. The companion CR could similarly speech-code common responses, dramatically reducing the need to talk. Such location updates may use far less battery power than speech. The commercial AACR could keep the radio transmitter powered off until speech or text traffic is offered, powering up in the appropriate mode, and powering down again immediately. An iCR learns such scenarios by observing the user's speech patterns, such as an increasing

TABLE 7-1 SDR Parameters—VHF

Software Radio Application	Sampling Rate (f_s)	Dynamic Range (dB)
VHF-UHF BB[a]	50–150 kHz	20–60
LVHF-IF (FH)[b]	12–200 MHz	66–108
VHF/UHF-IF	25–500 MHz	60–96
VHF RF	650 MHz	96–120

[a] BB = baseband. [b] IF = intermediate frequency.

average voice pitch when its user is under stress. The net effect of speech-code power management would be to enhance safety.

7.3.4.6 LVHF Radio Mode Parameters Knowledge Chunk

Propagation and air interface modes constrain the critical SDR parameters of Table 7-1. SDR ADCs and DACs sample radio bands at RF, IF, or baseband (BB). As the bandwidth sampled increases, the dynamic range requirements also increase.

Using That Knowledge: An AACR should model its own ADCs and DACs to manage performance and to characterize performance for nonexpert users in easy, medium, and dense RF environments. Such models and a priori knowledge assist the user in selecting the waveform appropriate for a given radio environment. The CR learns by monitoring radio resources (FH collisions and E_b/N_o or BER) over time. Thus, it could recommend parameter settings that avoiding interference or enhancing the QoS by optimizing the ADC, DAC, analog and digital filters, and roofing filters [144]. AACRs may adapt based on a priori knowledge, while iCRs learn from experience, tailoring ADC, DAC, and mode to the current situation.

7.3.4.7 Baseband SDR for LVHF Knowledge Chunk

Baseband digital processing accommodates single-channel voice and narrowband data communications.

Using That Knowledge for Backup: An AACR should know that if its wideband ADC or DAC fails but its narrowband ADC or DAC still works, then it can configure a narrowband waveform to minimize QoI loss. The autonomous generation of backup modes characterizes iCR. The iCR should learn such backup modes from expert user actions sharing techniques via CWN. The AAR cannot learn from this experience, but must have such backup modes preprogrammed.

7.3.4.8 LVHF IF SDR Knowledge Chunk

IF processing enables multiple channel radio relays, television, and other radio services. The IF dynamic range reflects the near–far ratio, noise variability, and interference background variations in VHF. RF dynamic range encompasses the entire band.

Using That Knowledge to Infer Causes of Near–Far Clipping: An iCR should go beyond notched filtering in the RF channel. The iCR should learn the space–time features of its local radio environment. It should identify those emissions that cause interference by reconciling space, time, identity, received signal strength, and estimated radiated power patterns against each other in the CR's dynamic ontological model of sources and propagation. It should then draw inferences about the sources of interference, making recommendations about how to deal with these sources operationally. These recommendations could be made to other CRs, to a CWN, or to operators of other equipment to enhance the radio environment that is under one's scope of control. For example, as radio users converge for a rendezvous, their radios interfere with each other. Today's solution may be to turn off some radios. The AACRs could autonomously manipulate network parameters to reduce transmit time while maintaining synchronization. Physically proximate AACRs could use a SHF ISM physical layer to maintain net membership and timing with low duty cycle LVHF transmissions to minimize interference.

Using That Knowledge for Radar Mitigation: In addition, LVHF interference could be mitigated operationally. For example, if a high powered radar's out-of-band emissions cause interference, the CR could recommend the movement of the radar or the pointing of antennas to optimize radar and communications QoI. Since metal structures exacerbate reflections, a CR with vision and propagation modeling could identify a barbed wire fence as a problem, resulting in the movement of the fence to mitigate interference.

7.3.4.9 LVHF Versus HF Knowledge Chunk

One benefit of operating in LVHF versus HF is the reduction in delay spread by three orders of magnitude from milliseconds to microseconds. In addition to improving the coherent bandwidth of the medium, it reduces the memory requirements and complexity of time-domain equalizer algorithms.

Using That Knowledge: An AACR should know the computational resources needed for each of its SDR bands and modes. It should be able to explain the benefits of one SDR mode over another to a nonexpert user. The CR should have sufficient knowledge, RF sensing, and explanation of RF alternatives to interact effectively with spectrum managers following the myriad rules of the local, regional, and national RAs, treaties, and agreements.

7.3.4.10 LVHF Noise Complexity Knowledge Chunk

The reduced noise complexity of LVHF enables constant false alarm rate (CFAR) squelch algorithms to reliably track the LVHF noise floor, while at HF, more complex algorithms are required.

Using That Knowledge: An AACR should adapt to noise and interference. Waveforms embed squelch algorithms sufficient for the typical radio environment. The iCR adapts to intense and statistically extreme interference by synthesizing mitigation such as prewhitening to remove non-Gaussian interference.

7.3.5 Exercises

7.3.1. LVHF knowledge is cumulative with HF knowledge. Extend the exercises of Section 7.2, Knowledge of the HF Radio Band, to LVHF.

7.3.2. The AACR should know whether it is wearable, portable, vehicular, or fixed. It should know which antennas are mounted and available for transmission and/or reception. Write RXML that expresses this to the <Self/>.

7.3.3. Write RXML that supplies the ontological knowledge for <Self/> to apply the radio horizon formula for a nominal $^4/_3$ earth. Write a stand-alone Java module to do this. Refer to the CD-ROM: can you teach CR1 to do this without modifying CR1? If so, then teach CR1 to do this, and if not, then add a PDA DL module that enables you to teach CR1 to do this.

7.3.4. An AACR should be able to list and explain each AM and FM SCPC mode to an inexperienced user. Search the Internet for AM and FM SCPC modes. If you can't find sufficient information elsewhere, look in the Canadian spectrum management web site. What are the benefits and the shortfalls of using this database? Look at the source files of the Canadian web site. Do you see any RXML-like semantic tags there? How would your research task be easier if RXML tags were compatible with the Semantic Web and the CRA <Self/>? Write RXML that patches this new knowledge chunk into <Abstractions/>. Is it better for AACR to incorporate the spectrum management knowledge in its own internal knowledge base, to keep this database in a CWN, or to query the Canadian database? Write a stand-alone program that incorporates the AM and FM SCPC modes at LVHF.

7.3.5. Teach CR1 about SCPC using the t.e.a.c.h primitive and the Hearsay AML component. What are the benefits of embedding knowledge into CR1 this way? The shortfalls? Does t.e.a.c.h scale up to hundreds of thousands of entries? Suppose the iCR only needs to know a few thousand such entries. Does this change the trade-off of learning versus database integration?

7.3.6. Write RXML that enables an AACR to query the Canadian database for an update each time it needs to know about LVHF SCPC modes. Suppose the database were queried a few hours ago. Should it be queried again, or should the AACR infer that the current data is definitive? What additional knowledge would the AACR need to know to decide whether to query or not?

7.3.7. Refer to the CD-ROM for a TETRA exercise.

7.3.8. Write RXML to enable your AACR to become vocoder aware. Program it to explain the voice coder to a nonexpert user, explaining a vocoder as a way of converting the signal from the analog microphone signal to a digital bitstream. Write RXML for the vocoder knowledge for a maintenance technician. Include semantic primitives that the maintenance technician employs to diagnose vocoder faults.

7.3.9. Write RXML to link an aircraft AACR to a radio engineer in an extended in-flight emergency (Mayday situation). Suppose the aircraft has plenty of power but is in the Bermuda triangle and is experiencing navigation faults to be conveyed by the AACR to the radio engineer. Write a program for the AACR to use this knowledge to compare dead reckoning, GPS, Glonass (not normally on the aircraft but available for download), and weak signals from AM and FM broadcast and radar to localize the aircraft, determine its heading, and guide it to the nearest major land mass.

7.3.10. The vocoder-aware AACR should be able to mix and match vocoders to the needs of the user in cooperation with a CWN. Suppose the user's native language is causing vocoder errors. Write RXML-Java to identify a more language-friendly vocoder for better QoI. The RXML should include A-law PCM, Mu-law PCM, RPE-LTP the GSM vocoder, and 16 kbps CVSD.

7.3.11. Consider the Digital TDM PCM transcoding knowledge chunk of the CD-ROM and complete the related exercise.

7.4 RADIO NOISE AND INTERFERENCE

The transition from HF to LVHF brings substantial changes in the statistical structure of radio noise and interference. Progression through the bands toward EHF reveals noise and interference environments less limited by interference and more limited by thermal noise. Over time and in popular bands, relatively clear RF environments abruptly transition from noise-limited to interference-limited. The iCR recognizes such changes and offers mitigating strategies.

7.4.1 Overview of Noise and Interference

Figure 7-6 illustrates HF, VHF, and UHF sources of radio noise and interference.

7.4.1.1 Low Band Noise Sources Knowledge Chunk

In the lower bands, atmospheric noise arises from the reception of lightning-induced electrical spikes from thunderstorms halfway around the world. Consequently, this noise component is stronger in summer than in winter. In addition, this noise has a large variance. The short-term (1 ms) narrowband (1 kHz) noise background varies at a rate of a few decibels per second over a range of from 10 to 30 dB, depending on the latitude, time of the year, and sunspot cycle. Interference below 100 MHz includes automobile ignitions, microwave ovens, power distribution systems, gaps in electric motors (e.g., elevators), and the like. In military settings, unavoidable interference results when tens of thousands of ground military personnel use their LVHF radios at the same time. Thus, high levels of interference characterize these congested low bands.

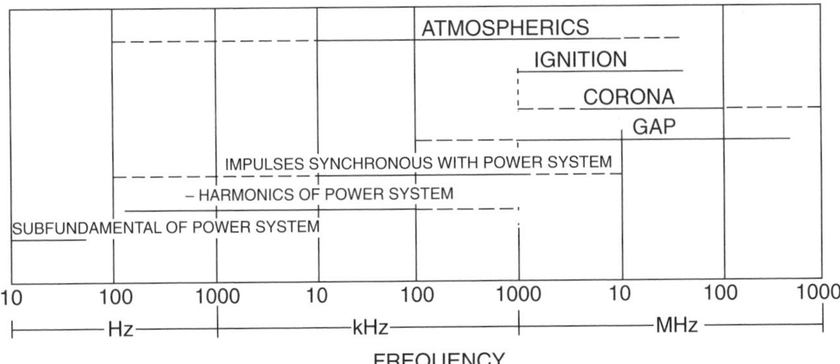

FIGURE 7-6 Frequency range of artificial noise. (Adapted from [174, p. 34-7, Fig. 6]; used with permission.)

Using That Knowledge: The AACR should be able to list the sources of radio noise and interference in each band as shown in Figure 7-7. It should be able to identify sources of radio noise and interference in its visual perception domain. For example, a truck present in the visual domain should be identified as a potential source of ignition noise currently experienced in LVHF. It should be able to advise the user on how to move to mitigate the effects. The iCR should provide feedback to the user regarding whether the mitigation worked or not. It should also be able to differentiate a few proximate sources of interference from many distant sources.

7.4.1.2 Midband Noise Sources Knowledge Chunk

Midband noise is stronger in urban than suburban <Scenes/> (Figure 7-7). Cellular bands from 450 MHz to above 1 GHz are dominated by the co-channel interference of other cellular users occupying the RF channel in distant cells.

Using That Knowledge: The AACR should be able to list those bands for which cellular service is authorized in a given area. It should be able to characterize which parts of the band have been built-out. A cellular-aware CR should be able to determine which cell systems are operating, which are busiest, and which are underused to pick the optimal cell system from among user accounts. The user may have authorized the CR to charge cellular service to a phone bill or credit card, for example, in anticipation of an upcoming trip to Europe, Asia, or North America. The AACR should be able to recognize underused ISM bands with low noise temperature as an opportunity for ad hoc networking.

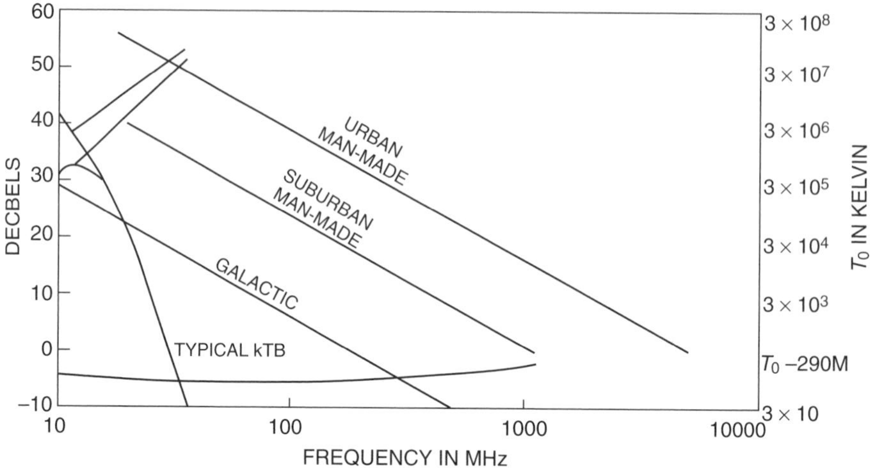

FIGURE 7-7 Median average noise power (omindirectional antenna near surface). (Adapted from [170, p. 34-7, Fig. 7]; used with permission.)

7.4.1.3 Thermal Noise Knowledge Chunk

Thermal noise is

$$P_n = kTB \tag{7-3}$$

where k is Boltzmann's constant, T is the system temperature (T_0 is the reference temperature of 273K), and B is the bandwidth (e.g., per Hz). In the microwave bands above 1GHz thermal noise approximates the noise background. In urban areas, however, incidental urban noise and interference dominate thermal noise until about 6GHz.

Using That Knowledge: The AACR should know the equation for thermal noise. It should know that as bandwidth B increases, total noise in the band also increases. It should know how noise temperature is different from ambient temperature. It should be able to autonomously estimate the total noise in a given band. The CR should determine from that estimate whether the band is dominated by thermal or other noise and diagnose the noise sources. Internal noise may indicate an incipient electronics failure. External noise may indicate suboptimal antenna placement. The CR should be able to compare noise levels across bands to assist the user in diagnosing communications problems caused by excessive noise.

7.4.2 Exercises

7.4.1. Write RXML that enables a mobile AACR (e.g., in a commercial aircraft) to recognize a transition from a noise-limited to an interference-limited situation, say, in crossing the Atlantic Ocean. Suppose the band has recently been liberalized to enable on-line web access from an aircraft that has 100 antenna elements embedded in its surface capable of achieving MIMO in the VHF–UHF bands. Write RXML that enables the AACR to reason about radio noise and interference as a factor in planning MIMO linkage to a ground–air MIMO site in the next big city.

7.4.2. Suppose the short-term (1 ms) narrowband (1 kHz) noise background in a HF band varies at a rate of a few decibels per second over a range of from 10 to 30 dB, depending on the latitude, time of the year, and sunspot cycle. Write RXML that enables an AACR to reason about these dependencies.

7.4.3. Write RXML that enables an AACR to know that the incidental and unavoidable interference below 100 MHz includes automobile ignitions, microwave ovens, power distribution systems, and gaps in electric motors (e.g., elevators).

7.4.4. How can an AACR learn about other sources of radio noise and interference below 100 MHz? What hardware, software, and ontological primitives are needed for AML to learn about noise versus interference?

7.4.5. Write RXML for a vision-capable AACR to identify sources of radio noise and interference in a visual scene given that a CAD-CAM model of a generic physical device can be recognized as belonging to a class such as automobile, power lines, and any noise sources in the ontology.

7.4.6. How could AACRs use knowledge of the physical presence and general locations of sources of radio noise and interference in a scene to assist a user who is having trouble joining a VHF network? An 802.11 network?

7.4.7. What hardware, software, and semantic knowledge and skill does an AACR need to determine that a large, old, unmaintained truck present in the visual domain is generating substantial ignition noise in the LVHF band? How would AACR use this knowledge to enable user connectivity if the user cannot move? Write RXML that enables your AACR to advise the user on how to mitigate the effects, such as moving away from the truck or moving the body between the truck and the radio antenna. See the CD-ROM for related exercises.

7.4.8. Assume that a CWN maintains a database of all RF emitters within a 70 mile radius. LVHF interference may originate from ignition noise or from 10,000 military radios of a multinational humanitarian relief group 50 miles to the west. Write RXML for your AACR to determine whether interference being experienced in the LVHF band is being generated by a few proximate sources or by many distant sources. Include the Query-Response behavior between your AACR and the CWN. Write Java code to use this RXML plus a spatial reasoning tool like Arc Explorer™ to simulate the function of a skilled radio operator in identifying sources of radio noise and interference. Train CR1 to respond to a particular scenario using Hearsay. Train it for the truck and present CR1 with a similar situation. Diagnose its learning faults. How often does it generalize appropriately? In what ways must you precondition the learning experience so that it produces the right result?

7.4.9. Write RXML for your AACR to know that in the microwave bands above 1 GHz thermal noise approximates the noise background. Augment this RXML so that it also knows that, in urban areas, incidental urban interference dominates thermal noise until about 6 GHz. Augment this RXML so that it estimates the total thermal noise. Teach CR1 to determine from simulated measurements whether a band is dominated by thermal noise or by interference.

7.5 KNOWLEDGE OF THE VHF RADIO BAND

By ITU and IEEE convention, the very high frequency (VHF) band extends from 30 to 300 MHz.

7.5.1 Alternative Definitions of VHF Bands

The convention ignores differences in propagation between the LVHF band and VHF above the commercial broadcast band (88–108 MHz).

7.5.1.1 VHF Bands Knowledge Chunk

The VHF band may be defined in any of several alternative ways. The ITU definition of HF, VHF, UHF, SHF, and EHF assigns each band exactly one decade of frequencies. The upper VHF band (100 to 300 MHz) approximates optical LoS with a $\frac{4}{3}$ earth radio horizon. Propagation between 100 and 800 MHz is well suited for air-to-ground communications. Thus, most regions of the world allocate a band above the 100 MHz commercial broadcast band for communications between pilots and ground controllers. In addition, the ICAO recognizes bands between 225 and 400 MHz for aeronautical operation. Thus, many commercial aircraft radios operate between 100 and 500 MHz, the VHF–UHF aeronautical band, which is another way of defining the band edges.

Using That Knowledge: The AACR should be able to list the alternative VHF radio bands, formal and pragmatic, including the aeronautical VHF–UHF band. It should be able to explain why band edges are defined differently between the ITU and other usage. The AACR should learn the local use patterns of VHF, such as pilot-to-tower communications. In an emergency, a CR might ask the assistance of a general aviation iCR or pilot on such a frequency by advising the pilot of the emergency (lost and had a heart attack while hiking). That pilot could render assistance by RF relay and by observing the scene from the air, guiding rescue to the victim. An iCR should use the $\frac{4}{3}$ earth radio horizon to estimate the range at which two aircraft may communicate on a VHF–UHF frequency.

7.5.2 VHF Physics-Related Knowledge

Frequency band: (30 MHz) 100–300 MHz
Wavelengths: (10 m) 3 m to 1 m

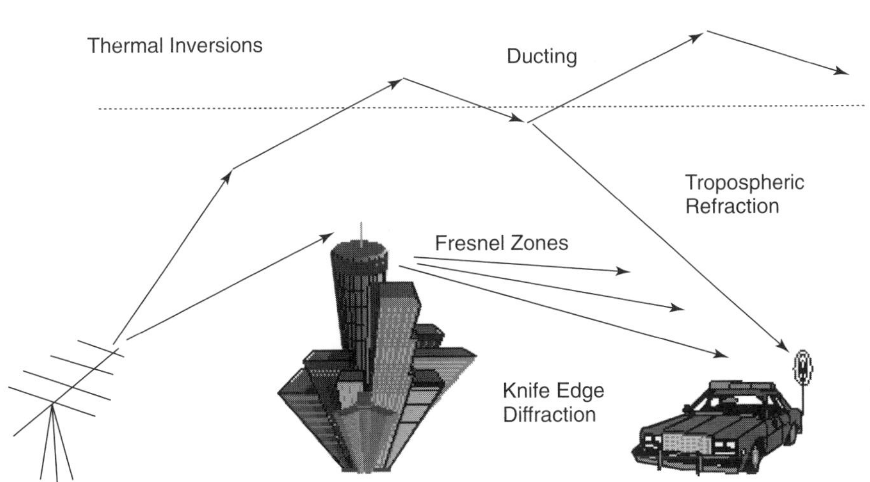

FIGURE 7-8 Spatial distribution of energy in VHF.

Propagation modes include skywave (radio line of sight (LoS), direct and multipath (e.g., air–ground), some ducting beyond LoS in the lower end of the band below 100 MHz, refracting or diffracting beyond LoS) and ground wave (short ranges, <10 km) with multipath delay spread of 1–10 μs.

7.5.2.1 Delay Spread Knowledge Chunk

The delay spread of 1–10 microseconds supports instantaneous modulation bandwidths of hundreds of kilohertz in simple receiver architectures (e.g., single-channel push-to-talk with AM conversion or FM discriminator receivers; or FSK mark/space filters for data signals).

Using That Knowledge: The AACR should know how to measure delay and Doppler spread from known broadcast stations with controlled time and pulsed emissions and by collaborating with peers. It should know by introspection (not by being told) whether a given waveform has an adaptive equalizer and how to compute the number of half-channel symbol period taps needed to compensate observed delay spread.

7.5.3 Spatial Distribution of VHF Energy

Typical spatial distributions of energy are illustrated in Figure 7-8.

7.5.3.1 VHF Fresnel Zones Knowledge Chunk

Upper VHF includes Fresnel zones, knife edge diffraction, ducting, and tropospheric refraction like LVHF. VHF has less filling of low lying and shadowed regions because the shorter wavelengths set up spatially smaller interference patterns. These patterns have smaller angles between successive constructive and destructive interference zones. This fine structure supplies less total power to the shadowed regions.

Using That Knowledge: The AACR should know how far away from a knife-edge obstruction the user should be to avoid shadowing. It should map constructive interference zones, to guide user mobility, and should share that knowledge with CWNs for HDR when in a Fresnel zone of high E_b/N_o.

7.5.3.2 Tropical Ducting Knowledge Chunk

Wavelengths from 1 to 3 meters typical of this band are readily trapped in thermal inversions in the atmosphere in subtropical climates, leading to beyond LoS ducting at the day–night boundary.

Using That Knowledge: The AACR should know how location in the tropics implies occasional BLoS propagation. It should search for reception of BLoS known broadcast stations to calibrate the ducting. It should enhance QoI by adapting HF ALE to VHF ducting.

7.5.4 Available VHF Communications Modes

VHF communications modes include AM (LSB, USB, VSB), FM (voice, fax), narrowband data (FSK, PSK, 75 bps to 9.6 kbps typical), multichannel radio relay (4–60 channels), and spread spectrum formats that include frequency hop (slow hop (<100 Hps); medium (100–1000 Hps), fast (>1000 Hps); wide hop (>10 MHz) and wideband BPSK with hop hybrids possible.

7.5.4.1 VHF Modes Knowledge Chunk

AM, FM, data modes, and FH spread spectrum such as the US/NATO HAVE QUICK I & II slow FH air interface are common in VHF. Wide hop separations are more practical in these bands than in HF because about 300 MHz less prior allocations could be available for VHF FH.

Using That Knowledge: The AACR should know about AM, FM, and FH and it should be able to observe the spectrum available for FH and explain this to a nonexpert user. It should be able to propose and validate FH style waveforms appropriate to current link conditions, QoI needs, and resource availability.

7.5.4.2 AM Voice Robustness Knowledge Chunk

The AM air interface waveform is particularly appropriate for emergency communications with aircraft. AM waveforms are intelligible at negative

SNR, just as speech is intelligible in low SNR. This feature extends the range and robustness of analog AM voice. FM voice, also a popular military mode, provides greater clarity of voice communications at channel SNR greater than 9dB. Below this SNR, the FM discriminator will not lock to the carrier, yielding only noise. Thus, AM voice is intelligible at total power levels well below those that render FM unintelligible. Consequently, AM voice is ideal for emergency reporting radio channels. Analog voice does not fully leverage modern signal processing technology. Recent research suggests extending analog modes via wavelet-based DSP [176].

Using That Knowledge: The AACR should know about AM and FM and should be able to explain to a nonexpert user why general aviation persists in using AM analog voice when so many other more modern channel coding methods are available. An iCR should know about wavelet signal processing and should be able to synthesize a wavelet-based air interface.

7.5.4.3 Channel Packing Knowledge Chunk

Improvements in components have reduced channel bandwidths from 100kHz or more in the early days of radio to typically 25–30kHz today, with $8^1/_3$ and $6^1/_4$kHz modes recommended by APCO. Congestion of air traffic control radio bands in Europe constrains AM/FM to $8^1/_3$kHz. This packs three SCPC subscribers into the 25kHz formerly occupied by one.

Using That Knowledge: The AACR should know why analog voice was designed with 25 or 30kHz channel spacing. An aircraft-aware AACR should know the new rules for denser spectrum packing. AACR should be able to set parameterized waveforms to achieve the denser spacing. It should cooperate with its peers to check air interfaces for spectrum mask conformance.

7.5.5 VHF Services and Systems

VHF service band allocations [170] include 87.5–108MHz broadcasting; 117.975–137MHz aeronautical mobile; 138–144MHz and 148–151MHz government (G); 151–162MHz nongovernment (NG); 162–174MHz and 220–222MHz G, NG mobile radio; 144–146MHz amateur satellite; and 156.7625–156.8735MHz maritime.

VHF antennas include log-periodic [172, p. 597] 20–220MHz; whip, blade, discone, corner reflectors, biconical horns ([172, p. 613], VHF through 960MHz).

Illustrative systems include AN/GRC-171(V) 20W, ECCM(HAVE QUICK), 36kg, 225–400MHz, AM voice, AM secure voice; Rhode & Schwarz Series 400: 15–300W, 12/40 channels. Emerging SDR products include MSRC prototypes (ITT 30–450MHz; Marconi VRC99; Rockwell ARC-220 0.002–2GHz).

7.5.5.1 VHF Services Knowledge Chunk

VHF includes commercial air traffic control (117.975–144 MHz), amateur satellite, and maritime mobile bands. Consequently, SDR access to VHF can provide services spanning air, ground, maritime, government, and amateur market segments.

Using That Knowledge: The AACR should know the services available in the bands, should list frequency allocations, and should summarize the general nature of the band. It should be able to read the standard text of any broadcast channel scanning services for topics of interest to the user to enhance QoI.

7.5.5.2 FM Broadcast Knowledge Chunk

VHF services include the 87.5–108 MHz commercial FM broadcast bands.

Using That Knowledge: The AACR should know the location of the FM broadcast band and tune to the user's favorite types of radio on voice request. It should be capable of proposing to play FM music matching the user's prior behavior patterns.

7.5.5.3 Air Traffic Control Knowledge Chunk

Air traffic control uses the 117.975–137 MHz aeronautical mobile band. This band is allocated to civilian air traffic control, while the companion UHF band is allocated to military air traffic control. Consequently, dual-band VHF/UHF avionics radios are common. There are also governmental applications in 138–144 MHz, 162–174 MHz, 220–222 MHz, and 148–151 MHz and nongovernmental bands from 151 to 162 MHz.

Using That Knowledge: The AACR should know the radio etiquette for air traffic control and government radio use. Specifically, it should know that it is not supposed to transmit in these bands signals that might interfere with aeronautical mobile user, air traffic control, or government applications. It should find policy broadcast and authoritative databases that specify the bands and modes used by the government. It should avoid interference when proposing modes of its own invention to operate in these bands.

7.5.5.4 Amateur Satellite Knowledge Chunk

The amateur satellite band extends from 144 to 146 MHz.

Using That Knowledge: The AACR should be able to explain Ham radio to the nonexpert. A Ham-aware CR should know about ARRL and the local Ham organization. It should download, test, and employ an AmSAT air interface with applications packages of interest to the owner.

7.5.5.5 VHF Antennas Knowledge Chunk

VHF antennas include whip, blade, discone, corner reflectors, passive network arrays, and biconical horns. The high gain horns, cavity-backed spirals, discones, and so on are relatively large because of the 3 meter wavelength at the low end of VHF. High gain military antennas are available for avionics and extensible antenna masts [172]. Some log-periodic antennas such as the Allgon Antenn 601 [172, p. 597] access the subset of VHF from 20 to 220 MHz. Others span VHF through 960 MHz [172, p. 613]. VHF/UHF operation is common for both antennas and discrete analog and programmable digital radios.

Using That Knowledge: The AACR should know about VHF antennas including those listed. For each of antenna type, it should explain their use for the nonexpert. The iCR should associate waveforms and antennas with the circumstances under which particular antenna types are especially useful. It should be able to nominate needed antennas from the supply system (e.g., if the current antenna is faulty).

7.5.5.6 HAVE QUICK Knowledge Chunk

The AN/GRC-171(V) general-purpose ground-based radio delivers 20 W from vehicular power. It includes the HAVE QUICK ECCM/EP (Electronic Protect) mode for interoperability with airborne radios. This radio weighs 36 kg, operates between 225 and 400 MHz, and supports AM voice, AM secure voice, and FM air interfaces.

Using That Knowledge: The VHF AACR should know about legacy VHF radios, SDRs, and other AACRs operating in this band. It should be able to summarize performance and choose waveforms to interoperate with the legacy radio. Since iCRs read natural language text and speech, they should recognize the mention by a user of a legacy VHF radio and request to interoperate, seting up the mode to match the waveforms perceived in the <RF-environment/>.

7.5.5.7 Multichannel Relay Knowledge Chunk

The Rhode & Schwarz Series 400 multichannel VHF radio relay produces 15–300 watts of power to relay from 12 to 40 channels. Each channel may have 25, 12.5, or 6.25 kHz bandwidth. This rack-mount radio is typical of military radio relays.

Using That Knowledge: The military AACR should know about rack-mounting to advise nonexpert users regarding configurations in coalition peacekeeping or humanitarian assistance. A configuration-aware AACR assists a nonexpert to configure such trunking radios into a field-expedient multichannel network. Since iCRs recognize speech, they route voice signals to appropriate users by name similar to the voice-activated phone book of a cell phone. A CR could create voice mail services on spare disk space. Multiple CRs should assist nonexpert users in configuring legacy radios for multichannel uses as well as configuring themselves with multichannel waveforms for ad hoc trunking networks.

7.5.5.8 Aviation Knowledge

Aviators have domain-specific information sources available by radio and Internet, including notices to aircraft (NOTAMS), weather facsimile broadcast, and air traffic control tower frequencies. A general aviation (GA) aircraft is typically small and low flying.

Using That Knowledge: The GA CR should know about GA-unique information and about how to get it on behalf of the user. It should also know about small, medium, and large aircraft. It should also be able to warn the GA pilot when rapidly approaching an obstacle detectable by radio or visual sensors for collision avoidance.

7.5.5.9 Commercial Trucking Knowledge Chunk

Commercial fleets (e.g., trucking) offer potential SDR insertion opportunities. Many truck fleets, for example, use a GPS-based location system coupled to a satellite-based fleet tracking system (e.g., Omintrax). In addition, the fleets use VHF or LVHF radio and commercial AM/FM broadcast for local traffic information.

Using That Knowledge: The trucker's AACR should acquire a model of a trucking user and should instantiate this model with expectations regarding services like location finding, CB, and AM/FM radio preferences as a function of space and time. It should use these patterns plus speech interaction to enhance QoI for the driver.

7.5.5.10 IVHs Knowledge Chunk

Local navigation, wireless on-line maps, and other intelligent vehicle highway systems (IVHs) are also emerging. Thus, commercial trucking fleets are evolving multiband multimode capabilities, potentially amenable to SDR insertion.

Using That Knowledge: The trucking-aware AACR should know what is important to a truck driver or motor freight operator. Some truckers may not be interested in location reporting if they routinely drive the same routes while others may benefit from location information. Markets have embraced GPS dedicated location-assisting devices. As other niche products emerge, such as IVHSs and automated toll payment systems, opportunities to simplify the electronics suite of a commercial vehicle may appear. The iCR MBMMR with cognition enables the integration of such helpful services focused on the needs of the trucking users.

7.5.5.11 Algorithm Complexity Knowledge Chunk

The algorithm complexity of VHF SDR is similar to LVHF. Many air interfaces use SCPC with narrow bandwidths. One potential benefit of SDR technology is the graceful introduction of the new narrowband modulation formats. Digital filtering, both on transmit and receive, can manage adjacent channel interference, even in $6^{1}/_{4}$ kHz bands. SDR with IF and baseband DSP also facilitate the introduction of vocoders and packet data in SCPC fleet networks.

Using That Knowledge: The AACR should know the computational complexity of its standard air interface personalities. It should know the incremental complexity of enhancing its standard air interfaces with more capable algorithms such as digital filtering. It should use this knowledge to enhance QoI within resources and to craft backup modes autonomously.

7.5.6 Exercises

7.5.1. Write RXML that enables your AACR to know that most regions of the world allocate a band above the 100 MHz commercial broadcast band for communications between pilots and ground controllers. Augment this knowledge to include the fact that the ICAO recognizes bands between 225 and 400 MHz for aeronautical operation. Augment this knowledge to include a method to find commercial aeronautical aircraft radios that operate between 100 and 500 MHz using VHF–UHF aeronautical modes. Write an algorithm to find details of such commercial products on the Web using a web crawler. Augment this knowledge to define the band edges in the aeronautical domain. Teach CR1 these facts. Offer a situation in which it retrieves one of the facts you taught it via the Hearsay model.

7.5.2. Write RXML that enables your AACR to list the alternative VHF radio bands including the aeronautical VHF–UHF band. Write a stand-alone Java simulation that enables an AACR to explain why band edges are defined differently between the ITU and other authorities. Simulate AACR behavior to learn the usage patterns of VHF in an area, such as whether pilot-to-tower communications are used. Postulate or use a speech recognition tool like ViaVoice.

7.5.3. Write RXML that enables your AACR to know how to measure delay and Doppler spread either from known broadcast stations with controlled time and pulsed emissions or by collaborating with local AACR peers. Write a stand-alone computer program that simulates the measurement of delay spread. Write a program that simulates introspection over a SDR-SCA or OMG-SRA regarding whether a given SDR waveform has an adaptive equalizer. Write RXML and a stand-alone computer program that enables it to compute the number of $T/2$ (half-channel symbol period) taps needed to compensate for typical and extreme delay spread measured.

7.5.4. Write RXML that enables your AACR to know about AM and FM aeronautical communications. Write a stand-alone computer program that uses this RXML to explain to a nonexpert user why general aviation persists in using

AM analog voice when so many other more modern channel coding methods are available. Teach CR1 about AM and FM analog voice.

7.5.5. Write RXML that enables your AACR to know the services available in the VHF and UHF bands without using a supporting database. The RXML should summarize the general nature of these bands.

7.5.6. Write RXML that enables your AACR to know the frequencies of the FM broadcast band. Write a stand-alone simulation of tuning to the user's five favorite radio channels that are already known upon request. Write RXML that enables your AACR to propose to play FM music at a time that matches the user's prior behavior patterns, assuming they are already known (e.g., news first in the morning then music for a half-hour until news just before arriving at work; music on the way home). Write RXML that enables your AACR to scan the channels and read the data tags on these (real or simulated) channels to find a class of FM broadcast similar to that known to be of interest to the owner.

7.5.7. Write RXML that enables your AACR to know about VHF antennas such as those listed above. For each of the general antenna types include the usees of such antennas. Write a stand-alone program to use that RXML to explain these antenna types to a nonexpert user. Enhance the RXML to include the association of a set of ten notional SCA/SRA waveforms defined in XML to antennas. Enhance the RXML to include the circumstances under which particular antenna types are particularly useful. Write a simulator to nominate specific antennas from the supply system it determined that the user needs, for example, assuming the current antenna is faulty. Train CR1 with this knowledge. Present it with a situation that evokes that knowledge in a way that would assist a user. Discuss the merits of training CR1 versus programming the stand-alone applications.

7.5.8. Write RXML that enables your AACR to know that commercial fleets (e.g., trucking) are interested in AACR insertion and that includes basic knowledge of the truck fleets; for example, use a GPS-based location system coupled to a satellite-based fleet tracking system (e.g., Omintrax). Include in your RXML that some fleets use VHF or LVHF radio and commercial AM/FM broadcast for local traffic information. Train CR1 to acquire a model of a trucking user that reflects the RXML of this exercise. Train CR1 to instantiate a Hearsay model with expectations, as a minimum regarding services in various bands and services such as location finding, CB, or AM/FM radio preferences as a function of space and time. Use the Hearsay model to offer advice to a trucking user regarding CB.

7.5.9. Write RXML that enables your AACR to estimate the algorithm complexity of VHF SDR. Include knowledge that resources for modulation formats using SCPC with narrow bandwidths are similar. Write RXML that enables the graceful introduction of the new narrowband modulation formats. Use this RXML to describe a specific arrangement of digital filtering, on both transmit and receive, to manage adjacent channel interference for a $6^1/_4$ kHz VHF SDR. Enhance this RXML so the AACR knows how to estimate the computational complexity of its standard air interface personalities. Augment it to know the incremental complexity of enhancing its standard air interfaces with more capable algorithms such as enhanced digital filtering.

7.6 KNOWLEDGE OF THE UHF RADIO BAND

UHF is clearly the most popular terrestrial commercial band. With the proliferation of cellular and personal communications systems (PCSs) between 400 and 2500 MHz, the most heavily used radio bands now are almost exactly the UHF band (300–3000 MHz). The following UHF radio knowledge chunks and micro use cases on the innovative use of that knowledge build on the ideas from bands discussed earlier. Most of the use-case ideas from bands discussed earlier are applicable to UHF, so although a few are repeated for emphasis and clarity, many are not.

7.6.1 UHF Physics-Related Knowledge

Frequency band: 300–3000 MHz

Wavelengths: 1 m to 0.1 m

UHF propagation modes include pure skywave (aircraft, square-law path loss), scattered (mobile radio, 2–4 path exponent), and pure ground wave (short ranges, <1 km). Impairments include multipath delay spread (2–5 ms [9]) and Doppler shift (75 Hz for 60 mph at 840 MHz), with Doppler spread of 2 × Doppler shift. Fast fading is distinct below 5 mph, noise-like above 5 mph.

7.6.1.1 General UHF Physics Knowledge

Pure skywave propagates between aircraft and the ground according to square-law path loss. Ground-based PCS channels scatter and attenuate the hybrid skywave/ground wave with path exponents between 2 (square law) and 4. Losses are nonuniform with range; loss exponents vary from square law near the antenna, to 2.8 in Rician zones, and 4th law in Rayleigh zones, with distance from the base station. Urban multipath delay spread typically is from 2 to 10 microseconds [177]. Doppler shift would be 75 Hz for a 60 mph vehicle and RF of 840 MHz, typical of cellular radio. The Doppler spread defines a range of frequency offsets that the receiver's carrier loop must track.

Using That Knowledge: CWNs use this knowledge for path prediction, assisting users with radio placement, and managing resources of the <Self/> as with HF, LVHF, and VHF. The cellular-aware AACR negotiates short-term rental with cellular service providers for the best deal for the <User/>. In situations requiring relatively minor changes to a SDR personality, the <Self/> may be told what to do rather than being offered a download. For example, the AACR could be advised to "Increase the Costas loop bandwidth limit by a factor of 1.2," which it could do for itself rather than accepting a download from the network. Such an approach reduces the download bandwidth and enhances the performance of nonnative radio equipment for which the CWN has no downloads.

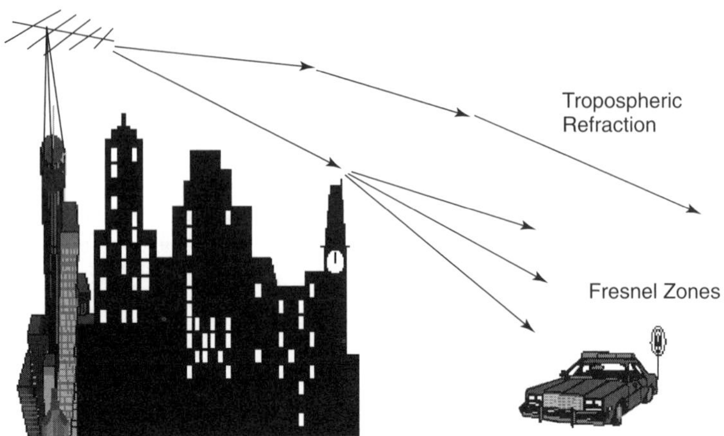

FIGURE 7-9 UHF propagation.

7.6.2 UHF Spatial Distribution of Energy

UHF propagation includes refraction and reflection as illustrated in Figure 7-9.

7.6.2.1 In-Depth LoS Propagation Knowledge

In UHF radio waves are approximated as traveling in straight lines to the radio horizon with sufficient beam spreading for multiple reflections at the typical receiver. This contrasts with HF, where skywave yields distant BLoS propagation, and with EHF, where antenna beams are often narrow enough to eliminate multipath.

Using That Knowledge: The AACR should be able to explain the differences in radio propagation from HF through EHF. The iCR applies that knowledge to autonomously choose the combination of bands, modes, and services that optimizes QoI within cost and resource constraints.

7.6.2.2 Multipath Reflector Knowledge Chunk

Since LoS waves reflect from large structures, more than one reflected UHF wave typically impinge on the receiver (Figure 7-10).

Using That Knowledge: The multipath-aware AACR can analyze the visual scene to determine the nature of proximate radio propagation in its <Scene/> that includes location, surroundings from a map, predicted propagation (RSSI), and probability of fading versus fade depth. The iCR identifies objects in the <Scene/> that affect multipath propagation.

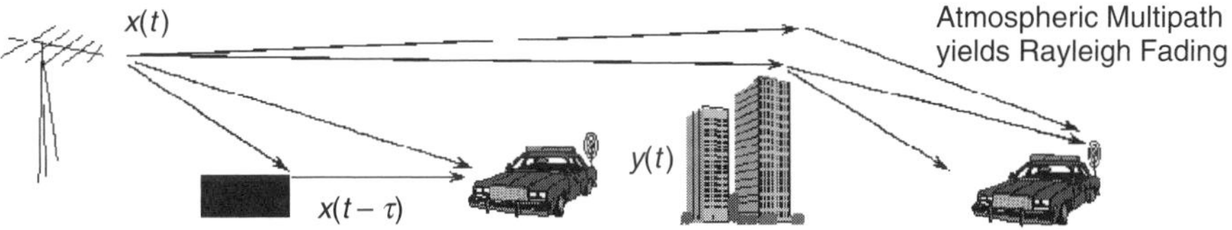

FIGURE 7-10 Physics of multipath propagation.

7.6.2.3 Analytic Model of Multipath Propagation

If the radiated wave is a cosine function of time, multipaths with direct and reflected paths have amplitudes α_1 and α_2 as illustrated in Equation 7-4:

$$y(t)=A(\alpha_1\cos(\omega_0 t)+\alpha_2\cos(\omega_0(t-\tau))=B(\tau)A\cos(\omega_0 t+\theta) \qquad (7\text{-}4)$$

where

$$B(\tau)=\left[\alpha_1^2+2\alpha_1\alpha_2\cos(\omega_0\tau)+\alpha_2^2\right]^{1/2}$$
$$\theta=-\tan^{-1}\left[\alpha_2\sin(\omega_0\tau)/(\alpha_1+\alpha_2\cos(\omega_0\tau))\right]$$

Therefore,

$$\text{Maximum amplitude}=A(\alpha_1+\alpha_2)$$
$$\text{Minimum amplitude}=A(\alpha_1-\alpha_2)$$

The amplitudes of the cosine waves differ with propagation. If the amplitudes are nearly identical, then the minimum amplitude ($\alpha_1 - \alpha_2$) is realized when the difference in path length is one-half wavelength (cosine waves that are approximately 180° out of phase), a condition known as cancellation or destructive interference. Extreme (30 dB) cancellation is known as deep fading.

Using That Knowledge: The multipath-expert CR applies the multipath equations to explain multipath to a nonexpert. It calculates the depth of a fade to reconcile RSSI to the visual <Scene/> to guide mitigation steps like choice of diversity-combining algorithm.

7.6.2.4 Flat Fading Knowledge Chunk

B varies as a function of differential path delay yielding constructive and destructive interference (Figure 7-11). The literature distinguishes flat fading from selective fading. If the signal bandwidth is an order of magnitude smaller than Δf, then as τ changes, the amplitude of the net received signal follows the curve in the figures, so the entire signal appears to have the amplitude of the point corresponding to Δf. Essentially the entire signal fades in and out

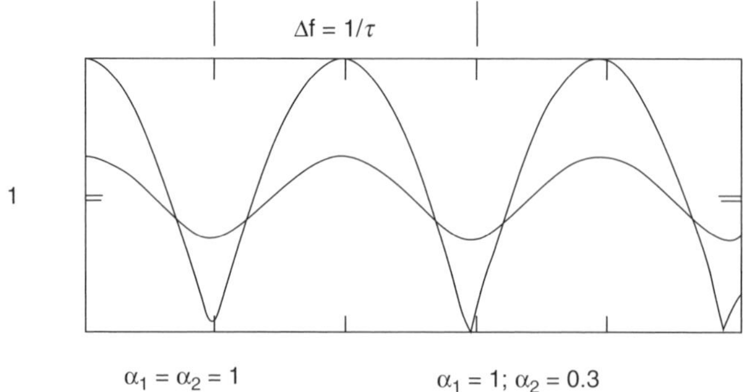

FIGURE 7-11 Zones of constructive and destructive interference.

at the same time. If τ is a microsecond, Δf is 1 MHz so, signals with a few kilohertz of bandwidth fade uniformly.

Using That Knowledge: The multipath-expert CR determines how an observed range of multipath delay spread results in flat fading to adjust signal bandwidth to mitigate flat fading.

7.6.2.5 Selective Fading Knowledge Chunk

For τ of 1 μs and signal bandwidth of 20 MHz, the faded signal has a deep null moving over time. Sinusoidal components that are nearly 180° out of phase fade deeply while most components remain unfaded for wideband selective fading. A tapped delay line equalizer overcomes selective fading with time delay exceeding the delay spread and with sufficient update rate.

Using That Knowledge: The multipath-expert CR can explain selective fading and can use multipath delay spread to adjust a tapped delay line equalizer to observed and projected delay spread. It learns the statistics of selective fading over time and space to share this knowledge with nonexpert users, with maintenance engineers, and CWNs.

7.6.2.6 Diversity Reception Knowledge Chunk

As the carrier frequency increases, changes in τ on the order of a fifth of a wavelength transition the received signal from deeply faded to moderately faded. Consequently, more than one antenna spaced appropriately receive independently faded signals. Diversity receivers select the antenna with highest RSSI. Digital diversity reception combines signals from diversity antennas more coherently than is practicable with analog techniques.

Using That Knowledge: The multipath-expert CR can explain diversity reception to nonexperts. It recognizes situations in which diversity reception enhances QoI to recommend diversity antenna parameters to the user. It adaptively combines diversity inputs. The diversity-aware CR learns new signal processing techniques to enhance diversity reception, synthesizing combining algorithms and collaborating with CWNs.

7.6.2.7 Diversity, Smart Antennas, and FH Mitigate Fading

The statistical structure of fading depends on reflection, refraction, and the speed of subscriber movement. At UHF, fades are distinct at speeds below 5 mph with regular deep fades observed as distinct events. Diversity antennas placed greater than a few wavelengths apart mitigate distinct fading because the nulls typically extend spatially less than a wavelength. Slow FH also mitigates slow fading by changing f and therefore t and x in the multipath fading model.

Using That Knowledge: The multipath-expert CR can explain how diversity, smart antennas, and FH mitigate slow fading. It can initiate slow FH with an AACR counterpart to overcome slow fading. It can also synthesize a FH scheme tailored to the fading, enhancing performance or conserving computational resources and thus battery life as the user situation dictates.

7.6.2.8 Coding Mitigates Fast Fading

As the speed of the subscriber increases above 5 mph the subscriber moves rapidly through multipath peaks and nulls, randomizing deep fade temporal structure. Such fades present a k-symbols erasure channel, so a code that can detect $d > k$ bit erasures and correct k bit errors removes the effects of the fading. As the speed increases so that the duration of erasures is on the order of a bit period, the deep fades have the structure of independent bit errors. Well known models specify BER, E_b/N_o channel symbol, and channel symbol demodulation method (e.g., hard decoding, soft decoding, trellis coding).

Using That Knowledge: The multipath-expert CR can explain the way fast fading changes from an erasure channel to a random bit error channel as the speed of movement increases. It can model E_b/N_o and use coding tables to design a forward error control (FEC) code tailored to mitigate the fast fading.

7.6.2.9 Using Propagation Models to Mitigate Fading

Sufficient received signal strength for communications in urban areas may require the equalization of a large number of reflections via space–time adaptive processing (STAP). Cellular propagation-modeling tools include WrAP [178] and RF-CAD [179]. By using 3D building plans accurate to less than 1 meter, calibrated tools predict RSSI to within 10 dB. Moving reflectors (e.g., trucks, aircraft, and other vehicles) complicate calibration and model validation. Calibrated models predict RSSI, equalizer, and STAP performance.

Using That Knowledge: The multipath-expert CR can explain cellular and PCS models. It can list current software tools and identify those available locally or via CWN. It can propose to use the tool appropriate to setting up a CWN (e.g., an 802.11 wireless LAN hot spot) or diagnosing CWN impairments. The iCR calibrates models via field measurements either individually or in collaboration with other CRs.

7.6.3 UHF Available Communications Modes

UHF includes AM (LSB, USB, VSB), FM (voice, fax), narrowband data (FSK, PSK, 75 bps to 9.6 kbps typical), multichannel radio relay (4–60 channels, PPM, FSK, PSK; fractional T1–E1), and spread spectrum (CDMA) mobile cellular and SATCOM.

7.6.3.1 UHF Modes Knowledge Chunk

Traditional narrowband air interfaces like AM (LSB, USB, VSB), FM (voice, fax), and narrowband data (FSK, PSK, 75 bps to 9.6 kbps typical) are common in UHF.

Using That Knowledge: A UHF AACR should be able to explain to a non-expert user the modes typically employed in UHF, including the bandwidth and channel spacing of legacy spectrum allocations. It recognizes narrowband UHF modes and can configure a waveform template to interoperate with legacy users. The iCRs configure <Self/> narrowband waveforms by recognizing and adapting to the counterpart air interface.

7.6.3.2 Multichannel UHF Knowledge Chunk

Multichannel UHF radio relays support 60–240 channels or more using FDM, PPM, FSK, PSK, and QAM channel symbols. Multichannel digital air interfaces include full and fractional T1 and E1, with protocols recognized by the channel symbol rate.

Using That Knowledge: The multichannel-expert CR should be able to explain multichannel radio relays, identifying channel symbols and protocols. It uses that knowledge to assist in setting up wireless backbone networks, establishing an air interface compatible with legacy radios.

7.6.3.3 JTIDS Knowledge Chunk

One of the most widely known DSSS hybrids, JTIDS, hops over 240 MHz in the 1.2 GHz RF band [174]. The U.S. Air Force and NATO publish the HAVE QUICK I and II slow FH air interface used by military aircraft.

Using That Knowledge: The JTIDS-aware CR should be able to explain the JTIDS air interface and protocol. It should differentiate between features that are in the public domain and features that are not. It should use that knowledge to detect JTIDS to avoid or interoperate.

7.6.3.4 CDMA Knowledge Chunk

The most widely deployed DSSS air interfaces today are the CDMA mobile cellular standard, IS-95, CDMA-2000 (2.5G), and WCDMA (3G). IS-95 has a 1.2288 MHz chip rate, supporting 64 subscribers plus signaling and control in a 1.25 MHz bandwidth. Smart antennas compensate for multipath [180]. The 3G CDMA standards offer spreading rates up to 20 MHz, alternate synchronization schemes, more efficient vocoding, time domain duplexing, and MIMO.

Using That Knowledge: The CDMA-expert CR should assist a CDMA technician in setting up and tuning a CDMA network.

7.6.3.5 GSM Knowledge Chunk

GSM's TDMA air interface predominates global cellular bands (450, 900, and 1800 MHz), with the transition from 2G-GPRS to EDGE and 3G on-going.

Using That Knowledge: The GSM-expert CR can diagnose the GSM air interface for services of interest, but for which the CR's SDR personalities are not suited, such as web browsing, displaying video clips, or tracking a location on a map. It autonomously seeks network services via GSM control channels, pattern matching to detect EDGE and 3G such as 16 QAM channel symbols, adapting its air interface for maximum QoI.

7.6.3.6 Law Enforcement Knowledge Chunk

Relatively simple FH and FH/DSSS hybrids are used in UHF for voice privacy, for example, by government and law enforcement organizations. APCO recommends air interfaces for law enforcement, such as APCO Study Group 25 standards for $12\frac{1}{2}$, $8\frac{1}{3}$, and $6\frac{1}{4}$ kHz in 25 kHz legacy channels.

Using That Knowledge: The law-enforcement CR should be capable of explaining the difference between an unrestricted law enforcement air interface such as UHF push to talk and a private air interface. It should apply APCO 25 to instantiate the waveform template for interoperability. It should determine from a trusted authority whether its user is a law enforcement official and should limit the use of security modes to authorized personnel.

7.6.4 UHF Services and Systems

UHF service band allocations [3] include mobile cellular (450, 850, 900, 1900, 2400 MHz), SATCOM, maritime satellite (1535 MHz), and aeronautical users. Antenna types include log-periodic, parabolic reflector, and discone array. Illustrative systems include the Marconi, Canada AN/GRC-103(V) Relay (220 MHz to 1.85 GHz, 15–30 W, 4–60 channels, PCM, DeltaMod FDM, 31 kg), and the FHM9104 Digital Radio Link Terminal (600–960 MHz and 1.35–2.1 GHz, 0.5 W, 10 channels, 45 kg) from SAT Paris [172]. SDR products have been developed by AirNet, MorphICs, enVia, and Toshiba.

7.6.4.1 UHF Services Knowledge Chunk

UHF service bands [5] include mobile cellular (450, 850, 900, 1900, 2400 MHz), fixed satellite communications, maritime satellites (1535 MHz), and aeronautical mobile satellite communications.

Using That Knowledge: A UHF AACR should be able to list UHF services and explain them in technically accurate radio terms. Specialized terminology like "channel" means different things in different air interfaces. The CR must employ technically accurate language when communicating with expert users and CWNs even if its user trains it with different jargon for informal communications.

7.6.4.2 UHF Antenna Knowledge Chunk

UHF antenna products include log-periodic, directional parabolic reflectors, and discone array antennas. The reduced wavelengths at UHF make the physical size of the antennas more compatible with avionics than VHF.

Using That Knowledge: The UHF antenna-aware CR should know the types and dimensions of common antennas for UHF. It should be able to perform as a technical advisor, diagnostics assistant, and/or supply clerk for UHF antennas.

7.6.4.3 Sectorized Antennas Knowledge Chunk

Cellular base stations use arrays of relatively high gain elements for diversity and gain. A sectorized cell site might employ an array of 3–4 ft tall antenna elements arranged in a triangle 30 ft on a side to provide 5–8 dB gain over isotropic (dBi) and diversity reception. The handset, on the other hand, might use a helical whip with less than 0 dBi gain.

Using That Knowledge: The base-station-expert CR assists in the deployment, maintenance, and evolution of fixed cellular infrastructure antenna systems.

7.6.4.4 Military UHF Radios Knowledge Chunk

Illustrative military systems in this band include digital multichannel radios such as the GRC-103 [181] and the FHM9104 [182].

Using That Knowledge: The military operations UHF CR knows the technical and operational parameters of legacy military radios. The visually capable CR recognizes the equipment to assist in configuring federated suites.

TABLE 7-2 SDR Applications Parameters—UHF

Application	Bandwidth (MHz)	Dynamic Range (dB)
UHF FDM-PCM	0.1–25	40–50
GSM cellular base station	0.200 (to 5.0)	85 (65)
CDMA cellular base station	1.5–20	30–45 (Power Managed)
UHF RF	300–3,000	48–90

7.6.4.5 UHF SDR Knowledge Chunk

Table 7-2 shows SDR applications parameters determined by the RF environment and air interface modes.

The Nyquist sampling rate is twice the bandwidth, but for practical SDRs, a ratio of 2.5 times the bandwidth is more appropriate. Dynamic range may be estimated in ADC bits by dividing decibels by 6 and adding 1 bit for the roofing filter [144]. DAC performance typically is better than the ADC for a given bandwidth and dynamic range.

Using That Knowledge: The SDR-aware iCR assists in defining and diagnosing failure modes in UHF SDRs. It recommends sampling rate and dynamic range for SDR signal acquisition and transmission.

7.6.5 Exercises

7.6.1. Write RXML that expresses the physical properties of the UHF band (frequency: 300–3000 MHz; fast fading is distinct below 5 mph and noise-like above 5 mph). Some RXML data looks like a database, while some looks more like a rule base. Define an embedded database that AACR could use to store all RF, bandwidth, and other tabular data. Discuss the merits of RXML versus MySQL schema versus web ontology languages RDF, RDF-Schema, DAML, OIL, and OWL for this task. Write a RXML model for fast fading.

7.6.2. Write the content of Exercise 7.6.1 as rules for a rule-based inference system. Describe how the AACR uses this knowledge to configure a waveform template for low or high processing demand (see [144]).

7.6.3. Write RXML for an AACR to advise users regarding the physical placement of UHF antennas.

7.6.4. Write RXML for an AACR to implement the advice "Increase the Costas loop bandwidth limit by a factor of 1.2." Show how it reasons from fundamentals rather than needing a network download. Quantify how this <Self/> modification reduces the download burden on a home CWN. Suppose an error in the Costas loop bandwidth precipitated a recall and the core image of each cell phone is 256 MB while the Costas loop bandwidth limit request takes 2 kB. Given a 10 million cell phone recall on a U.S. nationwide network, what is the cost/benefit ratio of the self-modification capability?

7.6.5. Write RXML for an AACR to be multipath aware. How does your RXML enable it to analyze the visual scene to hypothesize multipath causality? What classes of object does your RXML cast in the scene? How does it reason about distance to associate an equalizer delay tap response with an object in the scene?

7.6.6. See the companion CD-ROM for additional multipath and diversity exercises.

7.6.7. Write RXML for an AACR to explain to a nonexpert user the air interface modes typically of UHF, including the bandwidth and channel spacing for legacy SCPC spectrum allocations. Augment standard CRA <Self/> to do this. Discuss the merits of defining a new air interface such as an IEEE 802.11x that operates in UHF on unused TV channels in terms of existing <Self/> knowledge versus defining it as an <ISM/> capability parameterized for UHF TV channels (e.g., limited to 6 MHz).

7.6.8. Write RXML for an AACR to relate UHF waveform parameters of bandwidth and dynamic range to SDR platform and waveform parameters including automatic gain control (AGC); IF conversion; dynamic range; ADC and DAC bandwidth, resolution, and accuracy; digital engine precision and speed; and interconnect parallelism, clock speed, and bandwidth. Use the parameters of Table 7-2.

7.7 KNOWLEDGE OF THE SHF RADIO BAND

Although super high frequency (SHF), the microwave band, begins at 3 GHz according to international agreement, there is a transitional between 1 and 3 GHz in which LoS microwave characteristics are present.

7.7.1 SHF Physics-Related Knowledge

Frequency band: 3000–30,000 MHz
Wavelengths: 0.1 m to 0.01 m

"Microwave" propagation begins at ~1 GHz, based on beamforming with a high-gain antenna. Multipath delay spread of 0.5–1.5 ms is common with Doppler shift from low flying aircraft that may exceed 1 kHz at 15 GHz.

7.7.1.1 SHF Thermal Noise Knowledge Chunk

Microwave characteristics begin at about 1 GHz, including the transition of noise from externally generated to thermal noise. Since thermal noise is accurately modeled by an additive white gaussian noise (AWGN) process, thermal noise is sometimes called AWGN. Thermal noise is the physical process, while AWGN is a mathematical abstraction not identical to the physical process. The expected value of noise power (in joules) in a SHF band of bandwidth B is

$$P_N = kTB \tag{7-5}$$

where k is Boltzmann's constant (1.380658×10^{-23} J/K), T is temperature (K), and B is bandwidth (Hz).

At SHF, the noise temperature is established in the antenna feed and first stage amplifier—the low noise amplifier (LNA)—that establishes system sensitivity, the level of signal strength that the receiver can detect.

Using That Knowledge: The SHF-aware AACR should be able to explain that typical SHF channels are accurately modeled by AWGN. It can use the relationship among noise temperature, bandwidth, and noise power of the Boltzmann equation (7-5), converting joules to dBm or dBW as needed. The SHF-aware CR measures its own noise temperature, comparing internal SNR and E_b/N_o to predictions for <Self/>-diagnosis. It interacts with maintenance technicians, users, and the supply system for replacement parts.

7.7.1.2 SHF Ray Tracing Knowledge Chunk

SHF propagation is accurately characterized by LoS ray tracing, so the microwave bands are also called LoS bands. Propagation can include Doppler-shifted multipath similar to UHF multipath, but with fewer significant reflections. At UHF, 100 reflections can account for 99% of the received signal strength, with fewer than 10 (usually only one or two) at the high end of SHF.

Using That Knowledge: The SHF-aware AACR uses knowledge of multipath reflections to assist users positioning microwave antennas in a visual scene. See the CD-ROM for additional SHF multipath knowledge.

7.7.1.3 High Gain Antennas Knowledge Chunk

High gain antennas, easy to implement at SHF, reduce the number of multipath reflections with increasing carrier frequency. SHF pencil beams with spatial sidelobes 15 to 30 dB below the main beam are formed by parabolic dishes, horns, and lenses [170].

Using That Knowledge: The SHF-aware AACR relates the theory and practice of SHF antennas to its own antenna(s) to quantity antenna performance and to recommend antennas that meet user- or technician-specified needs, such as lower sidelobe reflections.

7.7.1.4 Atmospheric Absorption Knowledge Chunk

SHF energy is absorbed by the atmosphere better than UHF, reducing reflections from distant objects. Depending on climate, SHF absorption is about 0.007 dB per kilometer at frequencies below 10 GHz. Absorption peaks in the SHF band at the 21 GHz water absorption line at about 0.2 dB/km, dropping to 0.07 dB/km between 28 and 34 GHz.

Using That Knowledge: The SHF-aware AACR estimates absorption as a function of weather forecast for its SHF communications links using line of sight absorption in main beam and sidelobes. It advises the nonexpert user on the choice of frequencies in the SHF band based in part on atmospheric absorption.

7.7.1.5 Tapped Delay Line Knowledge Chunk

SHF multipath delay spread often ranges from 0.5 to 1.5 microseconds for LoS propagation, with fewer reflectors than VHF UHF. Time domain adaptive equalizers mitigate multipath. The number of taps in the $T/2$ time-domain adaptive equalizer delay line depends on BPSK data rate and delay spread:

$$N_{\text{taps}} = (\text{delay} - \text{spread}/\text{channel} - \text{symbol period}) * 2 = D * R_b * 2 \quad \text{(7-6a)}$$

$$\begin{aligned} N_{\text{taps}}(10\mu\text{s}) &= 10\mu\text{s} * 2.048\text{Mbps} * 1\ \text{bit}/\text{channel} - \text{symbol} * 2 \\ &= 10 * 2 * 2 \cong 40 \end{aligned} \quad \text{(7-6b)}$$

The computational cost of $T/2$ equalizer trading is proportional to the number of taps while the cost of rapidly initializing the equalizer is proportional to the square or cube of the number of taps because of a matrix inversion. Equalizers achieve BER as a function of E_b/N_o and multipath dynamics. Forward error control (FEC) improves BER at the expense of data rate overhead and MIPS.

Using That Knowledge: The AACR estimates and measures multipath reflections of a <Scene/> to configure its adaptive equalizer accordingly. It adjusts the computational burden QoI needs like enhanced data rate or constraints like the conservation of battery life. The AACR configures the <Self/> to combinations of data rate, equalizer settings, and FEC for observed link conditions.

7.7.1.6 Directionality Knowledge Chunk

SHF radio waves do not fill in shadowed areas well.

Using That Knowledge: The SHF-aware AACR can predict SHF shadowing and Fresnel zones and can employ this knowledge for CWN siting and for the mitigation of multipath and interference.

7.7.1.7 Radar Bands SHF Knowledge Chunks

Geometric LoS and high gain antennas at SHF are ideal for military and civilian radars that track aircraft and estimate weather severity. SHF radars use signals of large time–bandwidth (TW) product with nearly ideal ambiguity surfaces and high spatial resolution.

Using That Knowledge: Radar-aware AACRs analyze and mitigate interference with <Self/> and CWN communications signals. The radar-expert CR synthesizes SHF waveforms with good information density and ambiguity surfaces for simultaneous communications and radar functions.

7.7.1.8 WLAN Hot-Spot Knowledge Chunk

Cellular, PCS, and wireless local area networks use SHF spectrum near radar bands (e.g., the 5.4 GHz ISM bands) for small antennas, high directionality, and wide bandwidths for hot-spot applications. In hot spots, there are often many more radios than channels available (e.g., shopping malls and sporting events).

Using That Knowledge: The AACR learns the locations and incremental costs of hot-spots to shape traffic for user objectives of lowest cost or timely delivery. It also might learn the best location for a palmtop video teleconference at, say, 128 kbps.

7.7.1.9 Doppler Shift Knowledge Chunk

Doppler shift (e.g., from 400 mph aircraft) may exceed 1 kHz at 15 GHz. At frequency f,

$$f = \frac{c}{\lambda} \text{ so } \Delta f = -f\left(\frac{v}{c} - v\right) \cong -f\frac{v}{c} \text{ for } v << c \tag{7-7}$$

If a reflecting object moves away with velocity v, the distance is increasing, so the wavelength appears to be stretched and frequency is reduced. Table 7-3 shows Doppler shift for representative scenarios.

At HF, the plasmas in the ionosphere often have an apparent velocity of 2000 mph or more, inducing Doppler shifts of up to 5 Hz. Doppler shift is proportional to the cosine of the angle between an aircraft or satellite flight path and the RF LoS. Short-range EHF links experience Doppler shifts of over 1 kHz. In addition, the total Doppler shift between two moving platforms can be double that shown in Table 7-3. The Doppler shift between a ground-based receiver and a low Earth orbit (LEO) satellite at EHF decreases from +30 kHz to zero when the satellite is overhead and continues to decrease to

TABLE 7-3 Doppler Shift

Application	Frequency	Reflector	Velocity	Doppler Shift
HF	10 MHz	Ionosphere	2,000 mph	3.259 Hz
SHF ground	1 GHz	Aircraft	500 mph	81.48 Hz
EHF ground	15 GHz	Aircraft	500 mph	1.22 kHz
EHF LEO	21 GHz	Satellite	10,000 mph	34.2 kHz

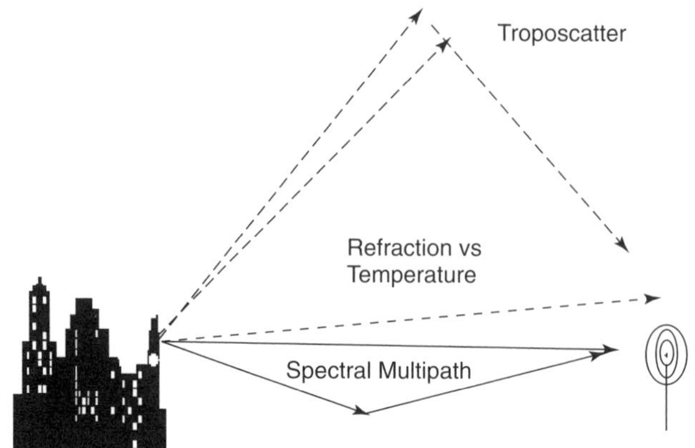

FIGURE 7-12 Spectral multipath causes Rayleigh–Rice fading.

–30 kHz as the satellite recedes. Doppler shifts, positive and negative, must be compensated by analog or digital carrier tracking loops.

Using That Knowledge: The Doppler-expert iCR knows the equations for Doppler shift and can identify those objects in a scene that affect Doppler shift. It also learns about frequency shift phenomena by observing the environment. The Doppler-expert iCR applies this knowledge to enhance QoI, for example, by pointing the antenna away from the nearby airfield to minimize Doppler spreading of a HDR SHF channel.

7.7.2 Spatial Distribution of SHF Energy

Spatial distribution of SHF energy includes troposcatter where a BLoS transmitter illuminates a point of refraction in the troposphere (Figure 7-12). LoS SHF communications over water encounter spectral reflections with Rayleigh or Rician probability distributions of multipath [183, 184].

7.7.2.1 Rayleigh Fading Knowledge Chunk

Rayleigh scattering induces hundreds to thousands of reflections of comparable signal strength with different time delay for apparently random phase of the received sinusoids. The Rayleigh fading model is a very good approximation for SHF scattering above 4 GHz.

Below 4 GHz, however, the probability that the signal level is less than the abscissa is not as high as the Rayleigh model predicts. FEC mitigates Rayleigh fading if the fade duration at erasure depth is less than the b-burst error correction capability of the code. If erasures exceed the FEC-correctable time delay, then packets may be corrected with automatic repeat request (ARQ) protocols.

Using That Knowledge: The SHF-aware iCR knows the Rayleigh distribution of signal strength and uses that to predict fade depth and duration in a <Scene/>. It configures FEC or ARQ protocols to accommodate the predicted fading, measuring actual fade depth (E_b/N_o during the fade) and duration to diagnose Rayleigh fading. The iCR recommends, synthesizes, and tests remedial actions in the field collaborating via CWNs.

7.7.2.2 Rician Fading Knowledge Chunk

Rice noted that the statistical structure of amplitude varies as a function of the number of strong multipath components, offering a model of amplitude distributions parameterized by the number of such strong paths. As the number of paths with approximately the same phase increases, the amplitude distribution becomes tighter and the variance of the amplitude distribution decreases.

Using That Knowledge: The Rician-aware AACR can describe the difference between Rayleigh and Rician fading to a nonexpert and can employ that knowledge to diagnose link conditions and to mitigate fading.

7.7.2.3 SDR Mitigation of Fades

SDR algorithms that mitigate Rayleigh–Rice fading include FEC, ARQ protocols, and bridging the data clock across deep fades. Coherently combining energy from diversity antennas reduces fade depth. Cyclostationary processing at the frame rate enhances E_b/N_o for synchronously framed data links. Because of the statistical structure of fades, the rate of convergence of such algorithms varies and processing demands also vary with fade depth. The Gamma function estimates the rate at which fade mitigation processing resources may be exceeded.

Using That Knowledge: The SHF-fade-aware AACR adapts the appropriate SDR waveform parameters to accommodate encountered fade statistics. It allocates SDR resources to remain within processing capacity and statistical limits. Collections of alternate mitigation algorithms gracefully adapt mitigation to the hardware platform.

7.7.3 SHF Available Communications Modes

SHF communications modes include:

High capacity microwave (FM/FDM, PPM, PWM, PCM/QAM (1.544, 2.048, . . . , 45, 90, 155, 622 Mbps))

Troposcatter (2 and 4.5 GHz bands, analog and digital)

Spread spectrum, satellite CDMA, FH-DS hybrids, and OFDM

Terrestrial SHF air interface modes include high capacity microwave, troposcatter, and spread spectrum communications.

7.7.3.1 High Capacity Microwave Knowledge Chunk

Point-to-point microwave radio was initially developed for high capacity backbone links of the Public Switched Telephone Network (PSTN). In the mid-1980s, digital microwave dominated this market but has largely been superseded by fiberoptics in developed economies. High capacity microwave retains niche applications in developing economies, in rugged terrain, for backup of primary fiber links, for cable television (CATV) distribution, and for microwave backhaul of cellular telephone traffic. Backhaul operates between cellular base transmission station (BTS) and the base station controller (BSC) or mobile telephone switching office (MTSO). To keep costs low, CATV and backhaul may use analog FM/FDM or commodity digital radios (e.g., T- or E-carrier).

Using That Knowledge: The SHF-aware AACR can explain historical and current applications of high capacity microwave radio to nonexperts and uses this knowledge to diagnose and mitigate interference in SHF <Scenes/>.

7.7.3.2 Mobile Microwave Radio Trunking Knowledge Chunk

Military markets and humanitarian relief employ mobile SHF microwave radio trunks for high capacity interconnect among deployable wireless base stations.

Using That Knowledge: The mobile SHF-adaptive AACR can configure itself for microwave radio trunking. It can advise the nonexpert user regarding antennas, siting, baseband switching, and RF interconnect for T- and E-carrier trunking.

7.7.3.3 T- and E-Carrier Air Interface Standards

The high capacity microwave air interface includes the legacy analog formats FM/FDM, pulse position modulation (PPM), and pulse width modulation (PWM). Modern systems use PCM with BPSK, QPSK, QAM, and partial response channel symbols. QAM amplitude–phase combinations range from 16 to 1024 in powers of 2, each requiring higher SNR. Data rates range from 128 kbps for military radios to T1 (1.544 Mbps), E1 (2.048 Mbps), and OC-1 (51.84 Mbps) multiples through the OC–N level of the synchronous digital hierarchy (SDH) [144]. Packing OC-12 into a 30 MHz spectrum allocation requires multicarrier–QAM hybrids like 4 × 256 QAM, needing 40 dB SNR, equalization, FEC, bit interleaving, and randomization for high computational complexity historically requiring ASIC hardware. Contemporary FPGAs and DSPs deliver equivalent GFLOPS, so low and medium data rate T- and E-carrier may be SDRs. The T- and E-carrier systems use Signaling System Seven (SS7) for dialing, call setup, switching, and PSTN interfaces.

Using That Knowledge: The SHF-SDH-aware AACR can explain the SDH to nonexperts. It can configure its processing, buffering, and interconnect for low to medium capacity microwave trunking with SS7 to cross-connect its wireless <User/> to the PSTN.

7.7.3.4 Troposcatter Knowledge Chunk

As illustrated in Figure 7-12, troposcatter is useful when transmitter and receiver are not within LoS of each other [170]. Each radio points at a scattering region in the troposphere where weak coupling mandates very large apertures (e.g., 10 meter dish), kilowatts of power, and diversity reception. Effective isotropic radiated power (EIRP) of 90 dBm provides sufficient SNR for multichannel relay but can interfere with nearby wireless systems. The military troposcatter networks connect headquarters to clusters of geographically dispersed units with lower acquisition cost than satellite communications and with the ability to operate at extreme northern and southern latitudes. Air interface modes include FM/FDM, PPM, PWM, and PCM [172].

Using That Knowledge: The troposcatter-aware AACR can explain troposcatter systems to nonexperts. It reconfigures RF, filtering, baseband, DSP, buffer memory, switching, and supervisory resources for troposcatter. With large antennas and powerful RF amplifiers, the troposcatter-adaptive CR replaces dedicated troposcatter IF, baseband, and supervisory radio functions with SDR. It also learns the best siting and configuration for QoI objectives while minimizing wireless interference.

7.7.3.5 Microwave Spread Spectrum Knowledge Chunk

The Joint Tactical Information Distribution System (JTIDS) occupies 250 MHz between 1 and 2 GHz [183]. JTIDS employs a 32 chip DSSS pseudonoise (PN) spreading sequence on each bit with instantaneous bandwidth of 3 MHz. Chip bursts hop across a 250 MHz agility band at over 1000 hops per second.

Using That Knowledge: The JTIDS-aware AACR characterizes JTIDS networks to avoid interference. The JTIDS-adaptive AACR instantiates a waveform template for interoperability, permission to join the network, identification of network services, role, and security.

7.7.3.6 SHF CDMA Knowledge Chunk

The U.S. Defense Science Board's panel on wideband communications [184] recommended the expansion of high capacity spectrum sharing technologies like CDMA. Instantaneous CDMA bandwidths of tens of megahertz are practical at SHF. Wireless LANs, for example, use 50 MHz CDMA to overcome multipath and provide asynchronous multiple access. Satellite communications also use CDMA with FH-DS hybrids for error mitigation and privacy.

Using That Knowledge: The CDMA-aware AACR uses its knowledge of SHF antennas, propagation, and the local <RF-environment/> to establish CDMA CWNs within local spectrum management constraints.

7.7.4 SHF Services and Systems

SHF service band allocations [172] include radio navigation, fixed satellite, maritime, MetSat, fixed/mobile satellite (X-band), fixed point-to-point links, and intersatellite communications. SHF antennas typically are horns or parabolic reflectors ("dishes"). Illustrative systems include the Alcatel TFH950S Digital Troposcatter (1.7–2.1 and 4.4–5 GHz, 10–1500 W, 2048 kbps, 2 DPSK, 37 kg [172, p. 181]), the Siemens FM15000 (15 GHz, 100 mW, 256–1024 kbps, 10 kg), and products from Nortel, AT&T, BT, and NTT. Emerging SDR products include SigTek [172, p. 70].

7.7.4.1 SHF Services Knowledge Chunk

SHF services leverage LoS, high gain antennas, and wide instantaneous bandwidths [170].

Using That Knowledge: The SHF-aware AACR can describe SHF services for nonexpert users. It continuously compares explicit and implicit user expressions of interest to its templates for SHF information services based on observed content.

7.7.4.2 SHF Antenna Knowledge Chunk

SHF antennas include pyramidal horns and parabolic reflectors, popular at 6 and 11 GHz. Antenna type, gain, transmit power capacity, noise figure, size, weight, power, and RF connector types characterize SHF antennas.

Using That Knowledge: The SHF-aware AACR can explain SHF antennas and technical characteristics, configuring the <Self/> for terrestrial trunking, ad hoc CDMA networks, and WLANs.

7.7.4.3 SHF Legacy Troposcatter Knowledge Chunk

The Alcatel TFH950S Digital Troposcatter system [185] operates from 1.7 to 2.1 GHz and from 4.4 to 5 GHz, radiating from 10 to 1500 watts of power with data rates to 2048 kbps. It weighs 37 kg, not including the antenna [172, p. 181].

Using That Knowledge: The SHF-troposcatter AACR can adjust a SDR waveform template to interoperate with legacy systems like the TFH950S.

7.7.4.4 Legacy SHF Digital Microwave

The Siemens FM15000 military digital microwave radio operates from 15.1 to 15.3 GHz, radiating 100 mW of power into a high gain antenna with data rates from 256 to 1024 kbps.

Using That Knowledge: The SHF-aware military or humanitarian relief iCRs can adjust SDR waveforms to interoperate with the FM15000 and other legacy digital microwave radios, verifying the air interface collaboratively with an expert user.

7.7.4.5 Advanced Topics

See the companion CD-ROM for additional knowledge chunks on legacy commercial microwave radios, SDR beam-pointing, and SATCOM.

7.7.5 Exercises

7.7.1. Write RXML that enables your SHF-aware AACR to explain that SHF channels are typically AWGN channels, and to identify those that may not be (e.g., ISM bands).

7.7.2. Write RXML that enables your AACR to know that it should measure its own noise temperature and compare internal signal strength (SNR or E_b/N_o) to expected signal in order to self-diagnose performance problems. Describe the functions of a self-diagnosis subsystem based on this knowledge. What other capabilities would the RF platform need to achieve built-in self-test? Discuss the merits of internally generated self-test signals versus the use of known external signals (e.g., broadcast signals). How does this trade-off change from HF through SHF?

7.7.3. Write RXML that enables your AACR to know the theory and practice of SHF antenna types. Describe the flow of logic as your AACR uses this RXML to explain its own antenna(s) in terms of that theory. Implement that logic for an AACR that knows about all of the SHF antennas in a standard reference such as *Reference Data for Engineers* [170]. Write Java for the automatic instantiation of the RXML from softcopy text and figures.

7.7.4. Find a SHF antenna manufacturer on the Web. Write RXML that integrates this manufacturer's product line into the answer to Exercise 7.7.3. Write a PERL script that finds the next antenna manufacturer on the Web and instantiates the RXML. Find a manufacturer that uses OWL for SHF antenna products. If there are none, write your own OWL for two products. If you find one, make your PERL work with OWL. If there are more than one, discuss the problems using OWL for RXML.

7.7.5. Write RXML for your AACR to explain the difference between SHF multipath and that of other bands such as UHF and EHF.

7.7.6. Write a program that estimates absorption for a given SHF communications link distance. Write RXML to describe this program to your AACR. Write a parser for NL text questions about absorption. Use this RXML plus the program to answer questions about SHF absorption.

7.7.7. Write RXML that enables your AACR to be radar aware. How can it use this knowledge of radar to analyze interference with communications signals? Define RXML and SDR functions so that the radar-expert AACR can synthesize CDMA codes with good ambiguity surfaces [186] to serve both communications and radar functions. Explain how the integration of ranging and communications enables cellular infrastructure to perform E911 emergency location.

7.7.8. Write RXML that enables your AACR to learn the locations and incremental costs of SHF hot-spot nodes in the user's environment. Write a PERL script or Java program that uses this RXML to shape traffic for lowest cost or most timely delivery of large files. Explain how to enable the AACR to learn the best location for a palmtop video teleconference at, say, 128 kbps via CBR. Describe the training of the CBR. Implement this in Java, C, or PERL instead. Measure the costs of CBR fraining versus coding.

7.7.9. Write RXML for a Doppler-expert AACR. Implement the equations for estimating Doppler shift in C or Java. Write RXML for a Doppler(s, f, W) function. Explain how <Self/> with vision could identify objects in a <Scene/> that will affect Doppler shift. Implement RXML and inference to identify a large track producing Doppler-shifted multipath.

7.7.10. Write RXML for an AACR to know the Rayleigh distribution of signal strength. Explain how it uses that RXML to predict fade depth and duration statistics. Write Java or PERL that uses the RXML to adjust an adaptive equalizer to deal with 90% of the fades predicted for a 5.5 GHz ISM band.

7.7.11. Find a rule-based expert system shell or an authoring facility to interpret RXML to explain historical and current applications of HDR microwave radio to nonexperts. Write RXML for AACR to explain the low and middle levels of the SDH to nonexperts using the expert system shell.

7.7.12. Write RXML that enables the <Self/> to reason about SHF trunking. Write Java, JESS, or C to use this knowledge to configure its processing, buffering, and interconnect for low capacity microwave trunking. Test the configuration in a simulated environment.

7.7.13. See the companion CD-ROM for exercises with microwave backbones and SHF antenna pointing.

7.8 KNOWLEDGE OF EHF, TERAHERTZ, AND FREE SPACE OPTICS

EHF, the millimeter wave band, spans from 30 to 300 GHz. Terahertz frequencies then extend from 300 GHz to 3 THz. The terahertz bands are of use in medical diagnosis but not telecommunications. Free space optics (FSO) frequencies of 300 THz correspond to wavelengths of 1000 nanometers. Communications applications of FSO include the ubiquitous infrared data access (IRDA) port and the TV remote control.

7.8.1 EHF Physics-Related Knowledge

Frequency band: 30,000–300,000 MHz

Wavelengths: 0.01 m to 1 mm (millimeter wave)

EHF propagation modes consist of direct and reflected LoS skywave with nearly complete attenuation from birds, clouds, rain, snow, and ice (except for nonlinear punch-through). Propagation can be very directional with little ionospheric scintillation.

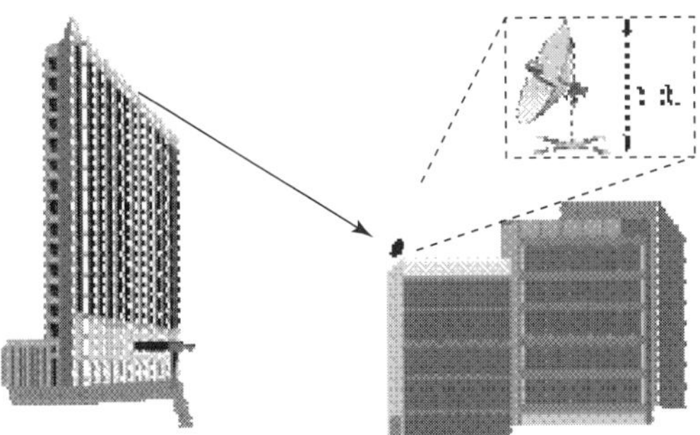

FIGURE 7-13 Propagation of EHF energy.

7.8.1.1 EHF Propagation Knowledge Chunk

Free space optics (FSO) and millimeter wave RF obey similar physical optics propagation characteristics with pronounced absorption by water and atmospheric gases. EHF is an ideal band for short-haul wideband communications systems via pencil-beams (Figure 7-13). Low data rate FSO reflects in small spaces for short range non-LoS links like TV remote control. Increased data rates require exact pointing with main beam tracking. EHF HDR have pencil beamwidths of fractions of a degree for 1 foot antenna apertures with pointing accuracy of milliradians over distances of kilometers. EHF carrier frequencies enable 100 Mbps to Gbps data rates with relatively low cost components.

Using That Knowledge: The EHF-aware AACR knows the link budget equations of EHF/FSO communications. It models low data rate reflection modes and HDR pointing and tracking modes. It applies these models to predict and diagnose EHF/FSO QoI enhancement opportunities.

7.8.2 EHF Spatial Distribution of Energy

The spatial distribution of EHF energy is illustrated in Figure 7-13.

7.8.2.1 EHF Attenuation

The atmosphere attenuates electromagnetic radiation through two distinct mechanisms: gaseous attenuation and absorption due to precipitation.

7.8.2.2 EHF Attenuation Knowledge Chunk

EHF signals are easily attenuated by clouds, rain, snow, and ice [170]. Anything solid from buildings to birds blocks the radiation. Gaseous attenuation loss is less than 0.01 dB per km to 12 GHz. Water absorption peaks at 0.2 dB per km at 22 GHz, a candidate for low probability of intercept (LPI) data links. For longer ranges, the "water hole" of minimum absorption of 0.06 dB per km at 34 GHz attracts military multichannel and commercial T1/E1 campus data links. Gaseous absorption increases from 0.2 dB/km (H_2O) at 60 GHz to 15 dB/km for O_2 at 120 GHz, and over 40 dB/km in the 190 GHz water absorption line. Low O_2 absorption (0.3 dB per km) at 90 GHz supports multichannel communications, radar, and so on.

Using That Knowledge: The EHF-aware AACR employs models of link attenuation to assist nonexpert users with siting, spectrum management, and diagnosis of link impairments.

7.8.2.3 EHF Precipitation Knowledge Chunk

In addition to gaseous water in the air due to humidity, precipitation increases specific attenuation. Precipitation intensities of 0.4–4 mm/h (nominal rainy day), attenuate frequencies below 5 GHz less than 0.01 dB per km. In the tropics, however, precipitation ranges from 10 to 50 mm/h to increase absorption to 0.2 at 6 GHz, 1 dB per km at 9 GHz and 10 dB/km at 20 GHz.

Using That Knowledge: The EHF-aware AACR employs its knowledge of precipitation to assist nonexpert users in planning EHF, terahertz, and free space optics communications.

7.8.3 EHF Communications Modes

EHF communications modes include high capacity microwave—FM/FDM, PPM, PWM, PCM/QAM (1.544, 2.048, 45, 90, 155, 622 Mbps)—and radar-related data links. Applications include "campus" data links, "up the hill" military links, and wireless PCS node interconnect ("backhaul").

7.8.3.1 EHF Modes Knowledge Chunk

EHF readily supports HDR microwave air interfaces, particularly PCM and QAM with data rates from T1/E1 (1.544 and 2.048 Mbps) to T3 (45 and 90 Mbps), and OC-3 (155 Mbps) to OC-12 (622 Mbps) and beyond. Applications like campus data links readily range from 100 m to 1.5 km at OC rates. The military "up the hill" radios link command centers to elevated radio-relay points over a range of a few kilometers. EHF for cellular and PCS markets connect BTS and BSC.

Using That Knowledge: The EHF-aware AACR can explain EHF modes to the nonexpert user and can assist the user in deploying EHF nodes to support use cases.

7.8.4 EHF Services and Systems

EHF service band allocations [3] include fixed/mobile satellite (39.5–40, 71–74, 134–142, 252–265 GHz), fixed point-to-point links, amateur satellite, radio-based location, and earth observation satellites. EHF antennas typically are lenses, horns, or parabolic reflectors. Illustrative systems include the Tadiran GRC-461 MM Wave Terminal (37.06–37.45 GHz and 38.5–38.89 GHz) that supports NATO, EUROCOM, and HDB-3 at 256–8446 kbps [172].

7.8.4.1 EHF Subbands Knowledge Chunk

EHF includes popular millimeter wave bands (39.5–40, 71–74, 134–142, and 252–265 GHz). Services include those listed above plus future 134 GHz EHF/FSO hot-spot application where the user's CPDA may point a narrow EHF beam at an information kiosk to obtain a Gbps download of a 10 Mbyte guide to the shopping mall, taking 100 ms for ephemeral HDR.

Using That Knowledge: The EHF-aware AACR employs this knowledge to predict, plan, and implement physical access to the bands and modes that enhance QoI. EHF/FSO links may require probing and cooperative pointing. The EHF CR recognizes access opportunities like EHF hot-spot kiosks, downloading necessary waveforms.

7.8.4.2 EHF Antennas Knowledge Chunk

Antennas employed in these bands include plates, horns, and parabolic reflectors.

Using That Knowledge: The EHF-antenna-aware AACR knows EHF antennas theory and practice, modeling its own EHF access to assist nonexpert users with EHF operations.

7.8.4.3 EHF Systems Knowledge Chunk

Illustrative EHF systems include the Tadiran GRC-461 [172] millimeter wave terminal with RF and direct-connect fiber-interoperability modes.

Using That Knowledge: The EHF-aware AACR assists the nonexpert user in locating EHF networks. For humanitarian relief, AACRs bridge low data rate voice and data nodes to the Tadiran legacy radio for E1 trunking among clusters of RF activity. It learns from experience to identify situations where EHF enhances QoI.

7.8.4.4 EHF SDR Knowledge Chunk

See the companion CD-ROM for EHF SDR knowledge.

7.8.5 Exercises

7.8.1. Write RXML for an AACR to know the link budget equations of EHF/FSO short-range LoS radios. Augment this RXML so that the <Self/> knows which radio is integrated into its own hardware platform and which others are integrated into peers. Augment this RXML to know that FSO is limited to low data rate reflection. Add high data rate pointing and tracking modes by adapting the SHF pointing and tracking exercise. Augment the RXML to predict available data rates. Write Java or C or use an inference engine to produce those inferences in Q/A interactions.

7.8.2. Write RXML to establish that the atmosphere attenuates EHF electromagnetic radiation via gaseous attenuation and absorption due to precipitation. Augment this RXML with a database model of link attenuation (e.g., digitize attenuation curves from the Web and place the RF–attenuation pairs in a database). Explain how this RXML could be used to assist a nonexpert user with siting of the AACRs for a humanitarian relief operation. With Java or Voice XML to do this via a structured dialog. Augment this RXML with EHF spectrum management knowledge. Explain how to use RXML to diagnose EHF link impairments similar to the SHF use cases.

7.8.3. Write RXML for an AACR to know about microwave HDR air interfaces including FM/FDM, PPM, and PWM. Augment this RXML to include PCM and QAM (T1/E1, T3, OC-3, and OC-12). Describe applications of these air interfaces for "campus" data links in RXML. Use this RXML to set up such an infrastructure in a simulated <Scene/>. Describe a military use of such links for "up the hill" communications via RXML to show that EHF links may be used between elevated radio relay points over ranges of a few kilometers. Write Java or C so that a simulated iCR discovers the radio relay application of EHF autonomously.

7.8.4. See the companion CD-ROM for additional EHF/FSO exercises.

7.9 SATELLITE COMMUNICATIONS KNOWLEDGE

Communications satellites [187] operate in the three orbital regimes illustrated in Figure 7-14.

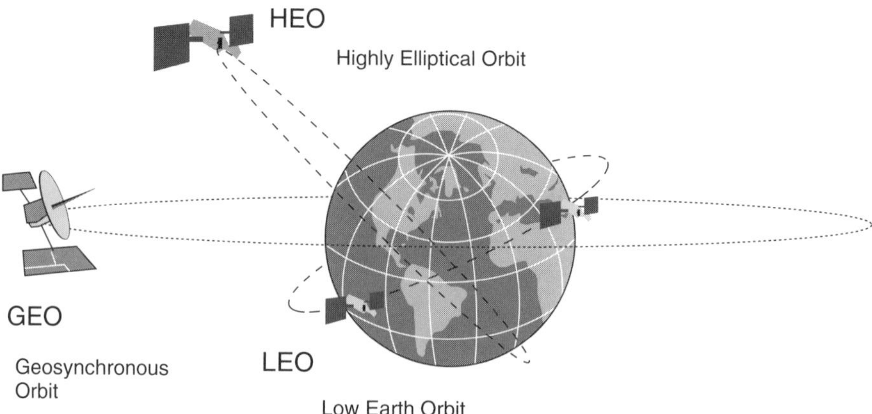

FIGURE 7-14 Satellite communications.

7.9.1 Satellite Knowledge Chunks

The following knowledge chunks and thoughts on the innovative use of that knowledge lay the foundation for the formal codification of knowledge of radio in the next chapter.

7.9.1.1 Geosynchronous Satellite Knowledge Chunk

Geosynchronous (GEO) satellites have an orbital period that is nearly identical to the Earth's rotational period, resulting in an apparent stationary position above the equator at an altitude of approximately 22,500 miles. A VHF SHF GEO satellite's tangential LoS visibility ranges between the arctic circles in an oblate elliptical footprint about 8000 miles across. Practical terminals require the GEO to rise a few degrees above the horizon, further limiting the useful footprint. Many GEO satellites support domestic satellite (DOMSAT) services, often through a directional domestic coverage footprint. The ITU and WRC GEO databases specify parking orbits and related frequency allocations. Adjacent-orbit interference results from excessive uplink antenna sidelobe radiation. One limitation of DOMSAT service is the 0.264 second round-trip delay between the GEO satellite and the Earth. Such long time delays annoy speakers, rendering GEO satellites better for video relay or data than voice.

Using That Knowledge: The SATCOM-aware AACR can explain GEO operation to the nonexpert user. The GEO-capable AACR assists the user in selecting pointing angles and frequency channels to establish and maintain links. The iCR learns these patterns from experience.

7.9.1.2 LEO/MEO Satellite Knowledge Chunk

Low Earth orbit (LEO) satellites orbit at between 150 and 1500 miles, reducing round-trip time delay to less than 20 ms. LEO satellites typically have 90 minute orbits with about 10 minutes of visibility above the horizon on each satellite pass. Medium Earth orbit (MEO) satellites operate at higher altitudes between 1500 and typically 8000 miles, yielding longer orbital periods. Motorola's Iridium [188, 189] satellite system provides LEO voice telephony and data services with global coverage via 66 satellites and on-orbit spares. LEO–MEO design parameters include the number of satellites, the time waiting for satellite visibility, or the number and duration of link outages per day. LEO satellites provide good coverage near the equator but longer and more frequent outages at more northern or more southern latitudes. Continuous coverage at higher latitudes requires a full (and thus relatively expensive) constellation. Ephemeris tables and orbital modeling tools enable one to accurately predict satellite orbits.

Using That Knowledge: The SATCOM-aware AACR can explain the differences among GEO, LEO, and MEO to nonexperts. Such AACRs can answer questions about specific orbits and types of satellites (e.g., using COTS modeling tools) [190].

7.9.1.3 HEO Satellite Knowledge Chunk

Highly elliptical orbit (HEO) satellites enable sustained SATCOM coverage at higher latitudes with only a few (typically three) satellites: one rising toward apogee, one setting from apogee, and the other in perigee. The apogee of such a modest constellation may be placed near the arctic circle for high continuity service to high latitudes such as Siberia (e.g., for the Russian Molnyia satellites).

Using That Knowledge: The SATCOM-aware AACR can explain the difference between HEO and other orbital regimes. The HEO-capable AACR employs modeling tools to predict pointing angles and to adaptively control antenna pointing to achieve link continuity. CRs learn over time to optimize pointing and tracking parameters. CRs also learn to adapt frequency usage and waveform parameters to enhance quality of service (QoS).

7.9.2 SATCOM Physics-Related Knowledge

Frequency bands: VHF, UHF, SHF, EHF.

Propagation modes include geosynchronous (low Doppler), highly elliptical (moderate Doppler), and low Earth orbit (high Doppler).

Propagation phenomena include Faraday rotation, ionospheric scintillation, ground station sidelobe control, and mobile satellite rain fades [8].

7.9.3 SATCOM Spatial Distribution of Energy

Satellite communications entail Doppler shift, Faraday rotation, sunspot-induced anomalies, the magnetosphere, and other channel impairments.

7.9.3.1 Satellite Doppler Shift Knowledge Chunk

Satellite frequency bands are traditionally the lettered band designations [170]. GEO satellites experience near zero Doppler shift while highly elliptical satellites near apogee exhibit moderate Doppler of from a few hundred hertz to upward of 20 kHz for ascending and descending operation. A C-band HEO satellite imparts 325 Hz of Doppler shift on a 2 GHz carrier for a nominal 1000 mph satellite velocity normal to radio LoS from the ground station ("radial velocity"). The same HEO spacecraft operating at 21 GHz imparts a 17.1 kHz Doppler shift on the carrier for an extreme radial velocity of 5000 mph. LEO satellites, on the other hand, impart high positive Doppler shifts as the satellite rises above the radio horizon, transitioning through zero and to high negative Doppler shift prior to setting on the opposite horizon. Iridium's 785 mi orbit imparts approximately 600 Hz of Doppler on its 3 GHz carrier, with a carrier tracking requirement of over 1200 Hz, or about 2 Hz/s.

Using That Knowledge: The SATCOM-aware AACR predicts, measures, and compensates for Doppler shift to achieve consistently high QoS.

7.9.3.2 Satellite Propagation Knowledge Chunk

Propagation between satellites and ground terminals must contend with Faraday rotation, ionospheric scintillation, ground station sidelobe control, mobility, and rain fades. Faraday rotation is the distortion of electromagnetic wave polarization as the wave transits the plasmas in the ionosphere, resulting in elliptical polarization at the receiver, with the attendant loss of received signal strength. Ionospheric scintillation is the equivalent of terrestrial multipath, induced by path length differences of multiples of a wavelength with nearly equal amplitudes for alternate constructive and destructive interference that present an erasure channel.

Using That Knowledge: The SATCOM-capable AACRs compensate for known path impairments. The iCRs discover path impairment phenomena independently, proposing to CWNs and testing mitigation strategies autonomously.

7.9.3.3 Sidelobe Knowledge Chunk

GEO ground stations control spatial sidelobes to limit sidelobe radiation to typically less than 40 dB below the radiated power. High GEO orbit packing density necessitates a large ground station for sidelobe control per international treaty. Active sidelobe cancellation imparts additional RF channels and processing requirements on the transmission segments of SATCOM SDR for spectrally and spatially pure transmitted waveforms.

Using That Knowledge: The SATCOM-aware AACR can explain sidelobe control to the nonexpert and can assist the user in establishing and maintaining sidelobe control.

7.9.4 Available Satellite Communications Modes

SATCOM modes include heavy Earth terminal (symmetrical), DOMSAT (multichannel, cable protection), mobile satellite terminals (INMARSAT—voice, LBR data), very small aperture terminal (VSAT), low data rate asymmetrical terminals, direct broadcast satellite (DBS), wideband asymmetrical terminals, spread spectrum, and military (SHF, EHF, LEO).

7.9.4.1 Fixed Heavy Earth Terminal Knowledge Chunk

Domestic satellites (DOMSATs) can connect distant cities via the PSTN in developing economies. In such symmetric applications all ground station antennas are nearly the same size (often tens of meters). With similar technology, the International Communications Satellite Corporation (INTELSAT) has operated transoceanic PSTN links for international gateways. This service includes legacy FM/FDM, T/E-carrier, and OC air interfaces. The deployment of extensive OC-3, -48, and -192 undersea fiber in the 1980s and 1990s curtailed the international satellite PSTN market. News and entertainment industries use international and DOMSATs for the delivery of program content.

Using That Knowledge: DOMSAT-aware AACRs can explain DOMSATs and assist nonexpert users with antenna selection and alignment. They can find, identify, and characterize SATCOM transponders and channels applying waveforms to enhance QoI.

7.9.4.2 Mobile Satellite Knowledge Chunk

Mobile satellite communications (MSC) services in wide use include the narrowband International Maritime Satellite (INMARSAT), with voice and low bit rate (LBR) data. Some terminals require a 3 foot directional antenna for acceptable signal quality. INMARSAT continues to invest in its satellites and terminal equipment to reduce the size, weight, and power of the terminals.

MSC services (Table 7-4) did not do well in 1999–2005. Iridium and Globalstar recovered from bankruptcy. Teledesic was delayed and ultimately cancelled.

The Orbcom MSC locates vehicles for commercial trucking companies. Odyssey has a medium Earth orbit (MEO) altitude at 10,354 km so that 12 satellites achieve global equatorial coverage.

TABLE 7-4 Mobile Satellite Services

Services	Globalstar Mobile Telephony & Data	Iridium Mobile Telephony & Data	Orbcom Store & Forward Messaging	Teledesic High Rate Fixed Service	Odyssey Mobile Telephony
Data rate	9.6 kbps	2.4 kbps	300 bps	16k-E1 + 1.2 G	N/A
Modulation	CDMA	TDMA	N/A	N/A [ATM]	CDMA
RF	1/3 GHz	1/3 GHz	148/137 MHz	30/20 GHz	1.6/2.5 GHz
Satellites	48	66	36	840	12
Altitude (km)	1400	785	775	700	10,354
Inclination	52	86.4	45	98 (Sun)	50
On-board processing?	No	Yes	No	Yes	No
Crosslinks	No	4 @ 25 Mbps	No	8 @ 155.52 Mbps	No
Mass (lb)	704	1100	85	747	4865
Partners		Motorola		McCaw & Gates	TRW
On orbit?	Operational	May 1997	Yes	Cancelled	Yes

Using That Knowledge: The mobile satellite-aware AACR knows the technical parameters of planned and deployed mobile satellite systems to explain the trade-offs among services, tariffs, antennas, SDR waveforms, and RF front-end hardware.

7.9.4.3 VSAT Knowledge Chunk

Very small aperture terminal (VSAT) satellite systems use large Earth terminals at a central hub and a high gain satellite with high power offsets to reduce user terminal antenna diameter to 0.5–1 meter. This asymmetrical arrangement significantly expanded the market for satellite data services because the subscriber Earth station is easy to deploy and maintain. VSAT applications include point of sales terminals with centralized control for large multinational corporations like McDonald's restaurants.

Using That Knowledge: The VSAT-aware AACR assists the user to obtain and employ VSAT modes.

7.9.4.4 DBS Knowledge Chunk

Direct broadcast satellites (DBSs) also use asymmetrical apertures with the heavy aperture in a central hub and VSAT or smaller terminals on customer premises. The Hughes 601 satellites, for example, deliver over 50 channels of digital television to a 24 inch aperture terminal via a 22.5 MHz satellite transponder. Leased DBS capacity can deliver large amounts of data to thousands of subscribers within the satellite footprint. DBS digital audio broadcast (DAB), XM service, and direct video broadcast (DVB) entered service between 2000 and 2004.

Using That Knowledge: The DBS-aware AACR recognizes DBS, satellite audio, and DVB in DBS bands to obtain air interface hardware, waveform downloads, and licenses if DBS services enhance user QoI.

7.9.4.5 Wideband SATCOM Knowledge Chunk

Between 1995 and 2000, DARPA and the U.S. National Aeronautics and Space Administration (NASA) cosponsored the Advanced Communications Technology Satellite (ACTS), which demonstrated 622 Mbps from a low cost geosynchronous satellite to a 10 meter parabolic aperture [191]. Japanese and European researchers have employed other satellites to study propagation phenomena (e.g., in the June 1997 Proceedings of the IEEE).

Using That Knowledge: The wideband SATCOM-expert AACR can explain wideband SATCOM technology, gathering current technical data from SATCOM web sites and autonomously evolving <Self/> SATCOM toward HDR as QoI dictates.

7.9.5 Satellite Services and Systems

SATCOM service band allocations [170] include C band (5.925–6.425 and 3.7–4.2 GHz), X Band, Ku (14–14.5 and 11.7–12.2 GHz), and Ka (27.5–31 and 17.2–21.2 GHz). Illustrative SATCOM systems include geosynchronous DOMSATs and INTELSATs; Molnyia, the Trans-Siberian HEO; and the Iridium LEO SATCOM systems. Emerging SDR products have been explored by COMSAT and SigTek [370].

7.9.5.1 SATCOM Services Knowledge Chunk

Spectrum use in the C, Ku, and Ka bands [170] has transitioned from large Earth terminal technologies for PSTN and governmental users to new VSAT, DBS, and DVB technologies with broad commercial services enabling new products, each a candidate for SDR.

Using That Knowledge: The SATCOM-aware AACR knows the capabilities, orbital parameters (ephemeris), spectrum occupancy, and information content of SATCOM services to enhance user QoI.

7.9.6 Exercises

7.9.1. The SATCOM-aware AACR can explain geosynchronous satellite operation to the nonexpert user. Using the knowledge chunks of this section, write an RXML description of GEO satellites. Extend this RXML to assist a user with SATCOM pointing angles and frequency selection. Include GEO ephemeris from the Web. Write a stand-alone GEO SATCOM-aware application (e.g., in Java) that uses RXML to explain to a nonexpert user about GEO satellites. Enable your AACR to learn about GEO satellite use patterns from experience scanning, pointing, and tracking via ephemeris. Enhance your AACR to learn about satellite services.

7.9.2. SATCOM-aware AACRs should be able to explain the differences among GEO, LEO, and HEO to nonexperts. Write RXML to explain these differences to include specific orbits and types of satellites by class and function. Evaluate the Satellite Took Kit™ (STK) modeling tool against a RXML embedded database for general knowledge of satellite behavior.

7.9.3. A SATCOM-aware AACR can explain the difference between HEO and other orbital regimes. Use RXML and a STK to predict pointing angles for a (simulated) electronically steered array on the roof of your car to track a HEO satellite. Add a LEO satellite. Enhance your software to learn the trade-off between pointing and tracking a LEO satellite, with an array of fixed pointing antennas using the strongest RSSI, versus a broad antenna beam. Describe how your AACR could learn to optimize RF, pointing, and tracking parameters for an amateur radio LEO satellite.

7.9.4. LEO satellites impart high positive Doppler shifts as the satellite rises above the radio horizon. Write RXML for LEO satellite Doppler signatures and write a stand-alone Q/A program to explain these signatures using the RXML.

7.9.5. SATCOM-capable AACRs compensate for known path impairments. Write RXML for path impairment phenomena and mitigation strategies.

7.9.6. DOMSAT-aware AACRs can explain DOMSATs to a nonexpert user. Write RXML describing three DOMSATs from the Web. Write a program to control a Ham scanner to find, identify, and characterize SATCOM transponders and channels for these DOMSATs. Write RXML and Java for this AACR to interact with a simulated CWN to download a waveform to access a DOMSAT soccer feed.

7.9.7. Write RXML for the technical parameters of planned and deployed MSC services. Write Java to use the RXML to explain the trade-offs among services; acquire tariff information from authoritative sources; recommend antenna and RF front-end hardware for SATCOM interoperability; and assist inexperienced users in acquiring hardware and waveforms for MSC information services.

7.9.8. DBSs also employ asymmetrical apertures with the heavy aperture in a central hub and VSAT or smaller terminals on customer premises. Write RXML for the Hughes 601 satellites, DBS capacity to deliver large amounts of data to thousands of subscribers within the satellite footprint, and the DAB and DVB satellites currently in service. Write a Java program to autonomously extend this RXML with information about an additional satellite from the (conventional) Web. Extend this Java to acquire RXML for additional satellites from the Semantic Web. Compare the two.

7.10 CROSS-BAND/MODE KNOWLEDGE

Table 7-5 summarizes band and mode advantages and disadvantages.

TABLE 7-5 Reliable Flexible Communications Via MBMMRs

Band[a]	Modes[a]	Key Technical Characteristics That Shape Software Radios
HF	Skywave, NVI, FH, burst	Long range, reliable in mountains/jungles, cheap, narrow BW, severe propagation, big antennas
LVHF	Beyond LoS, FH, burst	Low cost, general-purpose voice, data, and relay Rayleigh fading, Fresnel zones, interference
VHF	Quasi-LoS	Larger coherence bandwidths, cheap
UHF	Cellular	Spectrum auctions opening applications
	TDMA, CDMA	Spectrum crowding, beam forming, data rate
SHF	LOS TDMA	Ideal for space communications, large BWs, atmospheric and rain losses significant
EHF	Narrow beams, fiber protect	In-building, campus data links, spatial sharing, very short range but practical gigabit BWs

[a] No single band or mode delivers reliable, long-haul, high data rate, cheap, and convenient (unlicensed) service for mobile users.

Choice of band and mode balances GoS, QoS, QoI, and cost. Previously, there was no choice but to perform the trade-offs in advance, configuring a suite of discrete radios with computer-controlled mode selection for specific criteria. The confluence of SDR and computational intelligence offers the cognitive MBMMR alternative.

7.10.1 Cognitive MBMMR Knowledge Chunk

The maximum flexibility in the use of cross-band knowledge contemplates the ideal multiband multimode radio (iMBMMR) with bands from HF through EHF, including SATCOM. An iMBMMR may not be the size of a cell phone soon because of antenna constraints, but might be configured into a military or commercial SUV for humanitarian relief. Such an iMBMMR may be made computationally intelligent via RXML and AML for an ideal cognitive MBMMR (iCM). Each iCM uses a computational model of the <Self/> to reason over its own capabilities and limitations, dynamically choosing radio bands and modes to enhance QoI.

Using That Knowledge: The iCM can summarize the main features of each band and mode of Table 7-5. It knows its own ability for each of these modes. It relates the strengths, weaknesses, and content of these radio bands to QoI. Depending on the <Scene/>, one iCM might advise a remote <User/> that HF propagates readily over thousands of miles at low data rates. Another might configure itself for LVHF propagation for rough terrain. An iCM in a city could configure EHF pencil beams for low cost spatial sharing of the radio spectrum at gigabit per second data rates if it isn't raining.

7.10.2 Mode Selection Knowledge Chunk

Consistent mode selection behaviors may be preprogramed per Table 7-6. The evolution of SDR toward iCR includes such steps, but since an iCM can learn, its behavior may be enhanced by training it to weight QoI parameters according to the features of a <Scene/>.

Using That Knowledge: The iCM applies the critical mode selection decision parameters to enhance QoI. With reinforcement learning, multiple training instances from many information access experiences are aggregated into decision parameters. Algorithmic methods for analyzing such noisy and uncertain training include Q-learning and neural networks.

7.10.3 Mode Transition Planning

Seamlessness entails advanced planning of band/mode transitions as current modes deteriorate and as the scene changes. Information access gaps give the user a sense of seams in the network, while continuity of information access

TABLE 7-6 Critical Mode Selection Decision Parameters

Information Source	Parameter	Remarks
Source bitstream	Bit rate	Quality and coder complexity
Burstiness	Constant (CBR) or variable (VBR)	Minimum, maximum, sustained
Isochronism	None, real-time or near-real-time	Data transfers are not isochronous
Burst parameters	Maximum burst size	Large files create long bursts
Tolerance	Tolerance of parameter mismatch	See service quality parameters
Service quality	Error rates	Losses and delays
Bit error-related	Bit (BER), symbol (SER) errors	Requires error control
Delay-related	Transfer delay, variance, jitter	Excessive delays result in loss
Buffer-related	Packet or cell loss rate (CLR)	Overflows reduce rate
Availability	Probability (link), grade of service	Link and network provisioning
Quantity	Quantity of relevant valid information	QoI metric implies accuracy
Timeliness	Wall clock time delay for content	QoI metric reflects server in the network
Assuredness	Authentication, privacy	Significant for e-commerce
Cost	Peak, off-peak, service-related	Can be the critical factor

across discontinuities of physical access create a perception of seamlessness. Seamlessness may also require efficiency, balancing battery power, RF spectrum, data rate, and radiated power to the degree necessary for QoI.

Using That Knowledge: The MBMMR-expert AACR reasons about the differences among physical access, propagation path impairments, spectrum allocations, air interfaces, and information services from HF through EHF including SATCOM. It employs this knowledge to proactively enhance QoI via the appropriate band and mode, reacting appropriately to changes in the user's behavior. It should be able to differentiate exceptions such as working into lunch from the norm of entertainment during lunch. Cultural stereotypes reflect the prototypical behavior of many of the people some of the time, but rarely will cultural stereotypes be satisfactory to any given user. The iCM therefore learns user preferences in the timing and use of the bands and modes, removing from the user the burden of learning the radio nuances.

TABLE 7-7 Network Knowledge

Band	Network	Typical and Unique Knowledge
HF	ALE	Set up, tear down, TCP/IP interoperability
LVHF	CB, Ham	Voice interoperability
VHF	Air traffic control	User credentials, verbal protocol
UHF	Cellular bands	Legacy 1G, 2G, 3G services, licensing, tariffs
SHF	PSTN core	Link protocols, SS7, IP (MPLS)
EHF	Cellular backhaul	Ad hoc interoperability for humanitarian relief
SATCOM	DVB, DAB	Channels, licensing, security waveforms

7.10.4 Protocol Stack Knowledge

Although planning systems enable the smooth transition from one band and mode to another, they do not guarantee the delivery of the services that the user expects. Each of these bands and modes employs analog and/or digital network technologies for control and/or for traffic. Some of these networks are open to ad hoc participation and some of them are not. Once connected to a network, there may be a translation between the network's representation of the desired information and that of the <Self/>. For example, the <Self/> may express telephonic voice (not music) using DS0, 64 kbps Mu-Law encoded PCM. The GSM network represents the same voice as RPE-LTP coded 13 kbps bursts. Still other networks use any of the dozens of other voice coding methods. Digital networks, similarly, represent the same data in a variety of formats. SDR designers analyze the protocols at design time to create waveforms with a single internal representation. The iCM therefore also employs protocol stack knowledge as suggested in Table 7-7.

The first line of the table indicates that it takes a finite amount of time to set up and to tear down a HF ALE connection, but once established such a link interoperates at the network level via TCP/IP. This may be the preferred internal networking format of the iCM. Clearly, legacy 1 G and 2 G cellular systems don't use TCP/IP even though some offer CDPD or GPRS, so the iCM must know the mappings from the specialized radio access protocol to the more generic TCP/IP. In addition, there are many variations of TCP such as TCP-Reno and TCP-Jersey, each with different features for wireless connections [192]. Since most commercial services require licensing, the iCM must know both the licenses currently enabled on the <Self/> and the message protocols required to obtain short-, medium-, and long-term licenses as needed for QoI.

Using That Knowledge: The iCM employs its knowledge of the protocol stacks to obtain timely, relevant information for QoI. This behavior conforms to the user's value system for cost, timeliness, and other learned preferences. In addition, the iCM maps external networks to internal representations of voice, data, and multimedia streams. Some maps are available to the <Self/>,

TABLE 7-8 Typical Applications Layer Parameters

Applications Class	Characteristics and Parameters
All	Number of channels; underlying mode and bit rate; protocol profile (e.g., WAP over GSM and GPRS)
Location aware	Location accuracy, update rate, number of mobiles
Voice	Source code, bit rate, frame rate
Facsimile	Page-buffer space, number of channels proprietary protocols, beyond Group IV
Packet data	Protocol (e.g., V.xx, X.25, TCP/IP, MPLS [193], ATM), queue space
Email	Directory server, domain name server, gateways, proxies, security features, host application (e.g., Eurdora, Outlook, Netscape)
File transfer	Protocol (e.g., FTP), delay tolerance, maximum file size
Database	Size, query language (e.g., SQL), update rate and latency
Voice mail	Number of users, speech storage capacity, simultaneity
Multimedia	Mix of media (e.g., voice and shared whiteboard), delay tolerance, BER/FEC by class of service (e.g., line drawings versus voice)
VTC	Source coding, profile (e.g., capability within H.320)

some exist in the protocol stacks or SDR personalities, and others are available from CWNs, peer iCM, or third parties. The iCM may parameterize a mapping template to create the map autonomously.

7.10.5 Applications Layer Knowledge

Although the translation of representations is a necessary step in seamlessness, it may not be sufficient. Even simpler services like voice require applications layer support, for example, to bridge VHF push to talk to GSM, and VoIP on an IEEE 802.11 hot-spot bridge. The iCM conference call control module makes multiple voice connections via diverse protocols. Other such applications layer parameters over which the iCM operates are summarized in Table 7-8.

This table reflects applications interoperability parameters. Each of these applications domains includes dozens of kinds of services offered by large markets per region, with many more opportunities across national, regional, language, and other rapidly disappearing borders. The lists are illustrative rather than comprehensive, suggesting the scope of the challenges and opportunities. Many of these classes of application include interoperability features. Fax, for example, includes many interoperability features in Group IV. If one would like to send a fax to a multimedia database or into a video teleconference (VTC), interesting challenges and opportunities for new kinds of applications layer interoperability emerge.

Using That Knowledge: The iCM incorporates detailed knowledge of applications classes from location awareness to multimedia VTC. Its knowledge of <Self/> reflects its applications capabilities, while its knowledge of the <User/> reflects what it has learned about use of the application. The iCM also determines the user's need for cross-application interoperability, such as displaying a fax on a virtual whiteboard during a VTC. Such needs may be relatively ephemeral, so the iCM distinguishes itself by its ability to construct or obtain the necessary updates in real time from a trusted source.

7.10.6 Cross-Band/Mode <Self/>

The iCM with the ability to reason about the information structures, protocol stacks, and applications offered in various bands and modes of the MBMMR layer goes far toward removing the annoying seams that impede today's radio users. To organize the knowledge and skills implicit in the prior discussion of this section requires flexible interfaces among communications and information services (Table 7-9).

Reasoning about these layers of the <Self/> requires the organization of these abstractions (Expression 7-13).

Expression 7-13 RXML Scope of Layering

```
<Self>
<Information-services>
  <Communications-services> Voice facsimile
    <Data-services> Data-transfer Database-query email FTP
      </Data-services>
```

TABLE 7-9 Layered Software Radio Architecture

Layer	Applications/Services	Protocol Mappings
Communications services	*Applications*: voice, facsimile, data (email, file transfer), databases, voice mail, multimedia, VTC *Related services*: Location finding, Over-the-air downloads	WAP, CORBA, SQL, X.400, TCP/IP, MPLS, MIME
Radio applications	*Air interfaces ("waveforms")*, state machines, modems, *synchronization algebra*	Waveform/Air AVD, DS0, Group IV
Radio infrastructure	Data movement, memory management, *domain manager*	CORBA, ODP/X.900
Hardware platform	Antenna(s), analog RF hardware, ASICs, FPGAs, DSPs, microprocessors, instruction set architecture, operating systems	

```
        Voice-mail Multimedia VTC </Communications-services>
  Location-finding OTA-download </Information-services>
<Radio-applications>
  <SDR-application> <Air-interface/> <Protocol-stack/>
    </SDR-application>
  </Radio-applications>
<Radio-infrastructure>
  <Data-movement/> <Streaming/> <Timing/> <Synchronization/>
  <Memory-management/> <Middleware/> </Radio-infrastructure>
<Platform/> </Self>
```

With this framework, the RF band/mode knowledge chunks and cross-band/mode knowledge chunks and the exercises, one may address the synthesis of radio skills, the real-time flexible employment of radio knowledge to enhance QoI of the use cases.

7.10.7 Exercises

7.10.1. Define a process by which an iCM can summarize the main features of each band and mode suggested in Table 7-5.

7.10.2. Train CR1 to know that HF propagates readily over thousands of miles at low data rates given a suitable antenna; LVHF propagates well in rough terrain and wooded areas as well as suburban settings; VHF offers more coherent bandwidth but offers somewhat less propagation beyond LoS than LVHF; UHF is the TV, aircraft, and cellular radio band among other things; SHF marks the transition to the higher capacity more directional LoS bands; EHF pencil beams enable low cost spatial sharing of the radio spectrum while its high carrier frequencies support gigabit per second data rates if it isn't raining and if the beam is pointed in the right direction; and SATCOM can enable distant parties to communicate with each other and with the PSTN provided antenna pointing and the weather cooperate.

7.10.3. Explain the additional reasoning that CR1 must have to use each of these knowledge chunks in order to match a user's need to the right band/mode for Exercise 7.10.2.

7.10.4. Explain the reasoning steps required for an iCM to apply the critical mode selection decision parameters discussed above to supply information services via its iCM capabilities. Write each mode selection criterion of Table 7-6 in RXML. Write a stand-alone Java program to use this knowledge and the knowledge of Exercise 7.10.2 in transitioning from one band to another as each of the known bands becomes unavailable. Extend CR1 to use the knowledge of Exercise 7.10.2 and this exercise to implement a stand-alone Java program. Compare the RXML and CR1 implementations in terms of speed of implementation, likelihood of latent errors, and extensibility to new situations.

7.10.5. The MBMMR-expert AACR reasons about the differences among physical access, propagation path impairments, spectrum allocations, air interfaces, and information services from HF through EHF including SATCOM. Write

RXML to express the knowledge to proactively obtain information services via the bands and modes of Exercise 7.10.2 to meet user needs for voice, email, voice mail, and a VTC when located in an urban or a rural vacation setting. Write a stand-alone CBR application to differentiate exceptions such as working into lunch from the norm of using an entertainment service during lunch.

7.10.6. Write RXML to express the cultural stereotype of lunch at noon and departure for work at 5 pm. Do you personally conform to this prototypical behavior? Explain why and why not, and capture any exceptions in RXML. Ask three others whether they conform to the prototypical behavior and ask them to identify exceptions. Express all the exceptions in RXML. Implement the cultural stereotype in a CBR framework (stand-alone Java or CBR tool). Teach the CBR system the exceptions. Define a new exception that is close to one that has been taught already. Did it do the right thing? If so, then pick a new exception for which it will initially fail and train it to do the right thing. Discuss the trade-offs among stereotypical RXML and CBR exception handling.

7.10.7. The iCM employs its knowledge of protocol stacks for timely, relevant information for the <User/>. Write RXML to express the protocol stack knowledge of an IP/UDP protocol stack. Expand this RXML to reflect your personal value system regarding cost, timeliness, and relevance as defined in the QoI metric.

7.10.8. Write RXML for an iCM to map external networks to its own internal representations of voice, data, and multimedia streams. Let mappings among these classes of service be known and write RXML that makes them known to the <Self/>.

CHAPTER 8

IMPLEMENTING RADIO-DOMAIN SKILLS

This chapter develops software techniques for using the radio-domain knowledge of the last chapter by a suite of algorithms organized into a CRA for interactive and autonomous realization of the use cases.

The design principle for AACR, for practical engineering applications (i.e., without self-modification learning), is as follows:

> **AACR Radio Knowledge Exploitation Strategy**
>
> An AACR shall *optimize choice* of *its own radio band-mode and network for the user(s) fine-tuning* the air interface, network, and computational resources *better than the best human radio operator.*

Thus, the emphasis when referring to AACR is the optimization of existing SDR personalities with respect to the situation at hand reflected in the RF environment and the user's specific communications needs. The design principles for iCR research are more aggressive, focused on realizing more of the self-modification potential of AML in iCR:

Cognitive Radio Architecture: The Engineering Foundations of Radio XML
By Joseph Mitola III

iCR Knowledge Exploitation Strategy

An iCR exhibits *better performance* than a *radio engineer*, learning from experience by exploring alternatives both physically and via simulation, suggesting actions for users, cooperating with other iCRs and *creating new SDR personalities on the fly* that meet the needs of the use case *better* than any one band, mode, network, or parameter setting alone.

AACR may learn radio knowledge *by being told by a trusted authority*, just as a SDR can be enhanced by *download*. Subsequently, they will contribute to the CWN knowledge bases and thereafter may enhance their own skills autonomously.

8.1 COGNITIVE RADIO ARCHITECTURE STRUCTURES RADIO SKILLS

The CRA structures AACR into functional components with an inference hierarchy and cognition cycle. The CRA <Self/> ontological model of AACR includes <RF/> abstractions, related metalevel primitives, and examples, together referred to as the radio-domain ontology (RDO) summarized in Expression 8-1. This RXML expression captures the things that an AACR should know how to do with appropriately elaborated radio knowledge. In the ontological treatment, a RXML conceptual <Primitive/> has context-dependent semantics defined in the CRA <Self/> so that intelligent-agent class algorithms can realize the radio-domain use cases via radio-domain skill.

Expression 8-1 Radio-Domain Ontology (RDO)

<RF> <Radio-skills>
<Physical-access>
<Bands> <HF/> <LVHF/> <VHF/> <UHF/> <SHF/> <EHF/> <THz/>
</Bands> <Modes> <SDR-subsystem/>
<RF-sensing> <Signals/> <Noise/> </RF-sensing>
<SDR-waveforms/> </Modes> </Physical-access>
<Knowledge>
<Physical>
<Frequency/> <Wavelength/> <Bandwidth/>
<Canonical> <Transmitter/> <Channel/> <Receiver/> </Canonical>
<Transmitter> <Receiver>
<Antenna> <Efficiency/> <Gain/> <Noise/> <Direction/>
<Position/> <Array/> <MIMO/> </Antenna>
<RF-conversion> <Transmitter/> <Receiver/> </RF-conversion>

<Baseband/> <Bitstream/> <Security/> <Sources/> <Sinks/>
<MAC/>
</Receiver> </Transmitter>
<Channel/> <Path/> <Reflection/> <Refraction/> <Doppler/>
<Propagation> <Link-budget/> <2D/> <3D/> </Propagation>
<Signal-in-space> <Channel-symbol/>
<Multiple-access/> <Stream/> </Signal-in-space>
</Physical>
<Logical>
<Connectivity> <Grade-of-service/> <Quality-of-service/>
</Connectivity>
<Protocol-stack> <ISO OSI/> <Radio-access-protocols/>
</Protocol-stack>
</Logical> </Knowledge>
<Skill> <Radio-skills> <Connect> <Transmit/> <Receive/> </Connect>
<Bridge/>
<Transfer-data/> <Disconnect/> <Plan/> <Collaborate/>
</Skill> </Radio-skills> </RF>

The ontology shows that the AACR's radio-domain skills are based on physical access to the RF bands and modes, on knowledge about radio, and on the ability to use that knowledge to <Connect/>, <Transfer-data/>, <Disconnect/>, <Plan/>, and <Collaborate/>. Collaboration includes interaction with users, peer AACRs, legacy radios, CWNs, and RAs regarding the plans and actions of the <Self/>. The primary radio skills of connecting, transferring data, and disconnecting entail transmitting and receiving radio signals. The AACR knows what it is doing by using the RDO self-referentially. The dynamic RDO expresses what the AACR is doing, so it can plan and execute higher level goals, keeping the user connected in spite of impairments.

8.1.1 Functional Components Contribute to Radio Skills

The functional components synthesize radio skills as shown in Table 8-1. The functional component APIs support the information services API (ISAPI) interfaces 13–18, 21, 27, and 33, which implement the RDO exchanges.

The Cognition API (CogAPI) implements interfaces 25–30, 5, 11, 23, and 35 to accept RDO expressions from the other functional components and to assert control in RDO-consistent requests to those components.

8.1.2 Expressive Efficient RF Component Interfaces

Often, the SDR community becomes focused on computational efficiency when confronted with verbose XML in the SCA and SRA, CORBA middleware, and other structured information exchanges. The source code of modern computer languages such as Java and C++ can be written in a verbose style

TABLE 8-1 Roles of Functional Components in Radio Skills

Component	Role
User sensory perception (USP)	Identity of people in the audio-video scene who need radio services; convey radio-oriented beliefs, desires, and intent of users per RDO
Local environment sensory perception (LESP)	Senses RF, audio, video location, temperature, acceleration, and compass direction and extracts RF-related perceptions (e.g., of location) from USP
Information services (IS)	Expresses RF support needs to cognition functions necessary for current, historical, and future IS-provided services via the RDO
SDR subsystem (SDRS)	RF sensing and SCA/SRA compliant SDR applications
Cognition functions (CFs)	Control RF skills, SDR perception, planning, learning
Local effector (LE)	Speech synthesis; text, graphics, and multimedia displays

as well, but modern compilers transform verbose source code into efficient DLLs.

8.1.2.1 A Function for Verbosity

Similarly, RXML is excruciatingly verbose, so using RDO in the APIs at first seems inefficient. Posted on the Semantic Web, the verbosity assures that the computing agents refer to unique W3C, RDF, RDFS, DAML, OIL, XMLNS, and other authoritative web pages in "real time." During sleep modes, computing resources are available to employ the verbose versions of knowledge that express radio-domain knowledge most fully. Interaction with a new network might also entail initial RXML exchanges to mutually ground the ontologies for subsequent bit-efficient exchanges.

8.1.2.2 Efficient Frequent Internal Messages and Tasks

However, during real-time interactions among AACR functional components, there is little value in verbose formats. In particular, the ISAPIs and CogAPIs are compiled to efficient, Huffman-coded bitmaps so that the most frequent interchanges consume minimum interconnect and processing capacity per exchange. One of the major functions of the sleep cycle, then, is for the cognition component to examine the internal flows among the other components of the <Self/>, adjusting bit-level coding so that the AACR continually performs the radio tasks specific to this <User/> and as efficiently as practicable. Modern compiler technology can optimize register usage and message coding among functional components.

8.1.2.3 Verbose, Efficient, or Optimized Load Modules

The AACR functional components could have <Verbose/>, <Efficient/>, and <Optimized/> forms. The verbose form would be expressed in RXML as defined in (an evolved version of) CRA <Self/>. The functions might be interpreted rather than compiled for maximum flexibility in new situations with a mix of canonical knowledge and autonomous goal-directed experimentation. The efficient form typically would access bands and modes for the general user in general RF environments from rural vacationing to rush hour in an emergency. The efficient SDR personality would be balanced with functionality in DSP versus FPGA. The optimized form would be realized in ASICs and FPGAs to optimize power and capability per physical resource (interconnect bandwidth and MIPs). Optimized radio-domain capabilities may be described by a <Verbose/> version for flexible reasoning about the SDR waveform. This same strategy of verbose, efficient, and optimized implementations applies to the user sensory-perception component because of the computational burden of 3D visual perception.

8.1.3 Radio Skills and the Cognition Cycle

The top level relationships among the functional components, ISAPIs and CogAPIs, the Cognition cycle, and radio skills is summarized in Expression 8-2.

Expression 8-2 Radio Skills and the Cognition Cycle

<Cognition-cycle>
<Observe> <Sense> <User/> <RF> <SDR>
<Band/> <Mode/> <Environment/> </SDR> <Known/> <Novel/>
</RF> </Sense>
<Perceive> <User/> <RF>
<Connectivity> <GoS/> <QoS/> </Connectivity> <Backup/> </RF>
<Scene/> </Perceive> </Observe>
<Orient> <Self/> <User/> <Scene/>
<SDR-waveform> <State/> </SDR-waveform > </Orient>
<Plan> <Information-services/> <Game-theory/>
<RF> <Radio-skills> <Physical-access>
<Bands> <HF/> <LVHF/> <VHF/> <UHF/> <SHF/> <EHF/>
<THz> </Bands> <Modes> <SDR-subsystem>
<RF-sensing> <Signals/> <Noise/> </RF-sensing>
<SDR-waveforms/> </Modes> </SDR-subsystem> </Physical-access>
<Knowledge> <Physical>
<Frequency/> <Wavelength/> <Bandwidth/> <Canonical/>
<Self> <Transmitter> <Receiver>
<Antenna/> <RF-conversion/> <Baseband/>

<Bitstream/> <Security/> <Sources/> <Sinks/>
</Transmitter> </Receiver>
<Channel/> <Path/> <Reflection/> <Refraction/> <Doppler/>
<Propagation> <Link-budget/> <2D/> <3D/> <Propagation>
<Signal-in-space> <Channel-symbol/>
<Multiple-access/> <Stream/> </Signal-in-space> </Physical>
<Logical>
<Connectivity> <Grade-of-service/> <Quality-of-service/>
</Connectivity> <Protocol-stack/>
</Logical> </Self> </Knowledge>
</Radio-skills> <Collaborate/> </RF> </Plan>
<Decide> <User-criteria/> <Regulatory-criteria/> <Resources/>
</Decide>
<Act>
<Radio-skills>
<Skill> <Connect> <Transmit/> <Receive/> </Connect>
<Transfer-data/> <Disconnect/> <Bridge/>
<Skill/> </Radio-skills> </Act>
<Effectors/>
<Learn> <RF/> <User/>
</Learn>
</Cognition-cycle>

RF support functions of the Observe phase include sensing the radio's band(s), mode(s), and environment (e.g., for available backup channels) through the SDR functional component, supported by the CogAPI. The RF perception function of the Observe phase interprets GoS and QoS data from the SDR component, also identifying the available backup bands and modes to enhance QoI. XG stresses RF environment sensing via the Rockwell-Collins MBMMR XG sensors [194]. As ad hoc networks proliferate, the importance of RF sensing increases.

In the Orient phase immediate RF action mitigates catastrophic <State/> change in a current wireless connection. The Observe and Orient phases structure the behavior of current SDR systems, expanding RF sensing, but not fundamentally changing the nature of the radio. The Plan phase, however, implements a quantum leap in behavior since the <Self/> knowledge about <User/> QoI and the <RF/> <Scene/> drive <Goals/> in this phase. Today's radios don't plan. They do what they are told by the user and the host network. AACRs plan. They sense available spectrum, identify QoI enhancement opportunity, allocate resources, and enhance QoI <Goals/> within <User/> and <RF/> constraints. Instead of merely initiating a SDR personality with a fixed set of parameters, the AACR's radio skills are more refined, connecting to evaluate an available mode, but not necessarily using the mode. When transferring data, the AACR may adapt content to the changing RF environment, increasing error protection as fading parameters change. In addition to losing network connectivity, the AACR politely defers access to primary users when sharing spectrum. Bridging bands and modes yields uninterrupted conversation as the iCR switches from cellular to VoIP on the corporate wireless LAN. Collaboration among peer AACRs includes goal-oriented planning. Thus, the implementation of radio skills focuses mostly on the use of embedded knowledge via the Plan phase. The comprehensive scope of planning is reflected in the RXML of Expression 8-2.

8.1.4 Radio Skills Implementation Strategy

The strategy of radio skill implementation applies Occam's Razor to technology insertion. The simplest technology that accomplishes the use case is preferred. Thus, the chapter begins with the simplest methods of storing and applying radio knowledge and progresses to the more complex but more capable methods. The place to start embedding knowledge into AACR is the traditional database. Next are the rule bases and knowledge bases from eBusiness products, many of which have been assimilated into conventional database packages, like Oracle (e.g., 9i eBusiness rules are well beyond mySQL). In addition, conventional (nonintelligent) radio propagation modeling tools supply radio knowledge, particularly if calibrated to the <RF/> environment by the AACR or CWN. The format of radio knowledge differs with band, mode, and time through AACR evolution. The major conventional knowledge representations are databases, rule bases (including logic programming), object-oriented knowledge bases, agent systems, and domain-specific computational models. Each of these is a candidate for radio skill to observe, orient, plan, decide, act, and learn.

The inadequacies of the simpler approaches lead one to formal computational ontologies and the Semantic Web. In one approach, spectrum management is a multiplayer game [195] with heterogeneous knowledge representation. Microworlds organize scene-level knowledge and partition skills within specific task domains that are formally modeled, supported by axioms, a knowledge base, a rule base, and a domain-specific language [219]. <RF/> microworlds constrain AACR inferences to context for efficient, focused problem solving.

Finally, even these approaches do not fully support AML for iCR, which led to the specialized Radio Knowledge Representation Language (RKRL) [145], an extensible domain-specific language that is intentionally not Turing-computable and thus achieves finite introspection. This book realizes RKRL principles in RXML.

8.2 EMBEDDED DATABASES ENABLE SKILLS

SDRs and CRs are replete with databases. Efficient data structures like hash maps enable rapid access to internal databases. The CR challenge is the autonomous creation of properly organized perceptions and plans from the partially organized, incomplete, and errorful data of the external world. Database <Schema/> begin that process.

8.2.1 A Spectrum Management Database

Spectrum management databases are increasingly important with the FCC's support of DARPA's XG policy language for secondary users' spectrum. The following is an extract of a spectrum management database:

ILLUSTRATIVE RF SPECTRUM ALLOCATION

Frequency (MHz)	Allocated to
27.790	Forest products
27.900	U.S. Army
28.000–29.700	Amateur 10 meters
29.700–29.800	Forestry Service
29.800–29.890	Fixed service
29.890–29.910	Government
29.910–30.000	Fixed service
30.010	U.S. Government

The same spectrum point may be allocated to multiple users, and some entries express lower RF with an assumed per channel bandwidth while other entries specify both upper and lower edges, sometimes with different per channel bandwidths. The AACR database engineer brings consistent structure to these untidy differences for meaningful planning via the values in the above list. Storing this database for <Self/> spectrum management could use a simple schema like the following:

Expression 8-3 Spectrum-Allocation Schema in RXML

<Spectrum-use-database> <Spectrum-allocation> <Schema>
[<RF-low/> <RF-high/> <Allocated-to/>]* </Schema>
</Spectrum-allocation> </Spectrum-use-database>

In approach, a database engine in a CWN responds to queries. If the <User/> were an <Amateur-radio/> operator, then the AACR might embed such a database. The Ham might ask the AACR about frequencies allocated to forest products mapped to the CRA as illustrated in the following:

Sequence of CRA Activity for Frequency Allocations Database Access

Voice Stimulus: "Computer, what frequency is allocated to forest products?"

Perception Hierarchy
Observe Phase: NL processing

<User-addressing-Self>
<Question/>
<Frequency value = ??/> <Allocation to = "Forest products"/>
</Question> </User-addressing-self>

Observe phase presents <Question/> from NL to Orient phase.

Orient Phase: Detects no tasks conflicting with planning a response.

Plan Phase: Generates Plan to look up "forest products" in <Spectrum-allocation/> table of its *<Spectrum-use-database/>*.

Decision Phase: Initiates database action to look in frequency allocation table with forest products as the query and frequency as the result; and to say result.

Action Phase: Retrieves <Frequency>; (next cognition cycle) Says result.

The Action phase finds information in the internal database and yields the answer, which is spoken on the next cognition cycle. Actually, there are dozens of frequencies allocated to forest products, so dozens of frequencies would be reported. The planner might generate the response "The seventy-four frequencies are listed on the display," presumably held in short-term memory as "the [forest-products] frequencies" for further dialog.

Any PC-class database system from mySQL to Excel to Microsoft Access could perform this task. In addition to text and numbers, even the simpler embeddable databases accommodate large text fields, embedded objects, and hyperlinks. Predefined formats include dates and currency with field-definition APIs for application-specific <Schema/>. The database approach scales to millions of records, so a world-wide CWN could use Oracle for all of the policies and exceptions of all the spectrum management authorities in the world.

To facilitate machine reasoning, RXML asserts database features for knowledge representation, introducing needed conceptual primitives as follows.

Expression 8-4 Asserting Properties of Databases in RXML

<Abstractions> <Self> <Memory>
<Database>
<Definition> A database is a collection of tables, each consisting of records, each with a format defined by a <Schema/> that is identical for all records in the same table, and that is managed by a database system </Definition>

<Database-system> mySQL MS-Access Excel Oracle Sybase
</Database-system>
<Structure>
<Schema> <Definition> A schema describes the structure of data
with element names, data types, attributes and combinations
available for each element, including the rules of the data
document. Schemas model the database </Definition>
<Element-names/> <Data-types/> <Rules/> </Schema>
<Data-types> <Numeric/> <Text/> </Data-types/> <Delimiter> “,”
</Delimiter> </Structure>
<Example> <Database>
<Schema> <Name> Spectrum-use-database </Name>
<Element> <Name> RF-high </Name> <Numeric/> </Element>
<Element> <Name> RF-low </Name> <Numeric/> </Element>
<Element> <Name> Allocation </Name> <Text/> </Element>
</Schema>
<Record> <RF-low/> <RF-high/> <Allocation/> </Record>
<Spectrum-use-database>
<Record> 27.790, 27.790, “Forest products” </Record>
<Record> <End/> </Record> </Spectrum-use-database>
</Database> </Example>
<Scale> <Small> 100 </Small> <Big> 1,000,000 </Big> </Scale>
</Database> </Memory>
</Self> </Abstractions>

Expression 8-4 defines illustrative RXML conceptual primitives for AACR reasoning about database tools. The AACR resource planner needs to know about scaling, which is a conceptual primitive for iCR resource management, trading off learning, remembering, summarizing, and disposing of old information.

The AACR designer should realize that most of the hard work is the labor-intensive validation of database entries, including the removal of reasonable but inconsistent and illegal entries, like “TBD” in a numeric frequency field that is being revised. Although nearly all database tools can perform computations, the server programming style isolates the code from the associated data, sometimes requiring deep insight for desired behavior. Other candidate software tools include the spreadsheet.

8.2.2 Embedded Spreadsheets

Spreadsheets like Excel with Visual Basic for Applications (VBA) offer a step from databases toward the unstructured without sacrificing many of the benefits of the database such as the ability to scale up.

8.2.2.1 Spectrum Allocations Spreadsheet

For example, the source code of RKRL 0.3 is an Excel spreadsheet with VBA macros containing 47 microworlds [219], one of which contains a spectrum allocations database:

Band	Low	High	W_c
ISM	13.56	13.567	0.007
CB	26.9	27.4	0.5
ISM	27.195	27.205	0.01
Government, industry	29.7	50	20.3
VHF TV	54	88	34

W_c is the aggregate channel bandwidth of the allocation, readily computed in the spreadsheet from the low and high RF values as the numerical difference of the adjacent columns. This database occupies rows and columns dedicated to the database task, while other rows and columns describe spectrum management concepts, give examples, and supply context, citations, and other less structured knowledge of the spectrum microworld. Most of the ancillary data occurs just once and has little structure, so the spreadsheet asserts the supporting information conveniently. For example, the ad hoc fact that NFPT is the "Next-generation frequency planning tool" is included in the RKRL 0.3 spreadsheet to illustrate object-oriented association, relatively efficient storage, and the VBA retrieval tool.

Expression 8-5 Description of Spreadsheet in RXML

<Abstractions> <Self> <Memory>
<Spreadsheet>
<Definition> A spreadsheet is a collection of worksheets, each consisting of cells organized into rows and columns, each of which has its own format defined by usage, that employs macros for applications-specific programming, and that is managed by the spreadsheet system </Definition> <Database-system/>
<Structure> <Cell> <Definition> A cell consists of explicit data or a single computational expression defining the value of the cell </Definition> <Cell-format/> </Cell> <Macro-language/> </Structure>
<Example> <Spreadsheet> <Worksheet> <Name> Spectrum-use </Name>
<Cell> A 1 RF-high </Cell> <Cell> B 1 RF-low </Cell> <Cell> C 1 Allocation </Cell>
<Cell> A 2 27.790 </Cell>
<Cell> B 2 27.790 </Cell>
<Cell> C 2 "Forest products" </Cell>

TABLE 8-2 Radio Access Capabilities Matrix

Band	Mode	Capabilities	Entities
HF	ALE	:BLoS :Rb(2-9600) :Inet	:Ham :Web[OIL]
LVHF	FH	:BLoS :Rb(9.6-56)	:NationalGuard
VHF	NB-FM	:NLoS :Voice :Rental[]	:Taxi :Police
VHF	TV	:News :Sports :WX :Share1	:Broadcaster
UHF	GSM	:Tower :Voice :GPRS	:PSTN[SS7]
UHF	GPS	:SATCOM :Location	:Time-Std
UHF	WLAN	:LoS :ISM-Pwr :Rb (.2-2)	:WifeCR[RKRL]

```
        <Cell> <End/> </Cell> </Example>
        <Cell> A 5 NFPT </Cell>
        <Cell> B 5 "Next-generation frequency planning tool" </Cell>
    <Scale> <Small> 1 </Small> <Big> 100,000 </Big> </Scale>
       </Spreadsheet> </Memory>
</Self> </Abstractions>
```

The <Database/> and <Spreadsheet/> abstractions contain <Definitions/> for tutorial interaction with a user as well as related conceptual primitives (e.g., <Record/> and <Cell/>) that enable AACRs to share among CWNs.

8.2.2.2 The Radio Capability Matrix

Spreadsheets also help associate radio band and mode with access to entities the user enjoys (Table 8-2). Such a spreadsheet instantiated in advance acts like yellow pages, a generalized X.400 service. Over time, the AACR learns that the user's <Wife/> has a WLAN-capable cell phone to which the <Self/> may connect via ad hoc networking even though the OEM didn't program it to do so. Provided the AACRs don't violate regulatory rules, the <Wife/> LoS WLAN connectivity augments the prior knowledge. Similarly, learning that <News/>, <Sports/>, and <Weather/> originate with a <Broadcast/> enables the <Self/> to augment <User/>-specific interests with content.

The spreadsheet also readily reflects parameters of radio access, such as the available data rates ($R_b = 2$ to 9600 bps for HF and 0.2 to 2 Mbps for the WLAN). Since GPS is derived from satellites, an AACR that models radio propagation readily infers how GPS is limited in urban canyons. An AACR that reasons about movement of the user from both visual cues and GPS ignores GPS glitches with inconsistent optical flow of the visual scene. The more sophisticated uses of the radio access capabilities matrix are realized by the embedded spreadsheet and embeddable reasoning developed subsequently.

Comparing the RXML of the database and spreadsheet abstractions, one observes greater freedom of expression with the spreadsheet.

8.2.3 Spectrum Management Hash Maps

Hash tables are used in databases and spreadsheets to efficiently pack indexed data into finite memory. One Java library for embedded hash tables, ObjectSpace, defines a hash table as a HashMap:

1. HashMap serModel = new HashMap(); (8-1)
2. serModel.add("27.790","Forest Products"); (8-2)
3. serModel.add("27.790","Army"); /* Similar statements define a small database */ (8-3)
4. int responses = serModel.count("27.790"); /* This returns the integer 2 */ (8-4)
5. Object response = serModel.get("30.010"); (8-5)

The HashMap class (8-1) creates a stimulus–experience–response model (serModel) to which string constants may be added in pairs ((8-2) and (8-3)), the first of which serves as a key to the serModels microdatabase. HashMap.count reveals how many objects match the key object (8-4), while the get() function (8-5) returns a HashMapIterator over the objects that match a query.

Algorithm 8-1 HashMapIterator Processes Matching Objects

```
HashMapIterator Models = (HashMapIterator) serModel.start(); //iterates
   from beginning
   HashMapIterator end = (HashMapIterator)serModel.finish();// to end
   while(!Models.equals(end)){ // processes all the Models
      Pair current = ((Pair) Models.get()); //extracting the serModel pairs
      Pair next = new Pair(current.first, new String ("<Unallocated/>"); //
         and changing them
         if((String)current.first ! = new String("27.790")){ Models.
            put(next);} // if needed
      Models.advance();
   } //end while
```

The Java HashMaps of CR1 use memory efficiently for associative retrieval, an embeddable subset of database-spreadsheet technology.

Although consistent with the CRA, the above methods for remembering spectrum allocations do not take advantage of the built-in AML of even the CR1 prototype, which can be trained to remember spectrum allocations, in fact using HashMap as the internal representation. A vignette for training and using spectrum allocations with CR1 HashMaps is provided in the companion CD-ROM.

8.3 PRODUCTION SYSTEMS ENABLE SKILLS

This section briefly reviews production systems, the foundations for embedded rule-based expert system shells. Finite state machines are represented in XML.

8.3.1 State-Space Production Systems and Games in RXML

As is well known, a rule is a condition–action pair formulated so that when a condition is met, a rule is applied, and an associated action is taken. If the rules apply to a global memory, then the memory and associated rules are called a production system [196].

Expression 8-6 Defining a Production System to the <Self/> Via XML Tags

```
<Abstraction> <Reasoning> <Production-system>
  <Memory/>
  <Rule> <Condition/> <Action/> </Rule> </Production-system>
</Reasoning> </Abstraction>
```

The memory of a production system can reflect intermediate states in the application of rules and external stimuli. The memory may be discrete, finite, and small to represent the control states of a system or game board.

If each rule of the production system encodes one legal move, if all legal moves are represented, and if no illegal moves occur by applying the rules, then the production system spans the state space of the game. Moves not taken are hypotheses. Games like tic-tac-toe with small hypothesis spaces are played using such production systems. Games like chess are easy to encode but have state spaces on the order of 10^{120}, so one cannot exhaustively search the hypothesis space in a reasonable amount of time, even with heuristics, learning, and specialized hardware like Deep Junior [197].

Game theory's rich legacy has been applied to fairness in wireless protocols like the IEEE 802.11 Distributed Coordination Function Nash equilibrium backoff strategy [195]. Game theory applies to other planning and decision aspects of the CRA.

8.3.2 State Machines in RXML

If <Memory/> has a distinguished state (e.g., <Current-state/>), then rules may assert a next state and perform side effects, implementing a finite state machine as in Expression 8-7. This particular state machine would instruct an intelligent agent to say "Hello" once and then never to say anything else because the single <Rule/> sets <Current-state/> to 1, for which no actions are defined.

Expression 8-7 A State Machine Is a Production System

<Abstraction> <Reasoning> <Production-system> <State-machine>
<Memory> <States> 1 2 </States> <Current-state> 2 </Current-state>
</Memory>
<Rule> <Condition> <Current-state> 2 </Current-state> </Condition>
<Action> <Current-state> 1 </Current-state> <Say> "Hello" </Say>
</Action>
</Rule> </State-machine> </Production-system>
</Reasoning> </Abstraction>

In describing a state machine to the <Self/>, mentioning <Current-state/> in the <Action/> causes the <Current-state/> of <Memory/> to be set to the value in the rule, transitioning to the "Next-state" via the rule. The action <Say/> achieves the desired side effect when transitioning from one state to another. Such finite state machines can parse finite languages by starting in a distinguished state, <Start/>, comparing input symbols to values in rules that encode the structure of the language, and either ending in a distinguished state, <True/>, if the string is in the language, or transitioning to <Fail/> if an input symbol does not match any rule.

A state machine for using radio spectrum in accordance with the policy of a regulatory authority could have just a single state, the radio frequency, with a set of rules that encode the spectrum allocation database as suggested in Expression 8-8.

Expression 8-8 Sample Spectrum Management State Machine

<Abstraction> <Reasoning> <Production-system> <State-machine>
<Name> Spectrum Management Database </Name>
<Memory> <States> <RF/> </States> <Current-state> 24.0 </Current-state> </Memory>
<Rule> <Condition> <Current-state> 27.79 </Current-state>
</Condition>
<Action> <Say> Forest Products </Say> </Action> </Rule>
<Rule> <Condition> <Current-state> 27.900 </Current-state>
</Condition>
<Action> <Say> Army </Say> </Action> </Rule>
</State-machine> </Production-system>
</Reasoning> </Abstraction>

The rule base shows one strength of finite state machines—efficiency. Without the RXML markup that shows what it all means, the state machine has very compact content that could be represented in two variables as follows:

Current-state = 24.0 (8-6)

Array rules = new Array {(27.790, "Forest Products"),
(27.790, "Army")) (8-7)

Equation (8-7) shows a critical limitation of finite state machines and simple rule bases—limited expressive power. Since multiple rules apply, a control system that is not explicit in the RXML somehow picks which rule to apply. Is 27.79 MHz used by the Army or Forest Products or both? Database <Control/> aggregates all results into a vector of alternate hypotheses. Typically, state machines avoid such complexity, applicable in Orient, Plan, and Decide phases for fixed message exchange sequences. Such radio state machines are common in protocol stacks, typically well known to radio engineers expressed in SDL [44], which is easy for radio engineers to understand, but which is not as easy for AML algorithms to autonomously extend as RXML.

8.3.3 Push-Down Automata in RXML

If the state machine has a stack onto which it may place intermediate results, then the state machine is a push-down automaton (PDA).

Expression 8-9 Push-Down Automaton Has a Stack

```
<Abstraction> <Reasoning> <Production-system> <State-machine>
  <Push-down-automaton>
  <Memory> <States> 1 2 </States> <Current-state> 2 </Current-state>
    <Stack> </Stack> </Memory>
  <Rule> <Condition> <Current-state> 2 </Current-state> </Condition>
    <Action> <Current-state> 1 </Current-state> <Stack> FM 1200
      </Stack>
  </Action> </Rule> </Push-down-automaton> </State-machine>
    </Production-system> </Reasoning> </Abstraction>
```

The machine of Expression 8-9 pushes whatever occurs in the <Stack/> part of the <Rule/> onto the null <Stack/>. This rule pushes the string "FM 1200" that could represent a goal to turn on the FM receiver at noon. Such goal stacks popularized in the Blocks World planner [13, 89] are productized in the open procedural reasoning system (OPRS).

PDAs recognize context free grammars (CFGs) [286]. With side effects, such CFG recognizers parse and compile computer languages like C and Java, and analyze (structured) natural language [198]. Augmented transition networks (ATNs) [199] extend finite state and push-down automata to general computations during transitions, enabling side effects that make them Turing-computable and thus they may crash trying to reach unreachable states.

8.4 EMBEDDED INFERENCE ENABLES SKILLS

A production system that analyzes and asserts statements within a model of truth is called an inference engine. Axiomatic knowledge represented in Horn clauses uses first-order predicate calculus (FOPC) reasoning, typically via PROLOG. The less axiomatic forms, called expert system shells, have fewer constraints and are less predictable theoretically than the axiomatic forms.

8.4.1 JESS Embeddable Expert System Shell

The Java Expert System Shell (JESS)[1] classes implement the Jess language [200]. Jess is a LISP-like language [201, 202]. Atoms are the symbolic identifiers, with only nil, TRUE, and FALSE having special meaning. Jess parses floating point and integer numbers but does not accept scientific or engineering notation. Lists consist of an enclosing set of parentheses and zero or more atoms, numbers, strings, or other lists, such as:

```
(+ 3 2) (a b c) ("Hello, World") ()
  (deftemplate foo (slot bar))
```

As in LISP, Jess code has the form of a function call, a list with prefix notation for the function to be called, as in (+ `2 3`). Jess variables begin with a question mark (?), while multivariables begin with $ (e.g., `$?X`) to refer to multifield lists. The Jess `bind` assigns values to variables: `(bind ?x "The value")`. Control flow in `deffunction` is achieved via the control-flow functions `foreach`, `if`, and `while`. `Defadvice` executes advice code before or after the associated Jess function is called to wrap code around an existing function, for example, to extend Jess without changing Jess itself. One may create and manipulate Java objects directly from Jess, except for defining new classes.

Jess KBs allow ordered facts, unordered facts, and Java Bean-derived facts. Ordered facts are lists, where the head of the list is a category for the fact, such as (<Radios/> PRC-117 SINCGARS cell-phone). Ordered facts are added to the single global Jess KB via the `assert` function, such as `Jess> (assert (father-of Barb Joe))` that returns `<Fact-1>`. Unordered facts associate slots and values with an atomic identifier to create a Jess KB object such as Expression 8-10.

Expression 8-10 Defining a Spectrum Allocation Object in Jess

(<Allocation/>
  (<Frequency/> 27.790) (<Primary-use/> "Forest Products")
  (<Secondary-use/> "Army"))

[1] Jess 5.0 a6 is described here since it is representative of embedded rule-based shells.

```
<(deftemplate <Allocation/> "A frequency allocation."
  (slot <Frequency/> (type RU.FLOAT)) (slot <Primary-
  use/>) (slot <Secondary-use/>))>
```

Jess can represent Java Beans in the KB statically (changing infrequently, like a snapshot of the properties at one point in time) or dynamically (changing automatically whenever the Bean's properties change) via defclass and definstance functions.

Jess rules are defined via `defrule`: `(defrule <Rule-name/> <Equivalent-text/> <LHS/> => <RHS/>)`. In RXML, the LHS is a <Condition/> while the RHS is an <Action/>. For example, `(defrule now-switch-to-ALE "If all LoS links are unavailable, then switch to HF ALE." (LoS-down) => (queue-source-to-HF-ALE))`. Since Jess rules fire only when they are activated by matching facts, the RHS typically asserts something into the KB that causes other rules to fire, a process called forward chaining. Rule patterns include wildcards and predicates (comparisons and Boolean), where LHS variables match values in that position in the KB. Qualification tests, reference bindings, and other forward-chaining mechanisms are summarized in the companion CD-ROM.

Jess resolves rule conflicts via salience that is declared and manipulated by rules, all of which fire in order of salience, highest first. In addition, the depth conflict resolution strategy fires the most recently activated rules before others of the same salience, while breadth fires rules in the order in which they are matched to facts in the KB.

Failure is equivalent to negation in Jess, so if a fact is not in the KB, it is considered to be not true. The pattern `(not (<Allocation/> (<Frequency/> ? (27.790) (<Primary-use ?)))` searches for patterns related to a frequency of 27.790, so if none allocate that frequency, then the RHS negation fires.

In backward chaining the Jess inference engine identifies a sequence of rules whose preconditions are not met to identify steps toward a goal. The recursive factorial function can be implemented as a backward chaining rule instantiation process.

```
(do-backward-chaining factorial)
(defrule print-factorial-10
(factorial 10 ?r1) => (printout t "The factorial of 10 is "
  ?r1))
```

Jess asserts `(need-factorial 10 nil)` into the KB. The factorial rule that matches this need is:

```
(defrule do-factorial (need-factorial ?x ?)  =>
(bind ?r 1) (bind ?n ?x) (while (> ?n 1) (bind ?r (* ?r ?n))
  (bind ?n (- ?n 1)))
(assert (factorial ?x ?r)))
```

The rule compiler adds a negated match for the factorial pattern itself to the rule's LHS. Rules that match on (factorial), and need a (factorial) fact activate (need-factorial) so the needed facts appear, and the (factorial)-matching rules fire, achieving the Jess form of backward chaining.

Consider the VHF BLoS problem of the prior chapter. An AACR is to employ spatial reasoning about D-layer scattering via frequencies between 30 and 60 MHz. Jess can backward chain over distance and frequency by the backward chainable (do-backward-chaining <RF/>) and (do-backward-chaining distance). The following rules assert the frequency tuning constraints:

```
(defrule rule-1 (<RF/> ?A ?B) => (<Advise/> Set <RF/> between
  ?A and ?B))
(defrule create-<RF/> (need-<RF/> $?)
    (BLoS-distance ?X ?Y ?Z &:(> ?Z 40) & :(< ?Z 100))
    => (assert (<RF/> 30 60)))
(defrule create-BLoS-distance (need-BLoS-distance $?)
    => (assert (BLoS-distance <Self/> <Other-radio/> 63)))
```

Jess detects that rule-1 could be activated to get advice on RF if there were an appropriate <RF/> fact, so it asserts (need-<RF/> nil nil nil). This matches part of the LHS of rule create-<RF/> that cannot fire for want of a BLoS-distance fact. The constraint will check to see if the distance between ?X and ?Y is greater than a BLoS-distance threshold of 40 km and less than the upper limit of 100 km. Jess therefore asserts (need-BLoS-distance nil nil nil). This matches the LHS of the rule create-BLoS-distance, which fires to assert (BLoS-distance <Self/> <Other-radio/> 63), and 63 exceeds the range for LoS, but is within the range of D-layer scattering. This fact activates create-<RF/>, which fires, asserting (<RF/> 30 60), thereby activating rule-1, which then fires to advise the AACR to set RF between 30 and 60 MHz. Such rules could be embedded in the Plan component of the CRA.

JessAgent [203] shows how to implement reasoning in a JADE application via the JESS engine (JESS 5.1). When a message from another agent arrives, a new fact is asserted, so a rule defined in the JadeAgent.clp responds to the sender. The example also uses Jade classes Cyclic Behavior and ACL Message, illustrating techniques for linking the reasoning of one AACR to a CWN.

JESS and other embeddable expert system shells can implement radio-domain skills. The following rule includes advanced features, such as expressing uncertainty:

> If the style (attribute) of the radio-station (entity) is rap (value of attribute) => then it might (20%) be an-enjoyable-station (classification decision) (8-8)

The user's value judgments on radio stations expressed in Equation (8-8) may be organized into the following hierarchy:

"Radios" [Organic Wireless Connectivity Ports]
  "Cell Phone" [Mobile Cellular Network Services] . . .
  "Radio Stations" [AM/FM Broadcast]
    "Good radio stations" (3%)
      "Enjoyable radio stations" (20%)
      "Informative radio stations" (20%)
      "Cool radio stations" (60%)
    "Bad radio stations" (97%)
      If it is not "good radio station", then it is "bad radio station"

The hard part is the autonomous learning of these value judgments. The iCR should accurately infer when the user is expressing a preference. Users don't want to be a slave to the radio. Do you? No, of course not. Users do not willingly fill out user profiles, so the CRA envisions prototypical situations that can be reliably detected by an algorithm that reliably infers preferences, reasoning by analogy with what is already known, bootstrapping new knowledge in terms of old.

CR1 codifies knowledge in serModels that have features of embedded expert systems. Individual serModels have only shallow reasoning without supporting data structures or rule sets. Nevertheless, serModels organized into an associative perception hierarchy can learn responses to relatively complex stimulus sets, as shown in Table 8-3.

Since CR1 detects RSSI, BER, light level, and temperature from different sensors, and since it always aggregates contemporaneous stimuli as features of a scene, it can be taught to recognize the relatively complex situation in which the SDR waveform is reporting low RSSI and high BER while the light level is decreasing and the temperature is rising (when outside in ambient less than 70 °F). Those conditions trigger the radio to turn on the WLAN system, searching for an alternative to the cellular system that may soon be lost.

Rules are a mainstream commerce technology now widely accepted as a component of the Semantic Web. The Rule Markup Initiative took initial

TABLE 8-3 Stimulus–Response Models as Rules

Situation	Stimulus	Response
Accessing (cells)	Low RSSIs, high BER/FERs, light level decreasing, temperature rising	Search for WLAN (implicit "entering building" hypothesis)
{(lost cell, light, got hot)}	Found WLAN	Ask (permissive) user "Did we enter a building?"

steps toward a Rule Markup Language (RuleML), for forward and backward chaining in XML, but in 2004, there were an XML-only RuleML, an XML-RDF RuleML, and an RDF-only RuleML. Complementary efforts include Java-based rule engines such as Mandarax, RuleML, and XSB-RDF RuleML, with DTDs for basic RuleML sublanguages.

As shown from the examples, embeddable expert system shells like JESS enable forward and backward chaining within a KB for reasoning and goal-directed inference in AACR use cases. Backward chaining can be more efficient if rules are expressed as logical axioms, as in PROLOG.

8.4.2 PROLOG

PROLOG is the logic-programming language for reasoning with closed-world knowledge. In a closed world, the inability to find an assertion is treated as de facto proof that the assertion is not true. Horn clauses are an efficient form of FOPC. In Horn clause logic, facts are atomic and conditional inference is expressed as a conjunctive clause with a single negative conjunct [204]. In the PROLOG implementation of the Horn clause representation, facts are expressed as predicates, such as `Allocated (27.790, Army)`.

For the logic enthusiast, the following six expressions are equivalent. The last logic form is a Horn clause.

A implies B	B if A	$(\neg A \vee B) \wedge \ldots$
If A then B	Not (A) or B	$\neg (\vee \neg (A \wedge \neg B))$

Most practical logic can be expressed in FOPC as Horn clauses and interpreted by the PROLOG language [205]. As with Jess, PROLOG facts are atomic:

```
Allocated (27.790, "Forest Products") Allocated (27.790, Army)
```

As a logic programming language, PROLOG statements imply existential (∃) or universal (∀) quantification. The statement `Allocated (x, Army)` means there exists a frequency `x` that is allocated to the Army. Implications are universal, however. So the following statement means for any and all frequencies, if the frequency is busy, then we can't use it.

```
Cant-use (frequency) :- Busy (frequency)
```

(PROLOG variables traditionally are single capital letters like A or X. For clarity, verbose variables are in bold.) The statement is read "Can't use frequency if the frequency is busy." Placing the conclusion first is one of the features of Horn-clause logic that if–then rule programmers have to get used to. Conjunctive conditions are expressed on the right-hand side of the rule.

```
Use (<Self>, frequency) :- Allocated (frequency, user),
  equals(user, <Self/>))
```

PROLOG can backward-chain to establish a plan more efficiently than Jess. Some versions of PROLOG include built-in numerical functions that return the value in the last argument. For example, the statement "`?-*(10, 3, A)`" asserts A = 30, the product of 10 and 3. Since PROLOG is compact, one must read expressions carefully.

PROLOG engines from universities are designed to teach logic programming. PROLOG sometimes requires one to rethink a use case in ways not necessarily intuitive to a C, C++, or Java programmer. For example, it may take a Java programmer some practice to get used to the PROLOG cut operator (!), similar to the Jess "unique" flag for lists.

There are also industrial strength PROLOG engines like BinProlog™ [206], a high performance Internet-oriented PROLOG compiler with the ability to generate C/C++ code and stand-alone executables. BinProlog includes high level networking with remote predicate calls, blackboards, mobile code, multithreaded execution on Windows NT and Solaris and secure Internet programming with CGI scripting, multiuser server side databases, and rule-based reasoning. PROLOG utilities that make BinProlog more like conventional languages include dynamic clauses, a metainterpreter with tracing facility, sort, set-of, dynamic operators, floating point operations, and function definition. Its mobile code, user interfaces, 3D graphics, client/server, dynamic databases, and make-facility are summarized in the companion CD-ROM. Other commercial PROLOG systems include Quintus and SICstus PROLOG.

8.5 RADIO KNOWLEDGE OBJECTS (RKOs)

The previous sections surveyed current technology embeddable for the control of any SDR. To the degree that these techniques are expressed in a standard RXML, they facilitate AACR collaboration. This section sketches research ideas that may not be realized for some time.

During the 1980s powerful machine learning technologies were invented. One of the most interesting was AM [320]. Although many of the concepts of the early research were refined and improved, several pioneering contributions have yet to be fully realized, even in intelligent agent technology of 2005, yet are relevant to iCR. Specifically, AM pioneered the multifaceted **concept** autonomously evolved by rich heuristics, the inspiration for the radio knowledge object (RKO) of this section.

8.5.1 The AM Concept Data Structure

AM automated the discovery of data structures that express <Interesting/> mathematical expressions. Simple radio skills may be based on embedded

HashMaps and expert system shells, but the autonomous extension of these skills by iCR is a technology challenge not unlike that of autonomously deriving principles of mathematics from set theory. Radio-use rules defined by regulatory authorities are not unlike axioms of logic. Laws of physics are not unlike laws of mathematics.

Rules of the previous sections were expressed in RXML for autonomous extension by AM-class AML. AM employed 115 initial "concepts" like Set and Set-Union, Object, List, Compose, and Truth-value. From these plus heuristics for instantiating new concepts, it generated and evaluated as <Interesting/> ("discovered") data structures corresponding to Perfect-squares and Peano's axioms. AM used a very large state space of concepts and 242 heuristic rules for filling-in concept slots, checking intermediate results, suggesting search directions, and calculating interest. Mathematics is a big, open domain not unlike the radio and user domains of AACR. What Lenat called a "concept" might be called a knowledge template, or knowledge object (KO), a named data structure where knowledge is brought together (adapting Lenat's original language):

1. *Name(s)*: A string to which a person or agent may refer to the KO.
2. *Computational Definitions*: Metalevel methods for evaluating a concept.
   (a) Domain: List of sets over which the KO is defined.
   (b) Range: List of sets to which the KO can map or be mapped.
   (c) Lambda (λ) expressions: Anonymous functions attached to the KO, akin to methods of object-oriented systems. These functions are typically Boolean, testing instances for degrees of conformance to the KO.
   (d) Slots: Lambda expressions that contain data used by the KO.
3. *Algorithms*: Named expressions attached to the KO. These are KO-domain functions that implement some aspect of the KO, such as mapping domain to range.
4. *Generalizations*: More abstract KOs from which this KO may inherit properties. KOs form inheritance networks (heterarchy not hierarchy), rarely inheriting all properties, so rarely are generalizations strictly less constrained than a given KO.
5. *IsA*: KOs, the definitions of which this KO satisfies.
6. *Views*: A view of some other class of KO as if it were this KO.
   (a) Intuitions: An abstract analogy for this KO.
   (b) Analogies: Similarities drawn between this KO and other KOs.
   (c) Conjectures: Unproven theorems (hypotheses) about this KO.
7. *Specializations*
   (a) Derivative KOs: AM's heuristics could create virtually new KOs from one or more existing KOs, so that the new KO inherited very little from the base KOs.

   (b) Instances: More specific KOs that inherit directly from this KO.
8. *Examples*: Instances that are directly related to the KO.
   (a) Typical: Instances that conform to all the definitional lambda expressions.
   (b) Barely: Instances that conform to one or more but not to all of the definitional lambda expressions.
   (c) Not-quite: Does not conform to sufficient lambda expressions to be considered an example, but does conform to all but one or two requirements.
9. *Conjectures*: Untested hypotheses about the KO.
10. *Worth*: Some numerical, symbolic, or vector-space metric for value of the KO. A vector-space representation may define value with respect to different goals.
11. *Interestingness*: Numerical, symbolic, or vector metric for novelty, unexpected performance, important source, and so on.

The "typical" examples have no conceptual distance from the KO while "barely" examples have increasing Hamming distance from the ideal KO counting the number of facets observed in the example. Not-quite satisfies some, but not enough criteria to be that KO. Views guide the autonomous interpretation of data structures. Since the more aggressive use cases like Genie, Bert, and Ernie contemplate substantial autonomous extension of situation-specific RF knowledge, derivatives of the AM **concept** are worth considering.

The lambda expressions for AM **concepts** do most of the work, determining the type of object and invoking the appropriate function. Metalevel **concepts** like canonize and repeat are **concepts** with appropriate domain and range over other **concepts**. Heuristics guide **concept** evolution.

8.5.2 AM Heuristics

AM's 242 heuristics address four needs for autonomous discovery in a complex open-ended domain. Again, adapting Lenat's language, these are:

1. *Completion*: Rules for filling in facets of a KO.
2. *Refinement*: Rules for patching up facets for consistency.
3. *Suggesting*: Rules for generating new KOs, for example, to overcome obstacles.
4. *Value*: Rules for estimating worth and interestingness.

Completion heuristics develop a known concept further, while refinement heuristics adjust features of the concept. Since AM's control loop picks the next most interesting thing to work on, many of the heuristics adjust worth and interestingness similar to <Histogram/>. For example, the completion heuristic increases priority of pending tasks if they involve the KO most recently worked on. "Anything" is the most general concept and the root of the KO heterarchy. In addition, the heuristics that make suggestions define the direction of evolution of discovery, knowledge, and skill.

Adapting AM to AACR may seem to be a giant step backward, but a review of the current artificial intelligence, genetic algorithms, agent systems, Semantic Web, and artificial life literature shows that AM was perhaps decades ahead in defining data structures and procedures by which a reasoning system could autonomously evolve. Of course, Eurisko and CYC continued the evolution of AM toward the codification of commonsense knowledge, so elsewhere in the text CYC's relevance to AACR is considered but found wanting. As Lenat accurately points out, there is an inherent brittleness to autonomous evolution as the conceptual distance increases between the evolving system and its founding concepts. Therefore, adapting AM's concepts and heuristics to iCR requires RF or user validation of evolving radio KOs and heuristics. There are many opportunities and pitfalls of autonomous evolution in iCR.

8.5.3 Radio Knowledge Objects (RKOs)

Ultimate iCR applications will require the integration of substantial radio knowledge. This section suggests RKOs patterned after AM's concept KOs. RKOs represent static knowledge and enable autonomous extension of localized and network-generated knowledge.

8.5.3.1 RKO Template

The RKO is defined as follows.

Expression 8-11 Radio Knowledge Object

<Abstractions> <RKO>
<Name/> <Definitions/> <Slots/> <Membership/> <Domain/> <Range/>
<Functions/>
<Examples/> <Generalizations/> <Specializations/> <Views/>
<Hypotheses/> <Value/> </RKO>
<! . . . See the companion CD-ROM for the complete template . . . >

The definitions, slots, and methods should enable the RKO both to perform functions and to trace and/or explain the functions performed. The following sections suggest RKOs for the radio skills needed for the radio-domain use cases.

8.5.3.2 Abstract RF RKOs

<RF-RKO/>: Determines from its properties that an entity is a radio frequency entity (e.g., as distinct from IF, baseband, or bitstream).

Membership: Entity has a frequency in hertz within a known RF band. Derivative KOs include IF, baseband, and speech, with characteristic frequencies in other bands (e.g., speech has a frequency of ≤20 kHz) with other properties (e.g., IF is a property of a device inside the radio).

<Radio-skills-RKO/>: Identifies <Self/> radio skills, differentiating its own from those of others.

<Physical-access-RKO/>: Associates devices with RXML categories of physical access including RF, wireline, haptic, and tactile domains.

<Knowledge-RKO/>: Lists the categories of knowledge known to the <Self/>.

<Frequency-RKO/>: Explicates the **concept** of frequency; lists all frequencies the <Self/> has experienced; ancestor of Doppler-shift RKO.

<Wavelength-RKO/>: Explicates wavelength; derives frequency from wavelength; ancestor of antenna RKO.

<Bandwidth-RKO/>: Explicates bandwidth; measures bandwidth of signal; knows all bandwidths the <Self/> has experienced.

<Canonical-RKO/>: Explicates canonical **concept** of radio communications consisting of transmitter, channel, and receiver.

<Transmitter-RKO/>: Explicates canonical transmitter; ancestor of antenna, RF-conversion, and SDR RKOs.

<Channel-RKO/>: Explicates the radio channel as signal over space, time, and frequency; ancestor of propagation modeling RKOs.

<Receiver-RKO/>: Explicates canonical receiver; ancestor of antenna, RF-conversion, and SDR RKOs.

<Sources-RKO/>: Explicates the general notion of an information source in terms of entropy; ancestor of analog and voice sources RKOs and digital/bitstream sources like keyboards.

<Sinks-RKO/>: Explicates data streams consumed by entities to transfer information; ancestor of user RKOs, not just hardware (e.g., speakers and displays).

8.5.3.3 Band and Mode RKOs

<SDR-subsystem-RKO/>: Lists all SDR resources of the <Self/> including SDR hardware, software, applications factories, and configured waveforms.

<Bands-RKO/>: Lists all the RF bands, flagging bands to which <Self/> has access.

<HF-RKO/>: Explicates the properties of the HF band, each of which is elaborated by a related RKO, such as a <HF <Propagation-RKO/> /> or a <Morse-code-RKO/>. Each of the bands <LVHF/>, <VHF/>, <UHF/>, <SHF/>, <EHF/>, and <THz/> would have its own RKO.

<Modes-RKO/>: Explicates waveforms in the public domain, highlighting the modes of which the <Self/> is capable. Each <Mode/> has its own top-level RKO with relationships to GoS, QoS, and QoI RKOs.

8.5.3.4 RF Environment RKOs

<RF-sensing-RKO/>: Lists bands, modes, and accuracy of <Self/> RF sensing.

<Signals-RKO/>: Determines whether a space–time–frequency epoch is signal or noise.

<Noise-RKO/>: Verifies the noise hypothesis or characterizes interference.

Other abstractions in the RF environment readily represented in RKOs include the <Channel/>, <Path/>, <Reflection/>, <Refraction/>, and <Doppler/> shift. A <Propagation-RKO/> would be the ancestor of RKOs for <Link-budget/> and for <2D/> and <3D/> propagation modeling. The <Signal-in-space-RKO/> could explain the channel waveform in terms of <Channel-symbol/>, <Multiple-access/>, and <Stream/>, each of which would be derivative RKOs.

8.5.3.5 Hardware RKOs

Following the pattern of relatively narrow scope for RKOs, one may define hardware RKOs for <Antenna/> and its properties <Efficiency/>, <Gain/>, <Noise/>, <Direction/>, <Position/>, <Array/>, and <MIMO/>. Similarly, the <RF-conversion-RKOs/> for the transmitter and receiver could diagnose RF conversion faults. The <Baseband-RKO/> could characterize digitized analog signals like speech. ADCs, DACs, DSPs, shared memory, and interconnect each warrant RKOs. The <Bitstream-RKO/> would explicate the **concept** of a temporal sequence of bits, ancestry for the protocol stack RKOs. <Security/> RKO could be the ancestor for devices with IA properties.

8.5.3.6 Information Service RKOs

The <Services-RKO/> lists the abstract information services of wireless connectivity and information transfer. It is the ancestor of more specialized RKOs for <Connectivity/>, <Transfer-data/> and its parameterizations in <GoS/> and <QoS/> RKOs. Since information transfer is mediated by <Protocol-stack/> with <ISO-OSI/> and <Radio-access-protocols/>, these each would be RKOs. To provide services, the iCR needs RKOs for <Skill/> areas including the ability to <Connect/> and <Disconnect/>; to <Transmit/> and <Receive/> <Voice/> and <Data/> as well as to <Bridge/> among voice and data representations. RKOs to <Plan/> and <Collaborate/> explicate relationships to information services from simple voice, data, news reports, and web browsing to conference calls and real-time collaboration.

8.5.4 Radio-Domain Heuristics (RDHs)

The AM classes of heuristic may be adapted to define radio-domain heuristics (RDHs) that assist in the autonomous evolution of RKOs via AML. Heuristics are just RKOs, the domain and range of which are RKOs and which express a rule or functional, suggest a new RKO, fill in an existing RKO, maintain the consistency of an existing RKO, or estimate the value of a RKO. Stated in RXML in a slightly more general form so that a KO-based inference engine could find the right KO using pattern matching yields Expression 8-12.

Expression 8-12 RDH Defined as a Particular Type of RKO

<RDH> <RKO> <Name> Heuristic &? </Name> . . .
<Domain> KO </Domain> <Range> KO </Range>
<Functions>. . . <Rule> <Condition/> <Action/> </Rule> </Functions>
</RKO> </RDH>

Radio-domain heuristics include rules for generating new RKOs. Since radio is a structured domain, the content of many RKOs already exist in radio engineering. For example, if an AACR encounters a novel form of voice coding, say, RPE-LTP-8 for GSM, it can get the RKO from a GSM-aware CWN. This kind of RKO is a standardized machine-readable documentation package for the 8kbps voice codec, enabling the AACR to perform the autonomous bridging use case. As with AM, the heuristics for new RKOs determine the direction of reasoning—goal driven to overcome connectivity impairments to enhance QoI.

Completion heuristics fill in empty RKO facets with advice that is precise and domain dependent. For example, a Doppler RKO with a slot for vehicle speed can be filled in by RDH advice to ask the host vehicle how fast it is going. A MIMO RDH can fill in a RKO with a slot for number of transmit elements with a query to the SDR component. AM discovery ideas apply to plug-and-play properties where increasingly specific questions are asked via RDH between collaborating software components on the same iCRs among peers, and with CWNs.

Refinement heuristics independently examine the evolving RKOs for validity, correcting facets for consistency as needed. For example, a RKO for home RF accesses is updated by an address-change RDH when a family moves. While the suggestive and completion RDHs may create and initialize the RKO, keeping it current falls to refinement RDHs.

Value heuristics estimate worth and interestingness of RKOs, typically a vector of usefulness metrics for QoI. The related RDHs embed rules for computing and interpreting worth with respect to the associated goals. Increases and decreases in worth can mediate AML reinforcement, enabling the autonomous refinement of radio domain skills. Causal reasoning research has established the complexity and pitfalls of inferring cause from observations [37]. Interesting RKOs may be examined for causality by RDHs at length during sleep cycles.

8.6 EVOLVING SKILLS VIA RKO AND RDH

This section considers the potential contributions of RKO and RDH to seamlessness by applying per band experience to tailor cross-band connectivity to the user's situation.

8.6.1 Per Band RKOs

The iCR's experience in a given band can be aggregated into a collection of RKOs.

Expression 8-13 HF RKO of a Priori Knowledge

<RKO> <Name> <HF-Radio-Knowledge-Object/> </Name> . . .
  <Example>
<Typical> <Power> 100 <Units> W </Units> </Power>
   <Antenna> Long-wire </Antenna>
   <Mode> <Voice> <Single-sideband> Upper-sideband (USB)
     </Single-sideband>
   typical-HF-usb-voice.wav </Voice></Mode>
   <Range> 3300 <Units> miles </Units> </Range>
</Typical> </Example> . . . </RKO>

The CRA <Self/> ontology includes the a priori HF RKO of Expression 8-13. This RKO says that a typical HF radio can use voice over 3300 miles radiating 100 watts into a long wire antenna, typical for an experienced ARRL Ham radio expert [207].

8.6.2 Per Band Experience

A RDH to extend this RKO embeds experience with band X into a new band-X-RKO using <Typical/> <Example/> as a template.

Expression 8-14 Radio-Domain Heuristic Creates RKO

<RDH> <Name> Band-Radio-Domain-Heuristic </Name> . . .
   <Rule> <Condition> <RKO><Name> ?<Band/> . . . </RKO>
     </Condition>
     <Action> <Create/> <RKO><Name<Self/>/>-?<Band/> </Name>
     <Example Time = <Now/> Place = <Here/> >

```
    <Power/> <Antenna/><Mode/> <Range/> <QoS/>
    </RKO> <From <Condition/>/> </Create> </Action>
  </Rule> </RDH>
```

The simplified RDH (Expression 8-14) creates a HF RKO using <HF/> <Example/> the current power, mode, range, QoS, and QoI for its current use of the HF band. This new RKO captures experience, and a related Refinement-RDH updates the HF RKO each time HF is used. Representing RDOs and RDHs in RXML enables the CRA cognition cycle to use XML tools to update self knowledge. As the iCR uses other bands, other RKOs aggregate knowledge of attempts, successes, and QoI.

8.6.3 Network-Enabled Experience

The iCRs also learn associations between uses and the associated networks (Table 8-4).

Radio network use is readily expressed in RKOs, a core of which is shown in the RXML of Expression 8-15.

Expression 8-15 RKO for PSTN Access by Telephone Number

```
<RKO> <Name> Cellular </Name> . . . <Domain> <Dialed-number/>
  </Domain>
  <Range> <PSTN/> </Range> . . .
  <Function> <Accesses> <Dialed-number/><PSTN/> </Accesses>
    </Function> </RKO>
```

This simple RKO expresses the a priori fact that when the cellular network is dialed, it accesses the PSTN. A good RDH saves dialing experiences as in Expression 8-16.

TABLE 8-4 Network Knowledge for RKO

Network	Available Knowledge	CR Uses
Cellular	PSTN access	Dial a phone number
WLAN	Internet query	User entertainment
Broadcast	Spectrum rental policy	Obtain P_{TX} limits
KNN[a]	CNN in OWL format	Obtain news briefs
NOAA	WX broadcast	Scan for WX threats
. . .	?	?

[a] KNN is a Knowledge News Network.

Expression 8-16 RKO Evolved by RDH for Remembering Experience

<Example> <Accesses> <Dialed-number/>703-555-1215<Dialed-number/> <PSTN> "Hi, Wendell, this is Betsy. . . ."</PSTN> </Accesses>
</Example>

The RDH integrated Betsy's phone number and her conversation in the RKO. Several interesting things are readily associated via RKOs and RDHs. Since iCRs remember everything, memory is organized for ease of subsequent use. When a number is dialed, the iCR remembers the entire conversation, so the next time <Wendell/> wants to call <Betsy/> he just asks.

Since the iCR is driven to help Wendell, it has processed the audio segments resulting from PSTN access, extracting Wendell's and Betsy's names. Name-finding is reliably realized by commercial speech recognition algorithms so that <Wendell/> and <Betsy/> are grounded to known persons.

Wireless use cases also are functions of place and time, so AACRs need a priori spatial knowledge of the Universe (to do it right).

8.7 IMPLEMENTING SPATIAL SKILLS

As iCRs move, newly minted RKOs keep track of which bands work best in which locations and at which dates and times. For this knowledge to be most helpful, the iCR must have a high performance capability for navigation, location finding, and reasoning about physical space. Then it could reason about space and time as a coherent whole, relating location on the way to work to time to access the corporate wireless LAN, for example. Location knowledge for spatial reasoning about radio is more than GPS navigation, a radio-oriented model of space and time based on an ontological model of space for spatial KOs (SKOs).

8.7.1 Spatial Ontology

For radio communications, physical space may be organized into a hierarchy of logical planes: <Global/> (including outer space to know about sunspots and SATCOM), <Regional/>, <Metropolitan/>, <Local/>, <Outdoor-scene/>, and <Indoor-scene/> planes. Associated with each are spatial knowledge objects (SKOs), with spatial domain heuristics (SDHs) for instantiating, populating, and managing KOs to reflect experience with its own bands and mode in those specific places. Generic SKOs represent broad knowledge of radio in space, while heuristically evolved SKOs aggregate the experience of a specific user. Computational ontologies of space expressed in CRA <Self/> RXML and implemented in SKOs enable iCRs to autonomously employ maps, charts, GPS, radio propagation, and network-derived location knowledge. In

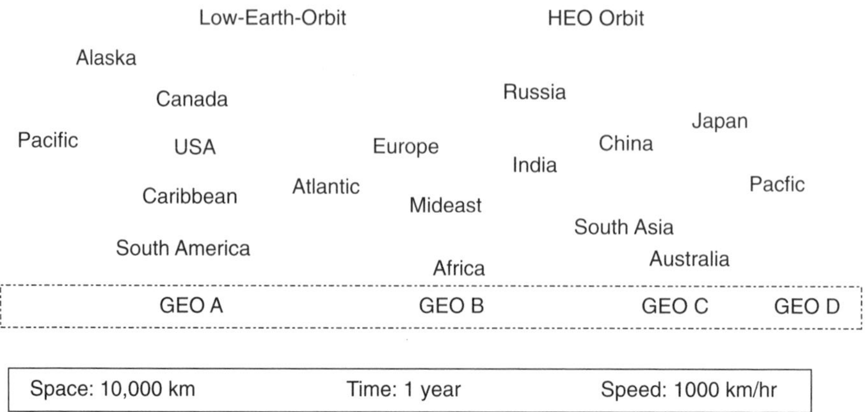

FIGURE 8-1 Spatial plane.

addition, each plane includes visual and acoustic features for <Scene/> detection and information space parameters across that plane to enhance the use of that space for QoI.

8.7.1.1 Global Plane

Figure 8-1 shows the top level of the physical-world spatial hierarchy, the global plane. Here are the telecommunications patterns that are global in scope. SKOs on this plane include global demographics like population, global connections like SATCOM, global traffic statistics, and global wireless patterns. The global plane SKO lists the regions into which this space is partitioned with the characteristic distances, speed typical when traversing this plane, and characteristic time over which patterns on this plane change significantly. These characteristics enable quick high level reasoning about the <Self/> situated in this plane, for example, making a trip that transits one or more regions.

Significant motion in this plane occurs at a rate of speed that can transit a substantial fraction (>10%) of the characteristic distance in an hour like a traveler moving at 1000 km/h (e.g., in an aircraft). In addition to the SKO's slots for aggregate characteristics, each plane describes interconnection mechanisms like global fiber and SATCOM; characteristic travel, like air, rail, or ship; annual temporal patterns, including population migration; and day–night zones. Planning SKOs include travel itinerary and critical information such as location, telephone number, Internet address, mobility, wireless access opportunities, path length, delay, data rate, QoS, and traffic density. Access to Iridium and other satellite telephones is driven by location in the global plane. The global SKO of Expression 8-17 highlights key features of this class of KO.

Expression 8-17 Global Plane SKO (Simplified)

<SKO> <Name> Global-Plane </Name>
<Definitions> <Spatial-extent/> > 1000 km </Definitions>
<Slots> <Members> <Sun/> <SATCOM/> <Alaska/> . . . </Members>
</Slots>
<Membership> <Indicators> Globe, Earth, Outer-space, Intercontinental
</Indicators>
</Membership> <Domain> Space Earth <Members/> </Domain>
<Range> <Members/> <Connectivity/> <Traffic-patterns/> <Statistics/>
<Characteristics> <Space Typical = 1000 km/> <Time Typical = 1 yr/>
<Speed Typical = 1000 km/h/> <Travel Typical = Air/>
<Interconnect Typical = SATCOM/> <Travel Typical = Business/>
<Travel Typical = Vacation/> <Space-time Typical = Itinerary/>
</Characteristics></Range>
<Examples> <Unique/> </Examples> . . . </SKO>

8.7.1.2 Regional Plane

Each partition of the global plane (e.g., Europe) has a corresponding regional plane that captures those features of a geospatial region necessary for AACR use cases (Figure 8-2).

Important properties of this region are as follows:

REGIONAL PLANE PARAMETERS

Interconnect	Fiber trunks, cellular roaming (terrestrial micro-wave or backhaul)
Travel	Commute by air (rail, ship, automobile)
Rhythm	Weekly (annual, daily), seasonal cycle of temporal patterns
Space–time	Itinerary, commuting habits, day–night boundary

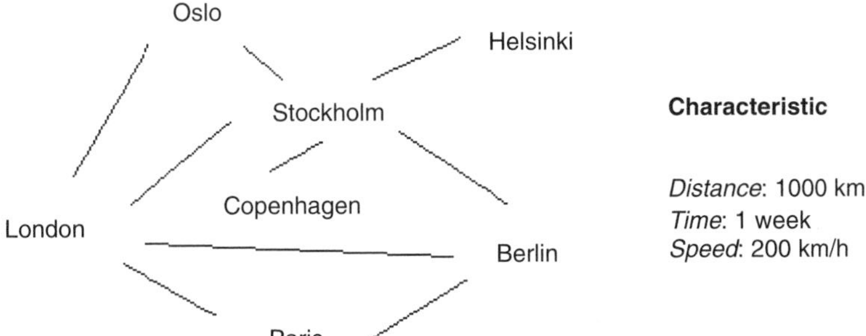

FIGURE 8-2 Regional plane SKO identifies links among metropolitan areas.

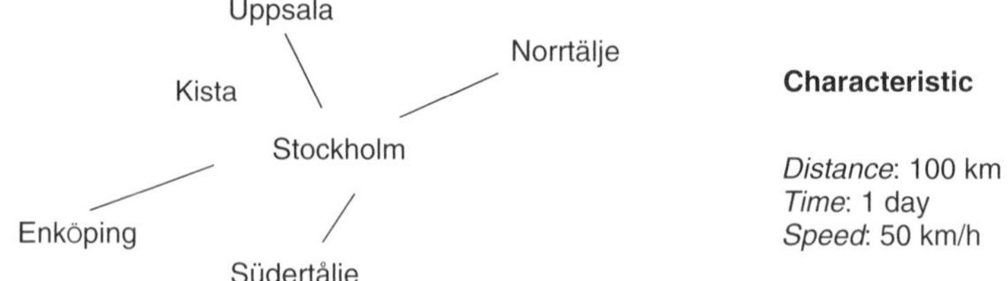

FIGURE 8-3 Metropolitan SKO describes geography and infrastructure.

Information space	Constraints imposed by geopolitical boundaries like national borders; physical barriers such as mountain ranges

8.7.1.3 Metropolitan Plane

Each region of the spatial hierarchy is partitioned into metropolitan planes that often have the greatest intensity of telecommunications infrastructure. Regions need not be metropolitan demographically: a large wilderness parkland, for example, could be a metropolitan plane. The criterion for the inclusion of a locale in one metropolitan plane is the relevance to QoI tasks of AACR, such as the inclusion of bedroom communities with the urban center. Commuters' daily patterns transverse these locations, typically by rail or automobile. Stockholm suburbs within about 200 km of the hub are included in the Stockholm SKO in Figure 8-3.

Other significant properties of this plane of the spatial hierarchy include:

METROPOLITAN PLANE PARAMETERS

Interconnect	Wireless, fiber trunks, cellular service, propagation, cell coverage
Travel	Commute by rail, automobile (air, ship)
Rhythm	Daily (weekly, annual, seasonal)
Space–time	Commuting habits, to-do list
Information space	Wide area access, best service provider

In this SKO, the quality of wireless coverage is important. Video data rates are not available everywhere and temporal patterns shift in a daily cycle driven by commuting and leisure pursuits, depending on the day of the week and the season. A commuter's normal pattern is shaped by the daily schedule, causing significant variations in the space–time pattern of demand for wireless. Visits to clients, luncheon engagements, and so on can shape the needs of wireless power users such as corporate executives. Since power users may be early adopters, their patterns inform use cases for early technology insertion.

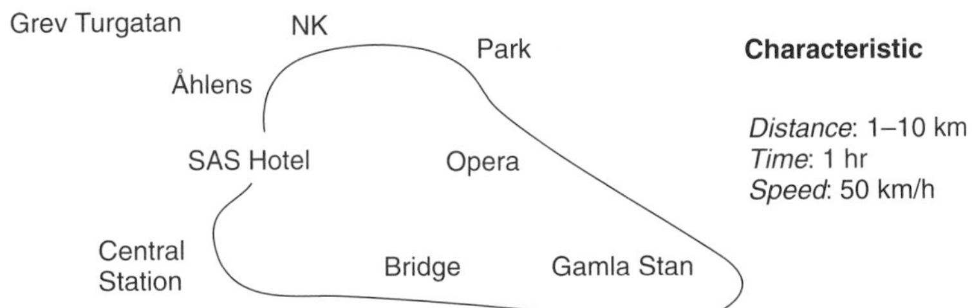

FIGURE 8-4 Local area SKO for a Stockholm experience.

8.7.1.4 Local Plane (<Locale/>)

The local SKO embraces the region that generates 90% of the radio multipath from PCS and cell towers serving typical users, spanned by a 5–10 minute walk (urban), subway, or automobile ride (rural). A subscriber's location varies over this area during the day. In Figure 8-4, a visitor to the SAS Royal Viking Hotel in downtown Stockholm takes a walk around town captured in a SKO.

The Åhlens and NK department stores within a few hundred meters of the hotel are adjacent to the large green Kungstradgarden park and walkway to the river. The Opera is adjacent to the old city of Gamla Stan with its historic restaurants and shops. The bridge connecting Gamla Stan via the Centralbron to the Central Station rail terminus offers convenient access to the SAS Royal Viking. These places are known via the local plane SKOs to contain important elements of information. To mutually ground the iCR and user in such small areas requires the correlation of visual scenes to user-intelligible maps, for example, by landmarks that are easy for the iCR video sensor to perceive reliably. Such perception assists the user in navigating an unfamiliar cityscape with unreliable GPS in Gamla Stan (Figure 8-5).

An initial SKO has a priori location of buildings, bridges, cell towers, and radio access points like Internet cafés and 802.11 hot spots. The SKO is the template from which the AACR builds a RKO and into which the AACR can integrate its own spatially registered radio experience, such as physical access to trustable hot spots. This plane contributes to radio access quality, small enough to be manageable yet large enough for normal daily and weekly patterns for a majority of users. Derivative SKOs record personal experience for introspection and QoI planning via local plane parameters:

LOCAL PLANE PARAMETERS

Interconnect	Wireless, cordless, satellite mobile; propagation—reflection, scattering, multipath; preferred bands—VHF, UHF, WLANs, hot spots
Travel	Foot, taxi, train (entry and exit points/stations) on board is a scene

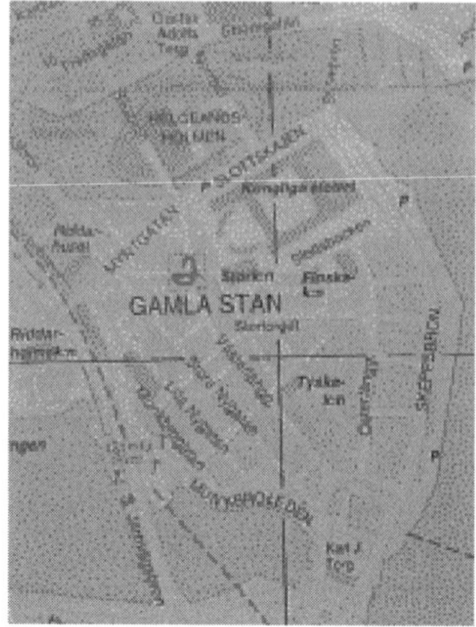

FIGURE 8-5 Spatially skilled CR understands a map of Gamla Stan.

Rhythms	Daily (weekly, seasonal)
Space–time	To-do list, meals
Information space	Local access, local coverage/gaps, kiosks

The ontological treatment of the local plane is the <Locale/>, a collection of <Scenes/>.

8.7.1.5 Outdoor Scenes

The immediate area plane has optical line of sight to objects resolved by a cell phone camera, not more than about 100 meters, as illustrated in Figure 8-6. Although radio propagation in the local plane can be modeled in two or $2\frac{1}{2}$ dimensions (x, y, and height in selected places), modeling propagation in urban areas requires 3D. A frame SDH spawns new SKOs whenever the iCR detects substantial change of visual scene. Dramatic changes of light, such as darkness and stormy weather, each spawn new SKOs that associate the visual scene with blobs, landmarks, and dead reckoning. The characteristic distance of the outdoor SKO ranges from a few meters (e.g., in an alley) to a hundred meters or more. The characteristic time scale is on the order of seconds to minutes, depending on motion on foot.

In addition, the information focus has changed as suggested by the following characteristics:

OUTDOOR SCENES ENTAIL MOVEMENT ON FOOT WITH B3G VIDEO EXPERIENCE

Interconnect	Voice, wireless, cordless
Propagation	Vehicular reflection, multipath
Preferred bands	VHF, UHF (EHF, HF)

Characteristic
Distance: 100 m
Time: 1 minute
Speed: 1 m/s

FIGURE 8-6 Outdoor <Scene/> of immediate area plane characterizes propagation in three dimensions.

Travel	Foot (taxi, train refer to inside of vehicle)
Rhythms	Hourly/momentary action needs
Space–time	Use of artifacts, dead reckoning and visual navigation
Information	Nearest infrastructure access point

Travel in this plane is limited to walking by definition. The outdoors immediate area plane is the user-centric plane. It is best described by a coordinate system that moves with the user, reflecting the immediate environment. Its SKO employs a mix of spatial models to identify opportunities to access RF LANs, future cordless telephones, and home networks when about in the yard, the "home" outdoor scene. The outdoor scene forms ontological primitive <Scene/> SKOs for the CRA <Self/>.

8.7.1.6 Indoor Scene

The indoor scene consists of the objects within 3 to 20 meters of the user as suggested in Figure 8-7. The iCR interacts with other objects in the environment such as the owner's personal computer (PC), or future smart TV, kitchen, or toaster via BlueTooth, WiFi, or an infrared port (IRDA).

Propagation effects in the indoor scene are those that change over fractions of a wavelength. This plane has the following additional characteristics:

INDOOR SCENE PARAMETERS

Interconnect	Physical contact
Propagation	Reflections from body parts, walls, furniture, appliances
Preferred bands	Infrared, optical, EHF (SHF), very low power ISM bands
Travel	Movement of body, artifacts

Characteristic
Distance: 10 m
Time: 1 μs
Speed: 0.1 m/s

FIGURE 8-7 The indoor scene SKO characterizes significant indoor objects.

Temporal patterns	Segment of momentary motion
Space–time	Fine-scale effects
Information	Very low power local exchange possible

Space–time patterns in the indoor scene are induced from the unit's own knowledge of its <Self/> approximate location (e.g., SAS Royal Viking Hotel) plus measurements continuously made by the device. If visual perception were completely reliable, then a new scene S1 is entered from S0 when all of the items of S0 are no longer observable. This is not a very good definition since the <Sky/> and the <Ground/> appear in most outdoor scenes and people move in groups to new scenes. Technical advances in perception technology are needed to recognize those things that distinguish one scene from another. In the CRA, <Scenes/> are to aggregate experience consistently. Thus, a <Locale/> consists of a set of spatially associated <Scenes/>.

Reliable definition of scene boundaries is a research area of computer vision and machine learning, so for the moment, less than ideal algorithms will identify scene boundaries with less than ideal partitioning.

This concludes the definition of the spatial inference hierarchy, enabling the exploration of the relationships among space, time, information access, and information content. The near-term use of SKOs for radio-domain competence may benefit from the idea of an information landscape.

8.7.2 The Information Landscape

Goodman [208] has been developing the concept of "The Geography of Information," referred to as an information plane. The spatial distribution of information has been a feature of human experience since the dawn of history. To see pictures of animals, cave dwellers visited a cave in France. To learn from books, ancients trekked to the library in Alexandria, Egypt. The loss of such information geography is a product of electronics that represents, transmits, receives, and stores information without the sophistication to create

TABLE 8-5 Goodman's Strategy for System Alignment to a Geography of Information

Plane	Criteria
Information	Urgency, localization, users
Environment	Signal propagation, network activity
Terminals	Location, motion, power supply
Network	Cell LAN, ad hoc, infrastructure

easily perceived landmarks for the new information geography. For AACRs to employ the information landscape autonomously requires the translation of artifacts from human accessible to computationally accessible forms. Let's relate Goodman's ideas to AACR through the integration of spatial and radio skills.

Goodman's strategy for expressing and exploiting the geography of information is based on the spatial location of signals, users, and information. He examines information, environment, terminals, and networks as summarized in Table 8-5.

Goodman's information plane characterizes the urgency of delivery, which is a social judgment, the localization of the information, which depends on caching, and the users, which are entities in the information plane. Social relationships imply needs and uses of information in an environment where faults or voids limit the geography of the information plane the way a river limits the geography of a city. Without a bridge, communities across the river from each other are isolated. Goodman seeks information, environment, terminals, and networks that enable users to perceive and manipulate the geography of information.

For an AACR to perceive a geography of information, that geography must be parameterized numerically and expressed in workable computational structures accessed by the ontological primitive <Information-landscape/>, a projection of features onto a spatial map, in RXML a projection from <Physical-space/> to <Abstractions/>.

Expression 8-18 The Information Landscape

```
<Information-landscape/>
  <Function> <Projection/>
    <Domain> <Physical-universe <Physical-space/> /> </Domain>
    <Range> <Abstractions> <User <Information/> />
      <RF-environment> <RF <Metrics/>/>
      <RF <Hardware <Devices <Terminals/>/>/>/>
      <RF <Systems <Networks/> /> /> </RF-environment>
    </Abstractions> </Range> </Projection> </Function>
    </Information-landscape>
```

<User <Information/> /> mapped against <Physical-space/> forms associative pairs like (<Here/>, <Owner <Information>> Income-tax-data </Owner </Information>). If <Here/> is a latitude–longitude pair, then the associative relationship (Where is the income tax data?) is not as helpful as if <Here/> refers to the <Desk/> drawer or file on the laptop. So there is a hierarchy of qualitative locations that makes the geography of information helpful. A particular latitude–longitude pair associated with a <Locale/> called <Home/> has both outdoor scenes where GPS can measure latitude and longitude and interior scenes like the basement where the <Desk/> is located. The presence and features of devices like <Desk/> that store information may be clustered into the <Home/>-SKO as a node in the user's information landscape, with devices and networks merely a means to an end. The income tax data could be accessed via WLAN from the laptop, so, the information landscape expresses the role of the devices, networks, and radio environment in accessing the income tax data (Expression 8-19).

Expression 8-19 Points in the Information Landscape Learned Via Experience

```
<Information-landscape> <Experience> <Now>
  1/28/2005 1:37:14 PM </Now>
  <Physical-space> <Here/> <Home><Basement><Desk/>
    </Basement><Home/> </Physical-space>
  <User <Information>>
  <Owner <Information>> Income-tax-data </Information </Owner>>
  </User </Information>>
<RF-environment>
  <Networks> IEEE-802.11-WLAN </Networks>
    <Metrics> <RSSI> <Strong/></RSSI> </Metrics>
    <Hardware> <Devices> <Terminals>
      <Laptop> Big-blue-Model-802 </Laptop>
      </Hardware> </Devices> </Terminals>
  </RF-environment> </Experience></Information-landscape>
```

The specific experience of Expression 8-19 formalized as a <KO/> could be an information landscape knowledge object, <ILKO/>. A proliferation of many kinds of knowledge object is confusing, so the sequel limits KOs to just three classes: RKO, SKO, and user KOs (UKOs). Other classes are hybrids of these orthogonal classes.

8.7.3 Constrained Information Landscapes

AACR navigation skills typically will include satellite location tools with electronic street maps of buildings, waterways, and points of interest sufficient for a user but insufficient for the <Information-landscape/>. The physical location must be augmented from the AACR's and user's viewpoints.

Expression 8-20 Gamla Stan SKO (Simplified)

```
<SKO> <Name> "Gamla Stan" </Name>
  <Definitions> <Physical-space><Physical-universe><Locale> Gamla-
    Stan </Locale>
    <Metropolitan> <Stockholm> <Metropolitan/> <Region><Europe/>
      </Region>
    <Global> <Earth/> <Global/>
    </Physical-universe> </Physical-space> </Definitions> </SKO>
```

The Gamla Stan section of Stockholm is part of Europe, which is a part of the globe in the physical model of the world. The tag <Physical-space> constrains Gamla Stan to be located in 3-space (i.e., it is not an abstraction like Santa Clause, which is real to the <Self/> as an abstraction of giving). The AACR encountering Gamla Stan for the first time may access as a priori SKO with landmarks, street names, the authority of the information, constraints on spatial relationships, and other characteristics that help relate its experience to features of the location. This section identifies the important constraints to the information landscape that must be known in order to enable well-grounded aggregation of experience for use of that landscape.

8.7.3.1 Constraining the Source

The source of the knowledge about Gamla Stan is not explicit in Expression 8-20, but in order for an AACR to reason about conflicting information, it must associate a source with each KO. For example, the AACR might learn that there are no public WLAN access points from the following source:

```
<Source>
  <Place> "Fairfax, VA" </Place>
  <Time> 199905121605 </Time>
  <Author> "J. Mitola III" <Author/> </Source>
```

Some might trust this source, while others might want a more current, more authoritative source. If another source says that there are indeed public WLAN access points on Gamla Stan, dated 200501041610, then which would you trust? Suppose another source later says that there are none because the network is down. Yet another source said five minutes ago that the WLANs are all up. Now which one do you trust? Reasoning over such conflicts requires the codification of the idea that the most recent relevant experience is the one to be most trusted. This idea isn't hard to implement computationally, so there is a SDH, advising that the most recent status report be believed:

```
<SDH> <Name> <RF <Network <Status/>/>/> </Name>
  <Rule> <Condition>
    <Status> S1 <Time> X </Time> <Place> L </Place> </Status> &
      <Status> S2 <Time> Y </Time> <Place> L </Place> </Status> &
      Y>X
       </Condition>
    <Action> <Believe> <Status> S2 </Status> </Believe> </Action>
```

If location L is Gamla Stan, then the heuristic guides the radio to believe S2 (e.g. WLAN is up) in Gamla Stan. If later AACRs report the WLAN to be down, then the <Self/> believes the new status.

8.7.3.2 Collaboratively Exploiting Spatial Knowledge

Of course, GPS receivers enable the rote services of displaying the location to the user and plotting the position on a map. An AACR autonomously uses location, for example, to check its equalizer with a rule such as the following PROLOG rule from Kokar's ontology-based radio (OBR) [88]:

```
checkPerformance(X)  :-equalizerError(E),
  pv(obr8upperPerformanceThreshold,
  object8MonitorServiceDispatch, UPt),
  pv(obr8lowerPerformanceThreshold,
  object8MonitorServiceDispatch, LPt), compare(‘>’, E, LPt),
  compare(‘<’, E, UPt), assign(‘Continue ’, X).
```

This PROLOG rule advises the OBR that if its equalizer error E is out of bounds, it should do something about it. Since this rule may fire at any time, the reaction of negotiating equalizer training sequences with the host wireless network may be initiated at any place or time. In Kokar's scheme, the OBR may try to fix the performance problem by switching to its most computationally expensive and best performing equalizer, but if that fails, it tries to negotiate a longer training sequence with the host network. Suppose the host wireless network knows how to embed a longer training sequence in a GSM burst but has not recently been asked to do this. It could take minutes before the enhanced sequence mode is available, rendering the strategy ineffective. Even if the network has such a mode ready to initiate when asked, a few frames of delay may render the call disconnected, defeating the strategy.

Now suppose that the OBR also has spatial knowledge. When it queries the CWN for directions from the Royal Viking to dinner in Gamla Stan, it detects a bridge crossing plus water as Rayleigh fade problems across the water [170]. It may find that other AACRs experience dropped calls unless they use the extended training sequence—via SKOs that reflect this experience. If the AACR installs the extended training sequence at the Royal Viking, then it wastes bandwidth and power for no benefit. Space–time planning over the SKOs can infer this implication of changing to the extended training sequence too soon. Although Kokar's OBR did not express the space–time constraints needed to generate such a plan, an embedded power budget instantiated with the cost per bit of the normal equalizer and the extended equalizer could estimate the power drain and loss of bandwidth associated with early employment of the extended mode. In addition, the CWN would not want to pay the penalty of lost network bandwidth unless the specific circumstances demand it.

The location-adaptive AACR can generate a collaborative plan with the CWN to enable the long training sequence when the AACR approaches the bridge to Gamla Stan, the Old City. When it finds itself actually in transit to Gamla Stan and on the bridge, then the CWN is ready to initiate the remedy proactively, avoiding dropped calls. If the network isn't particularly busy, then the resources could well be available, and the incremental revenue certainly would be welcome. The ability to reason about space, time, and radio performance with collaborative planning enabled by SKOs differentiates AACR from reactive OBR, which is a great step in the right direction.

8.7.3.3 Refining Location Skills in Outdoor Scenes

Suppose the AACR is in the Washington, DC area. The CWN has generated the maps of RSSI as a function of distance from a particular cell tower, as in Figure 8-8a. The <SKO> for the Washington, DC area includes the underlying data set, an array, and the visual representation, the <Graphic/>. The slice of RSSI represents a path that the AACR could take through the city. If the RSSI map were generated initially by a radio propagation tool, then the validity of the values might not be known. Values updated based on the experience of AACRs in the environment retain their validity subject to major changes in the scene, such as the construction of buildings, changing air traffic patterns, and diurnal variations in vehicular traffic. These dynamics are typical of the <Outdoor/> <Scene/>. As the user rides in a vehicle, the location and

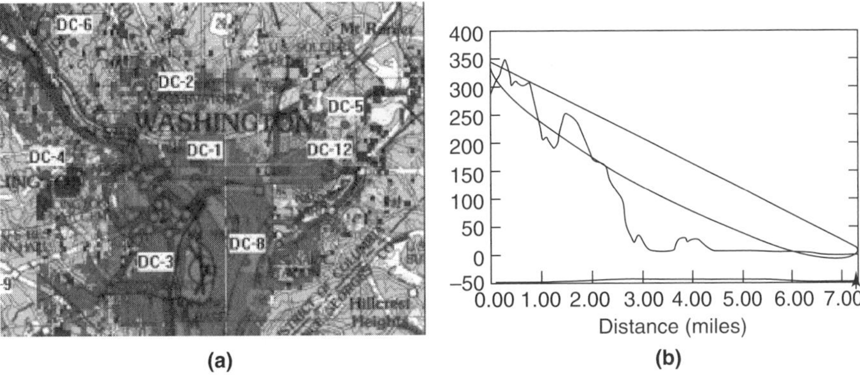

FIGURE 8-8 Spatial map of RSSI. (a) Two-dimensional representation of RSSI array. (b) Slice representing a path. Courtesy WrAP@used with permission.

FIGURE 8-9 Royal Viking Hotel entrance.

average RSSI change quickly, while when not moving, RSSI changes slowly because of RF dynamics. A path through a <Locale/> like a commute consists of an associated sequence of <Outdoor/> <Scene/>s. The <Locale/> SKO associates RF experience with these <Scenes/>.

8.7.3.4 Refining Radio Skills in Indoor Scenes

In addition to Rayleigh fading over water, other types of locations suffer from chronic radio channel impairments. Physically constrained spaces like Figure 8-9 set up interference patterns that distort radio reception.

The AACR that uses its cell phone camera to perceive such situations can associate its radio experience with the visual scene, such as the broad, wide, and low hotel entrance. A radio entering such a scene may know from sharing SKOs with other AACRs that there will be no reception during the transit under that entrance. It could ask a nearby AACR to act as relay during that transit, possibly using an unused TV channel as an ad hoc band. The <SKO/> needed for this function includes three scenes. One <Scene/> is the one shown with the AACR worn by its user looking at the entrance, but not yet severely affected by the propagation anomalies under the overhang. Another <Scene/> <SKO/> would be on the exterior. Both of these would show strong RSSI. The third would be the <Typical/> <SKO/> recorded actually in position under the overhang, where the RSSI is not zero, but only an extended mode of the extended equalizer can sustain information transfer.

8.8 GENERALIZED <INFORMATION-LANDSCAPE/>

SKOs and SDHs could aggregate experience to characterize the <RF/> aspects of an information landscape, but they should also associate <User/> <Information/> for which a <Need/> is expressed on the <Information/>

plane. Distance between user and information may be measured in time to access the information, one of the QoI parameters defined above. Reducing information (content) access time often enhances QoI. QoI may be low because of a region near a building where the cell towers are shadowed. One approach to improving the information landscape (an adaptation of Goodman's ideas) might be the bridging of faults in the access geography. The iCR empowered to assist nearby cell phones in need could relay information to shadowed users, making the landscape flatten. If the price to be paid by the user doing the assistance is workable, then all benefit by helping each other. In such an interaction, the signal propagation of the environment plane, the power available in the terminal, and the limitations of the networking plane all play a part.

8.8.1 Shannon Information

One challenge in discovering the connected information landscape may be to transition from a focus on data transfer to a focus on information transfer. Shannon considered strings of symbols from alphabets. If there were S different strings, then each of those S possibilities could be represented in $\log(S)$ bits, with base 2 logarithms. If a source produces these strings, distributed with probability $P(s_i)$, then observing symbol s_i supplies information $I(s_i)$ that is the number of bits required to express the inverse of that probability:

$$I(s_i) = \log(1/P(s_i)) \tag{8-9}$$

Shannon then went on to publish his noisy-channel coding theorem and many other distinguished works [209]. The noisy channel can transfer information without error if the speed is consistent with the channel capacity. Communications systems engineers know all the rest, but rarely employ Equation (8-9) to enhance information as content, not just as data. The noisy channel conveys information, not just data. An image to be transferred from one cell phone to another contains information. A JPEG-coded image can't be meaningfully decoded until the entire image is present. The image as a block of coded data has no explicit information content since it has been encapsulated into opaque bits. When decoded into a display, it becomes meaningful information to the user. The same wavelet-coded information conveys the underlying shape of the image in the early block transfers, and this may be decoded incrementally, shape first; and details may be filled in subsequently. Some forms of communication thus exhibit the strong statistical structure (information) of the content of a message payload while others do not. One generalization of <Information-landscape/> is to make the statistical structure of information content explicit in sharable SKOs.

The strong statistical structure of information expresses itself not just as the probability density of bit patterns, but also as the discoverable nonuniformities that shape the human and iCR experience of information. The

<Histogram/> was selected as the algorithm for the introduction to radio-oriented machine learning because the <Histogram/> estimates the information content as a probability density function in the simplest possible way—just count occurrences. The <Typical-example/> in Lenat's AM **concepts** and thus in the RKO tableau is defined simply in terms of counting. Specifically, the example that has the most counts has occurred the most in the entity's experience and therefore is the <Typical/> example. In the absence of traumatic shock to sensitize the memory, we remember things as typical that occur many times.

8.8.2 Information, Memory, Experience, and Skill

There are two kinds of information to which the information landscape must facilitate access: the typical and the atypical. Human memory seems intriguingly nonlinear. We often remember a single experience vividly, but sometimes cannot remember regular experiences very well (What did you have for breakfast today, anyway?). Neurophysiologists now know much about how the neural pathways of memory are wired. The molecular basis for memory and learning includes the chemistry of neuronal development and adult plasticity [210]. In traumatic situations, plasticity is a function of the concentration of biochemicals like adrenaline and their cocatalytic associates [211]. The fine structure of biological memory is reflected in the macroproperties of the mind. Jerry Fodor's classic monograph *Modularity of Mind* [212] formulates memory as belief. He also posits mental skill development as the work of repetition that sublimates steps of conscious awareness into fluid, modular, unconscious skills. The biochemistry seems to justify this philosophical view with neuronal plasticity and response to repetition. Athletes subsume the tedious details of training ("Keep your eye on the ball") with experience (Stance, pitch, swing, hit—or miss) into modular skills that their own introspection does not penetrate well. The information landscape, then, should extract spatial structure from experiences to SKOs and sublimate procedural details away from the user and the <Self/> with repetition.

We seem to naturally gravitate toward doing the most rewarding things most often, doing repeated things better than things we have never attempted before. Some of us also like novelty, like the 10% of ants in the ant hill that don't follow the pheromone trail and thereby save the colony from extinction walking down the snout of the anteater. Life seems to thrive on such entropy peaks and valleys. Thus, the domain heuristics of CRA <Self/> focus on remembering all experiences in a way that makes it easy to retrieve both the most typical and the extreme atypical examples. So the most common experiences are flagged by domain heuristics as <Typical/> for indexed access. In addition, the CRA postulates embedded associative memory exemplified by the HashMaps of CR1 that include not just serModels, but also the counting of repetitions of stimuli and their hierarchical associations into SKO scenes, computational frameworks for information landscapes.

8.8.3 The Statistical Structure of Content

Zipf counted words in *The New York Times* in the 1940s [11, 213, 214]. He found that a single word—the word "the"—accounted for nearly a tenth of the words in the corpus, with the next most common word occurring 1/10th as often and 1/10th of the words occurring only once, supporting the hypothesis that the product of the number of occurrences n times the square of the relative frequency of occurrence f^2 is nearly constant. In addition, the product of the rank order r and the relative frequency of occurrence f of the word is nearly constant [215]. The product form is a fractal distribution, a distribution in which the product of size and an exponential of size is approximately constant [216]. Although there are exceptions, the pattern is so consistent that it can be used like <Histogram/> to detect <Interesting/> information. The fractal structure of language is expressed on three levels. The structure of language dominates, comprising one fractal with "the" as the most frequently used English word. The second fractal distribution may be associated with a domain. The term "radio" occurs with dramatically more regularity in communications about radio than in general communications, for example, in *The New York Times*. Finally, each component of communications such as a book, technical paper, or paragraph each exhibit a topic, the keywords of which occur with Zipf-fractal relative frequency compared to those of the domain. Text retrieval systems use such statistical properties of text to improve precision and recall, and to estimate the relevance of a document to a query. Thus, the AACR may use <Histogram/> to extract content <Domain/> or <Topic/> to characterize the information content of a <Scene/>.

The CRA <Self/> offers <Histogram/> and domain heuristics to estimate those statistical trends. Suppose then that the SKO for the scene accumulates a <Histogram/> of the words of the dialog, less the <Structure-of-language/> words. Let <Dialog/> be the histogram of the content words. The following domain heuristic then applies.

Expression 8-21 DH Flags the Most Frequent Word as the Topic of Conversation

```
<DH> <Name> Topic-clustering </Name>
  <Function> <Rule>
    <Condition> <Histogram> <Domain> <Dialog/> </Domain>
      <Range> D </Range> </Histogram> </Condition>
    <Action> <Topic> <Extract> <Most-frequent> D </Most-frequent>
      </Extract> </Topic> </Action> </Rule> </Function> </DH>
```

The DH simply flags the most frequent non-language-structure word of the dialog as the <Topic/>. Computational linguists and information retrieval experts improve this result via metrics and rules for identifying topics [217], but the topic-clustering heuristic in Expression 8-21 is surprisingly effective.

Huffman coding assigns the shortest codes to the most common strings of a communication [218]. So although the statistical structure of text communications often is not made available to the control system of the radio, it often is computed in the process of coding communications for efficient transmission. <Histogram/> operations on the speech phrases in a scene reveal the statistical structure of <Topic/>, enabling the iCR to associate dialog topic with location, building the information landscape in SKOs.

8.8.4 Associative Information Landscape

On the global plane of the spatial hierarchy, the user is one of millions of others moving from place to place. The global patterns are represented in statistical aggregates. A given AACR and user interact with the global level of the information landscape in unique ways. For example, a travel itinerary may define sequence of <Scenes/> across regions of the global plane such as airports, hotels, car-rental lots, and the like. Some AACRs may experience the global plane regularly because of the travel patterns of the owners, while others may not experience them at all. The SKOs reflect these differences as the AACRs extract information content, associate it with radio tasks, and employ it in a timely way.

Table 8-6 shows a flow of associations made through the heuristic evolution of SKOs, populated with associations of the <Information-landscape/> discovered while interacting with <Global/> plane <Scenes/> at Dulles Airport. Initially, the AACR knows only the most generic facts about airports. When approaching Dulles for the first time, SDH retrieves a new SKO for <Dulles/>, which makes the transition from a conceptual <Abstraction/> to the <Self/> to an experiential SKO. Entering the airport, the AACR detects "Check-in" as a topic of conversation, so it seeks wireless and check-in, finding new wireless check-in kiosks similar to today's mechanical kiosks where passengers must physically enter credit cards and physically push buttons. The AACR advises the user that WLAN check-in is available with directions to the WLAN kiosk. One check-in topic is seating, so the AACR generates a message on the WLAN to change the seats so the owner and spouse can sit together

TABLE 8-6 Illustrative Information Landscape Associations

SKO	Scene	Topic	Association	Employment
<Airport/>	Dulles	Check-in	Kiosk via WLAN	Direction to counter
<Tickets/>	Ticket counter	Seating	Seat-change message	Change seat
<Tram/>	Tram	Directions	Route plan from WLAN	Direction to gate
<Gate/>	Gate	Lunch	Wireless credit card	Pay for lunch

in an exit row. The AACR notes physical location and information function of the WLAN, of a new wireless credit card that enables it to pay for the owner's lunch, engaging in a visual dialog on the screen about a 15% tip. Since lunch conversation had included "Great Food," the CR proposed a 15% tip. Had it heard "Awful" or "Yuk," it might have proposed less.

So the SKOs and SDHs suggested above could enable the AACR to address the when, where, how, what, and who of Goodman's geography of information in a personal way. SKOs store experience with time tags to isolate current, recent, and historical aspects of the information landscape. In addition, the CRA's <Plan/> phase enables the AACR to project experience into hypothetical SKOs for plans.

With the associative skills suggested above, AACRs will readily populate SKOs with metrics such as availability, GoS, QoS, and QoI. Relevant research includes Goodman's work, the CAHAN project of NJIT for efficient peer routing to overcome dead spots, and the 7DS project of Columbia University (peer-to-peer local data sharing optimizing power conservation).

8.9 MICROWORLDS

The SKOs and associated SDHs and experience must be organized so that algorithms can access relevant information quickly and can respond appropriately to stimuli from either the user or the radio environment. Where those responses are preprogrammed into protocol stack or graphical user interfaces, they may be used as they are, but where the situation is in some ways novel, the generation of responses may be combinatorially explosive. The microworld offers one way of managing the combinatorial explosion. A microworld is an association of objects that integrate SKO, RKO, and UKO by <Scene/> and other relationships.

8.9.1 Defining the Microworld

This section provides an overview of microworlds [219].

A microworld consists of those computational components necessary for a computational entity to perform tasks in a narrow task domain such as updating a schedule, planning a trip, or buying groceries. Microworld components (Figure 8-10) include collections of formal and informal inferences drawn from models contained in the microworld.

8.9.2 Skill Components

The knowledge codified in a microworld includes preprogrammed and learned aspects aggregated into KOs by DHs. Table 8-7 contrasts types of domain knowledge to be formalized alternatively via an air interface or in the related microworld.

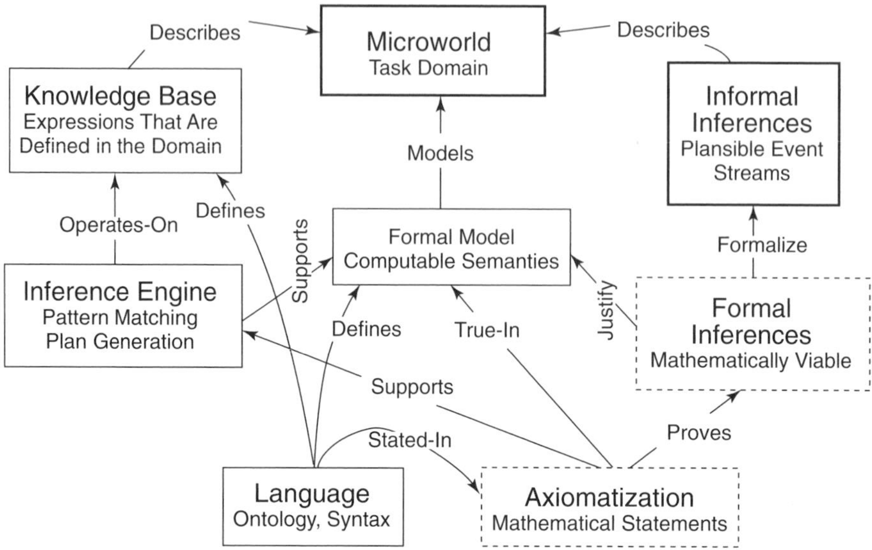

FIGURE 8-10 Skill may be organized in microworlds.

TABLE 8-7 Microworld Knowledge and Skill

Knowledge	Air Interface Component	Microworld Contribution
HF has unreliable channel	ALE probe signals	When to use HF
SHF has narrow beams	Pointing/tracking algorithm	Use SHF for wideband LoS
GPS provides location	Mapping signal to coordinates	GPS is unreliable in urban canyons
FM broadcast digital labels	How to decode the label	What "news" <USER/> likes

The fact that HF has an unreliable, fading, and dispersive channel is well known to radio engineers. The ALE protocol contains probe signals to find HF channels connectable between two points through the ionospheric skywave. A military SHF data link may track an unmanned aerial vehicle (UAV) for battlefield surveillance. The soldier using the system lacking prior SHF line of sight (LoS) pointing and tracking experience may not be aware that the link will be lost when the UAV flies behind the hill. This knowledge may be embedded in the UAV mission planning system, but if the soldier is not a regular user with no authority to plan the UAV mission, that software may not be available. An iCR could explain why the link was lost so that he could drive his Humvee (HMWWV) up the hill to reacquire the link. Table 8-7 also suggests similar examples from GPS and FM broadcast.

8.9.3 Radio Microworld

The microworld of radio consists of the following:

1. *Task Domain*: Radio services.
2. *Informal Inferences*: Plausible event streams that describe the services domain.
3. *Formal Models*: RKOs with computable radio-domain semantics of the CRA <Self/>.
4. *Formal Inferences*: Mathematically viable statements that formalize informal inferences such as RDHs fired during the cognition cycle.
5. *Axiomatization*: Mathematical statements not necessarily in FOPC that substantiate formal inferences and are <True/> in formal models.
6. *Knowledge Base*: Including RXML with embedded rule bases (If–then or PROLOG) that are computable in support of behavior.
7. *Inference Engine*: With novelty bindings when it encounters new things and with phase operations for the cognition cycle.
8. *Language*: Ontology, syntax, and semantics that define formal models and knowledge base.

Fully developed use cases capture the informal inferences that underlie the plausible event sequences of the tasks defined in the use case. The formal model uses the RKO structure and the cognition cycle. The formal inferences of the radio microworlds are implemented in the CRA and the <Self/> via RXML and exemplified in the PDANodes of CR1 of the companion CD-ROM.

8.10 RADIO SKILLS CONCLUSIONS

A cognitive MBMMR incorporates RF-band microworld knowledge, employs that knowledge to sustain connectivity, and resolves knowledge and skill conflicts to enhance QoI without annoying the user. The classical radio engineer's view formulates the radio as a bit pipe. The AACR view offers continuing evolution toward self-aware communications assistant with the AACR that knows that it is a radio, knows its user(s), and learns to continually enhance the user's QoI experience via the CRA, the <Self/>, RXML, KOs, and DHs introduced in the current and previous chapters.

One near-term implication is that the FCC vision of cognitive radio for secondary spectrum use can be built with conventional SDR augmented with appropriate RKO, SKO, and DH rules. RKO–SKO pairs could express both the broad permissions and constraints of the XG policy and the implementation constraints in a given region or locale. A regulatory policy microworld configured from such KOs could address the entire breadth of regulatory policy configured in a CWN policy server.

To continue evolution of AACR toward the user-aware RF assistant requires further attention to sensory-perception for autonomously inferring the user's needs. User knowledge objects (UKOs) with UDHs facilitate the evolution to a user domain with greater degrees of uncertainty that require additional techniques for managing complexity.

8.11 EXERCISES

8.1. The use of a spectrum management database for CR applications requires <Self/> knowledge. Write RXML that enables an AACR to know that it is a Forestry radio. Suppose the AACR were not initialized with knowledge of Forestry. Define a method by which it could find out. Train CR1 with the first 10 elements of the Canadian spectrum management database from the Web. Teach it that it is a Forestry radio. What kind of internal association facility does CR1 need to reply that the <Self/> is authorized to use one of the Forestry radio frequencies if you ask it "Can we use 27.790 MHz?" Expand CR1 using the PDL DL to respond accurately to the question about Forestry RF.

8.2. Can you find anything on the Web regarding bootstrapped heuristic learning of the kind explored by AM? Start by trying to find citations to AM and threading the trail through its successors. What about Eurisko? What were you able to find? Explain its implications to AACR. To what degree have the foundations of AM's *concepts* been reflected in open CYC?

8.3. Find references on the Web to Rule XML or RuleML, SWRL, and WebRL (WRL). What standards have the best support? Use a RuleML to generate your own RDH.

8.4. Write a stand-alone program to summarize the strengths and weaknesses of radio bands from RKOs. Write the following RKOs from which to generate the summary in response to either a general query or a specific query beginning with the phrase "Which radio band . . ." explaining that HF propagates readily over thousands of miles at low data rates given a suitable antenna, that LVHF propagates well in rough terrain and wooded areas as well as suburban settings, or that VHF offers more coherent bandwidth but offers somewhat less propagation beyond LoS than LVHF. Develop RKOs from AML RDH that reflect UHF is the TV, aircraft, and cellular radio band among other things; that SHF marks the transition to the higher capacity, more directional LoS bands; and that EHF pencil beams enable low cost spatial sharing of the radio spectrum while its high carrier frequencies support gigabit per second data rates if it isn't raining and if the beam is pointed in the right direction. Alternatively, postulate a SATCOM RKO that enables distant parties to communicate with each other and with the PSTN provided antenna pointing and the weather cooperate.

8.5. Complete the per band RDH that instantiates a personalized RKO per band if one does not exist. It must check that the <Self/> has no personalized RKO for that band. What else must it do? Write a stand-alone program to interpret this RDH to generate a new RKO when a simulated iCR first tunes to that band.

Use it to generate HF, UHF, and SATCOM RKOs from the generic ones of CRA <Self/>. Augment this program to update the appropriate RKO when the radio tunes to the band subsequently. Augment this program to flag the most common example as <Typical/>. Test it by using several different modes more than the others to confirm that the <Typical/> flag is removed from the previously typical usage. Under what circumstances would you not want to remove the flag?

8.6. Complete the definition of a global plane SKO that enables a complete international air travel thread from the planning stage to showing the pictures to the family and distributing a few gifts upon return. Define SKOs for travel planning, travel to the airport, airport, on-board the outbound leg of the trip, arrival at a foreign airport in a country that does not speak the native tongue, travel by taxi to a resort hotel, staying at the resort, taking a side trip to a local museum, attending a local sporting event, buying souvenirs, returning by taxi to the same airport, traveling on the return leg to the departure airport, picking up the car, driving home, sleeping, and meeting the family at home at noon on the next day for show and tell. Postulate the radio network aspects of the information landscape of these SKOs. Define a SDH to autonomously acquire and evolve SKOs and RKOs to aggregate the activities of the trip into a coherent memory. Write a stand-alone application to apply the SDH to the SKOs to autonomously evolve them from simulated stimuli observed during the hypothetical trip. What additional kinds of machine learning would reduce the amount of RXML needed in the initial SKOs? What additional kinds of machine learning would reduce the SDH? Is it possible to eliminate the SDH with AML? Implement this AML in Java.

8.7. Consider the U.S. FCC's plan for cognitive radio to facilitate markets for the secondary use of allocated radio spectrum. Find the FCC's NOI, NPRM, and other documentation on the Web. Find at least three associated briefings or technical papers supportive of the FCC plan. Define a RKO that expresses the top-level strategy of permitting ad hoc networks to be formed in unused TV channels. Write skeleton SKOs for a <Locale/> and for some simulated <Scenes/> that describe a priori knowledge about the physical space in a metro area. Define a sequence of RKOs by which an AACR can aggregate experience regarding the use of TV bands in a <Metropolitan/> plane. Write a SDH by which the AACR adapts these <Metro/>, <Locale/>, and <Scene/> SKOs to use the RKO to implement the FCC policy. Identify the challenges of networking in such an arrangement. Find technology on the Web that addresses these challenges. Describe how to define, develop, and configure radio and spatial domain software to address the challenges sufficiently to initiate secondary use services.

8.8. Define the microworlds of Exercise 8.6. Suppose the user travels a lot. How would the definition of microworlds change? Suppose the user takes a trip overseas every month. Describe how repeated exposure to the same experiences creates Zipf-like indications of <Topics/>. What classes of unique events should be readily accessible for future use? Implement a stand-alone CBR program to bind new experiences to frequent, infrequent, and rare experiences of the global plane's hypothetical information landscape. Enhance this program to reflect wireless information kiosks at hotels, restaurants, banks, and museums.

8.9. Using the KO template of CRA <Self/>, define a user KO, your own idea of UKO. Write UKOs for the Genie use case. Write UKOs for the Bert-and-Ernie use case. Write a UDH for acquiring user preferences for travel on foot and by vehicle. Explain how the UKOs would be evolved by this UDH. Write a stand-alone application to evolve initial UKOs into ontological skills via the UDH.

8.10. What complexities of real-world KOs and DHs have been ignored or brushed over in this chapter? Address them. ☺

PART III

USER-DOMAIN COMPETENCE

CHAPTER 9

USER-DOMAIN USE CASES

The use cases of this chapter motivate the inclusion of visual and audio-visual sensor perception into the cognitive radio architecture (CRA). The diversity of user templates motivates further AML and uncertainty management technologies in the CRA. The iCR should save its user substantial work by performing increasingly complex home, office, and leisure tasks via trustable wireless services as directed by the user, but mostly autonomously. It should make life more interesting by autonomously looking up that flower in the garden of Hampton Court with interesting horticulture lineage. The chapter begins with a study of a near-term AACR assisting in maintaining personal safety in an emergency situation and concludes with a conceptual treatment of far-term iCR functions.

9.1 EMERGENCY COMPANION USE CASE

Consider an AACR that monitors the owner's vital signs. Niche products already monitor and report blood pressure, heart rate, and respiration. Postulate a wardrobe in which the user wears the sensors and the wearable AACR itself autonomously monitors the readings for conformance to the user's historic norms and excursions. This AACR might use PROLOG rules to monitor these vital signs analogous to Kokar's rules for monitoring the adaptive equalizer. When bounds are out of range, the AACR takes appropriate action. Suppose now that an earthquake has left the owner injured in the rubble. The AACR detects the owner's elevated heart rate and labored breathing out of normal ranges, so it attempts to report the emergency but finds that the cellular network is down. The conventional emergency radio just keeps trying, draining the battery before help arrives.

The iCR explores alternative behaviors. First, the iCR's audio and video sensory perception will register the sights and sounds of an earthquake. Suppose the user was at work in a midrise office building when the earthquake struck. Previously, the iCR had developed and reinforced many SKOs in the daily work routine. Suddenly, the parameters of the visual scene become chaotic as the iCR encounters its first acoustic experience of a <Scream/>. In the CRA cognition cycles following the <Scream/> the iCR instantiates a <Personal-emergency/> UKO hypothesis for the chaotic audio-visual scene. The a priori UKO advises the iCR to use ASR to monitor police, fire, and rescue radio channels for indications of a larger emergency. This UKO enables the iCR to reliably determine whether the user's vital signs are abnormal because of personal trauma like falling down stairs, or an event that includes other people, like an earthquake, fire, or flood.

This proactive iCR monitors the <RF-environment/> to discover changes in the <Information-landscape/> in its amateur radio, police, weather, sports, cellular, and WLAN bands. Although the cellular network is down police calls penetrate the rubble pile. The iCR's speech-to-text recognizer and <Histogram/> detects "501" as the <Topic/> which in the dictionary of local police terms downloaded weeks earlier means "earthquake." Its <Speaker-ID/> algorithm detects two speakers on the police channels, triggering an RDH to instantiate UKOs for a <Police-cruiser/> and <Police-dispatcher/>. The cruiser is close by. Instantiating the <Earthquake/> UKO, the AACR powers down its own cell phone to conserve power. It then autonomously sends a formatted distress signal to the nearby police car on its VHF band, and the police acknowledge the message. It then advises the police AACR that it will sleep for 10 minutes to save power unless the vital signs turn for the worse. The message included the GPS coordinates of the building, the name of the user, company, floor, and suite number. It reports to the police AACR that the elevator's BlueTooth channel is reporting mechanical problems and heat. Police rescue the victim before the building is engulfed in flames. If I'm in LA and worried about earthquakes, I want to be wearing *this* iCR.

Implementing the Use Case: This iCR learned critical information about the user's situation by combining its native ability as a SDR with ASR and Zipf's Law <Topic/> spotting. Thus, autonomous reallocation of MIPS and radio resources enhances the user's likelihood of getting help quickly and thus of surviving the disaster. Technology challenges include reliably interpreting the chaotic scene. Social challenges include police acceptance of maritime safety protocols and standardization of message formats.

9.2 OFFICE ASSISTANT USE CASE

Why spend a whole chapter on use cases? Developing and using use cases is hard work. Do you like your MS Office Assistant™? I'm sure some people do, but I don't. Why not? I have been programming computers since high school, which is common now but was not all that common in the 1960s. Does the MS Office Assistant know this? No. Does it seem to care? No. It doesn't ask, and I don't tell. If it did, it would get out of my face without requiring me to train its silly voice system. I don't need any help with most of what it offers to help with. Some people like it. Some who don't like it do not know how to turn it off. I do, so that's the first thing I do in setting up an environment. I give it a chance to see if it has improved since the last one. Then I bury that little dog like many of you. Cell phones don't have little dogs—yet.

So the first CR needs ways of finding out about the user that are unobtrusive and natural. Otherwise, like the MS Office Assistant, expert users will turn it off, and some not-so-expert users will wish they knew how to get rid of the little dog. But the future is not so bleak. No doubt Microsoft is working on the Assistant-After-Next that will unobtrusively find out about your preferences before launching something that you gotta admit is cute to a user who just wants control-meta-H (a Symbolics LISP Machine help function) to come back. The first Toyota with a voice didn't usher in an era of cars that speak. Yet today, if your luxury car's GPS navigation system says, "Turn left at next signal," you are glad to hear it rather than take your eyes off the road to get that advice from a blinking display.

How can one structure user interactions with iCR so that the user actually wants to interact when needed? This is hard work, but computationally intelligent toys are being accepted by kids. Smart toy dogs, dolls, and now Asimo, a household robot, are creating new markets. The first successful iCR may have not one user interface, but a set of user interface skills from which inviting interactions can be tailored to the user, doing less for expert users and doing more for the less expert.

Recently, a friend bought a smart toy dog. After teaching the dog its name, the grandson got tired of its obviously rigidly timed interactions. Every so many minutes it wanted attention. Go away dog. Go to sleep. Don't bother me. I'm eating. I have to do my homework. The first toy dogs didn't understand that feedback. The first commercially successful iCR may understand all that and more.

Implementing the Use Case: ASR technology is rapidly maturing. Most 800 directory assistance is accomplished without people. The task domain of 800 number directory assistance concerns the names of only a few tens of thousands of companies with 800 numbers. Although some names sound alike, and users interject nonmeaningful expressions like "Ummm," and the speech recognizers aren't very tolerant of background noise, it still works more than 99% of the time. The trick isn't software with deep understanding of 800 numbers, but rather the structuring of the domain so that the task is within reach of the technology.

The microworlds construct of the last chapter assists with this necessary systems engineering trickery. A wireless office assistant microworld consists of prototypical UKOs for address book, calendar, to-do list, formatting, and printing documents with UDHs to instantiate, constrain, and expand the UKOs per experience with the <User/>. Even unreliable speech recognition and extraction technology (e.g., name finders) with statistical <Topic/> spotting may enable an iCR to recognize user states needing the wireless office assistant, such as discovering BlueTooth from the user's laptop to synchronize address books autonomously ("Is the laptop turned on?"). Problems like thrashing from one microworld to another (laptop to address book) need to be characterized and solved. Discovery of user intent remains a chronic problem. It will take not just natural language processing and computer vision, but also substantially evolved AML to gracefully and passively acquire relevant knowledge from users and to learn, again passively, what to do to actually help the user in the way the user wants to be helped.

9.3 COGNITIVE ASSISTANTS FOR WIRELESS

A cognitive assistant should focus on the following wireless (not office) tasks:

1. *Perform wireless access tasks described at the <Information/> level of abstraction*, such as "Keep track of the weather" or "What's the weather going to be like in Boston?"
2. *Notify* the user of problems, such as inability to access or interpret <RF/> in a new scene.
3. *Suggest alternate plans* for addressing in a <Situation/> (<Scene/> plus user <Needs/>).
    (a) Explain steps required to execute a plan.
    (b) Justify plans in terms of QoI value added.
4. *Competent mixed-initiative* interaction:
    (a) Initiate dialog for autonomously sensed situation changes.
    (b) Explain reasoning for <Self/>-initiated actions (not stilted).
    (c) Answer informal questions about the wireless devices and networks.
    (d) Pose planning questions ("Can we move 500 meters that way?").
5. *Reliably detect and predict user intent*:
    (a) Continuously track stereotype scenario templates.

   (b) Rectify decisions and actions against templates ("cases").
   (c) Obey meta level commands ("Knock it off").
6. *Adapt to users*:
   (a) Who regularly take apparently inappropriate actions.
   (b) Agent's scene perception must adapt to user actions.

These very challenging prototypical tasks are common to the following cognitive wireless assistant use cases.

9.3.1 Zeus in a Box

A start-up company called Zeus Wireless offers *Zeus in a Box* (ZB), an autonomous wireless agent in a set top box. ZB focuses on wireless network management in the home. Zeus marketing wants ZB to detect all home wireless networks—WiFi, 802.11, cordless telephone, Home RF, power line communications, TV-band ad hoc networks, and other ISM wireless. The consumer turns on the new wireless microwave oven; it talks to the wireless refrigerator, to the UPC scanning kitchen cabinets, and so on. It thus cooks your microwavable snack without requiring you to read the directions. (Like it so far?) A sensor on the front door detects the <Owner's> BT trust codes and connects it to the home network. Visual sensor reports "Sensor 2 front door—Suzie is home now." ZB is security aware, with aggressive personal firewalls changing long pseudorandom passwords to protect the home network from identity theft.

Implementing the Use Case: ZB uses a subset of the CRA <Self/> consisting of RKOs for the networks, SKOs for apartment buildings, suburban neighborhoods, and rural settings, and UKOs for prototypical users from <Computer-illiterate/> to <Accomplished-hacker/>. The Zeus CWN retains the master KOs, uploading the appropriate ones at point of sale. The Zeus CWN uploads RKO-SKO-UKO instances from customers via the Internet, solves home networking problems with Google-class MIPS and storage, and downloads solution KOs and DHs respecting consumer privacy and greatly enhancing per-customer QoI. Although there are many engineering and market challenges, few technology challenges preclude this use case.

9.3.2 ZB Space–Time Dialog About Home Networks

In its first employment, ZB has created an ad hoc LAN in unused TV channels for the home wireless network. The air interface delivers as much capacity as possible within the spectrum allocation and cost envelope of the hardware. It deals with fading and interference dynamically, implementing Kokar's OBR among other features. Even with such great technology, network setup and radio resource management can be challenging.

Suppose the family likes to watch the evening news via wireless relay of a cable channel to the new HDTV display outside on the lanai. Others are using the same TV channels as wireless LANs in this suburban neighborhood at the same time, creating interference. In addition, the placement of items in the RF scene causes fine-scale multipath fading. The iCR models the RF-related entities over space and time in order to generate methods to mitigate interference.

ZB offers alternative interfaces from a cute dog with step-by-step instructions to a bare bones but much faster command-line interface for experts. The dog can rescue the "experts" without appearing too cute. The average purchaser doesn't read directions or follow the steps. The components are poured out of the box and plugged in. Nothing is showing on the HDTV. Now what? The voice of ZB speaks from the ZB set top box:

"Is everything plugged in or supplied with batteries and turned on?" User says yes.

"Thank you for welcoming me into your home. I see a NTSC TV set, a DECT cordless telephone, two CDMA cell phones, a BlueTooth enabled laptop, but I don't see the remote control or the HDTV display that you purchased with this system. Did you turn on the remote control?" User says yes.

"Could you set the remote control into the cradle on top of this unit?" User says no.

"Without the test widget, I can't help you figure out what is wrong with the HDTV display." User says something unintelligible about a two year old. Unable to recognize the reply, ZB says: "If you leave this unit powered on, then I will start helping again when you set the remote control in the cradle on top of this unit." After a few minutes, the user inserts the remote control into the set top box. ZB clears and reprograms the remote control via the physical connection of the cradle. The controller memory had been corrupted by a magnet in the motor of a zip-zap toy car purchased at the same time.

"Thank you. Do you have the time to continue?" User says yes, OK, hurry up. "Could you pick up the remote control from the cradle and walk toward the HDTV?" ZB records digital snapshots of the scene every 2 seconds as the user walks out of the house onto the lanai. From its video blob detector, it recognizes the visual signature of an overhang (probability 80%).

The ZB client in the handheld remote-controller says, "Sorry. I just lost the wireless connection to the set top box. Could you please back up?" The user goes back into the house.

"OK, thank you. Your home seems to have walls and an overhang outside that are causing problems with the radio link. You may have metal studs in the walls. If you know this to be the case, please say 'yes'." The user says, "What do you think I am, an architect?" ZB says, "If you don't know, that is OK. With your help I will work around it. Is that OK?" User says OK.

"If the HDTV is still powered on, then please walk around near the HDTV as much as you can without getting hurt. I will record the radio network conditions as you walk so that we can figure out how to make your HDTV display work." The user walks around the lanai.

The ZB remote control says, "Please stop. This is a good place to put your HDTV for wireless network reception. Could you place the HDTV here or close by?" The user says something audible but unintelligible. ZB replies, "Sorry. If you continue to walk around, we can probably find some other places where the HDTV will work. Would you like to do that?" User says OK.

ZB says, "How about here?" User says, "OK" and moves the HDTV, which works in this particular location. ZB says, "Where is the HDTV?" and the user says, "The lanai." Meanwhile it has reallocated processing resources to enable a pseudo-MIMO mode. ZB tells the homeowner, "Your system will work now, but you can't use the high performance mode of your video games at the same time as watching HDTV. If you would like to do both at once, then purchase the MIMO antenna to play interactive games through the HDTV. Alternatively, you could move the set top box and HDTV much closer together so I don't have to work so hard to reach the HDTV from the set top box." User says something unintelligible again, but the <Topic/> is "Finally," which it interprets to be positive.

Implementing the Use Case: The use case needs a very strong ASR dialog manager as well as calibrated RSSI in a handheld client. Its SKOs for suburban neighborhood must include features of local construction, and this would be expensive to knowledge engineer, so the CWN uploads data from building permits databases from the county seat. It then <Topic/> spots with <Histogram/> to identify features of the buildings that cause problems with RF and downloads the resulting KOs to the point of sale terminals in that <Locale/>. Its dialog planner should classify utterances as male, female, infant, child, and adolescent, with 20% or less confusion matrix per utterance to adopt its dialog to the <Scene/>. The iCR could detect television speech by correlating the speech background of the room with the audio of the TV channel in use. The cross-correlation characterizes the impulse response of the room, enabling ZB to suppress the TV broadcast from the room acoustics to enhance ASR quality. Finally, the dialog manager switches among standard, careful, and quick modes of interaction depending on the features of the <Scene/>.

9.3.3 CHERISH: Beyond ZB

"The VCR should set its own bloody date and time, I'm too busy."

ZB was a success in the marketplace, so Zeus launches the Cognitive Home Entertainment Radio In the Smart Home (CHERISH). CHERISH goes beyond ZB to manage all home, travel, and office wireless connectivity and information access, driven by QoI.

1. *Entertainment*: CHERISH plays CDs, DVDs, and so on. When asked to "keep us up to date on the weather," it collects weather broadcasts into a personal weather-radar web site viewable on the home TV. The user no longer must tune to Bay News 99 for the latest weather radar since CHERISH updates the home TV weather web site from all available feeds automatically.
2. *Wearable Awareness*: The CHERISH Wearable Wireless (W2) PDA fits every member of the household, including the infant and the dog. Parents know the location and health of all family members at all times. If a W2 detects a threatening condition, like a child approaching a street, it warns the wearer and advises the parent autonomously. W2 also respects privacy, turning off video during private moments.
3. *Location Alerting*: The dog's CPDA records audio, video, and location, alerting the owner to threatening situations. Theft of the beloved family pooch is met with digital snapshots of the perpetrator and GPS location of the dog so that the police can rescue the dog and pick up the perpetrators.
4. *Range Extension*: CHERISH-in-the-Car extends the range of W2s via car radios.
5. *Wireless-for-Free*: CHERISH uses WLANs in the car, at home, and at the office, avoiding expenses of cellular networks when possible.

CHERISH reduces tedium and amplifies the fun. When the family decides to take vacation next month, CHERISH at home tells CHERISH at work avoiding the tedium of synchronizing calendars. Each wearable CHERISH may be manufactured by a different supplier since they all use OWL and RXML to normalize semantics. The home client submits a Vacation Request to the employer with the necessary details reducing the paperwork burden. CHERISH also knows about entertainment and travel. It finds interesting things to do, interesting places to go, and interesting information about almost any place in and out of the network. When users talk about a place, CHERISH detects it as a <Topic/>, looks it up on the Web, and proactively displays information instantly when asked for high QoI. It suppresses Web pop-ups and TV commercials, obtaining needed information without the advertising, refreshing personal yellow pages from the pop-ups. When and if the user wants to know about special events or prices, it already has the latest information.

Consider Lenny, the W2 PDA, and Charlie, Lenny's owner. Charlie should get information services without reading the user's manual even once. This is hard. It costs less to put the burden of using a complicated system on the user. The product gets to market quicker and with more features. Only the most honest users admit defeat at the hands of the VCR.

Implementing the Use Case: Writing a user's manual is hard work. Even with good authoring tools, the organization of the material and the decisions of what to include can be challenging. Error code listings and diagnostics were standard for computer systems of the 1970s and 1980s, dwindling in the 1990s as complexity increased and the benefit of explaining all that to the user decreased. Just reboot. Take two aspirin and call me in the morning. Could an AACR read its own RXML to the user? That would be the strategy with CHERISH. Instead of writing manuals, why not just write a rule interpreter for RXML? After all, the KOs both contain and express the strategies, tactics, and data structures on which CHERISH is based. Although such an approach may be out of reach at present, technology for systems that know what they are doing is emerging.

9.3.4 Wearable Awareness

A person might not mind wearing a cell phone, but the family dog might. So would the newborn. Yet wearable awareness calls for the integration of the W2 into the wardrobe. RF identification tag for the dog; a nice belt for the young boy; and a true fashion statement for super-Mom (Yes, that's a Gucci design). Soon, every cell phone will have long battery life, great graphics, and discrete voice effects. But discriminating customers will cherish their CHERISH fashion statements. Wearable systems must learn to be discrete, a technologically daunting task. In addition to the mores of segments of society implied by gender, age, religion, and other types of orientation, quirky individual preferences must be learned. There may be an Italian UKO (I'm an Italian so I'm allowed to make politically incorrect remarks about my passions). The default religion might be Roman Catholic. The AACR could observe its Italian wearer going to the cathedral for Mass. St. Mark's might take Wireless Credit Corp.'s PKI login, while St. Luke's might not accept wireless donations (yet). One parishioner might think it in poor taste for the W2 to process prayers, while another might want to be reminded of who to pray for, welcoming the W2 into the inner circle.

Implementing the Use Case: The iCR sales force may need a recommender system [220] that analyzes the purchaser profile to download stereotypical UKOs with tailored UDH to learn details, nuances, exceptions, and rules from Charlie. Lenny's UKOs should relate to Charlie's preferences (Expression 9-1).

Expression 9-1 User Privacy Stereotype

```
<UKO/><Name> US-stereotype Married Male Middle-aged </Name>
  <Slots> <Private-scenes> Bathroom Church </Private-scenes>
    </Slots> . . . </UKO>
```

Expression 9-2 Verifying Stereotypes

```
<UDH> <Name> Verify-stereotype </Name>
  <Rule> <Condition>
    <Detect> <Scene> <Private> S </Private></Scene> </Detect>
      </Condition>
  <Action> <Dialog> <Verify> <Private> S </Private> </Dialog>
    </Action> </Rule> </UDH>
```

W2 units disable video systems when they detect private scenes. A <Profanity/> feature could suppress the owner's voice if profanity is detected, remembering <Expletive/> deleted. When Lenny with Expressions 9-1 and 9-2 detects the entrance of St. Rocco's Catholic church, a UDH fires to ask, "Charlie, should I turn off the audio and video for greater privacy while you are attending church?" Charlie might say, "No. Leave it on in case of an emergency." To which Lenny replies, "My sensory-perception always detects emergency situations, but nothing is remembered unless you give me permission." Charlie says, "Can you still make a wireless donation to light a candle?" W2 says, "If you like." Since it is labor intensive to define fixed dialogs, VoiceXML popular for electronic call centers provides flexible synthesis from RXML for <Self/> defined dialog. Charlie may tell Lenny to turn off all that stuff and just take a digital snapshot when told to do so, turning Lenny back into a cell phone, which is one way of using the inherent flexibility of the UKOs.

9.3.5 Location Awareness

Location alerting tasks enhance personal safety. Lenny can help turn any neighborhood into a safer neighborhood. Visit Capitol Hill. This is an interesting neighborhood between the U.S. Capitol and some challenging neighborhoods. Which places are open all night? Which ones tend to be safer than others? If you walk, which way is the most scenic at a given time of day at a given time of year, given the police reports for the last six months? How much of this should Lenny process for alerting and display? Should Lenny offer a suggestion when Charlie heads out to get a loaf of bread? "Why don't we go the other way, Charlie? Last week there was a mugging the way we usually go and the street lights are still out." How did Lenny know that? There might be 802.11 enabled streetlights. A wireless logistics network could broadcast street light status data to the Washington, DC government. With CHERISH, Lenny reads it too and advises Charlie to take the lighted route home.

Implementing the Use Case: Lenny needs neighborhood SKOs that describe every street within walking distance. As Charlie wears Lenny, the GPS coordinates are correlated to Lenny's audio and visual <Scene/> perception. Each closed loop from home and back spawns a new SKO. Each such path is analyzed by introspection during a sleep cycle. Lenny recognizes that Charlie didn't walk at 800 mph across the street and back, but that GPS tends to glitch near the high rise building. Comparisons of loops reveal semantic path equivalence, such as two different paths that originate at <Home/> and reach <Giant-food-store/> without stopping anywhere else. Although no path matching is without error, semantic path matching relates the grounded semantic features, rendering correlations at a level of abstraction where errors can be diagnosed interactively. "Why did we use West 19th Street to go to Giant?" "Just for fun." Lenny's UKOs and UDHs accept vacuous reasons as a social skill.

If Lenny has to be programmed for access to the streetlight status in DC, then there must be an expensive group of programmers to keep such services current. On the other hand, if the FCC or some other public body were to adopt, endorse, and promote an open plug-and-play RXML for public wireless networks, then wireless assistants like Lenny can autonomously discover the presence, function, and QoI value of such wireless networks. The Internet grew on open TCP/IP and http markup, enabling syntactic interoperability. Semantic interoperability needs RXML. Although the IEEE Standard Upper Ontology (SUO) seems to have taken a back seat to DAML/ OIL/ OWL, open architecture semantics too is growing. Public WLAN standards, TCP/IP, and http need OWL and RXML for Lenny's semantic interoperability.

9.3.6 Spectrum-Cash Range Extension

Lenny connects via WLAN to Charlie's cognitive automobile, Sam. Sam's higher power transmitters and maybe 80 radio antennas (don't laugh, the 2000 Mercedes had over 60 antenna elements) can extend radio range for revenue generation as each relay operation on behalf of another AACR accrues spectrum-cash, barter for relay services. Consuming a few watts of power for a day or two is no big deal. In Europe, however, a car may be parked for a month while the family is on vacation. In the United States, that is less likely. How do Lenny and Sam discover the use pattern of this particular family? Question and answer during a setup phase is not user-friendly; neither is a dead battery after a month in Switzerland. Lenny could read the family's travel itinerary and thus put Sam to sleep for a month.

Radio engineering questions abound for Lenny and Sam range extension. Store-and-forward relay introduces time delays. Packet structure and content differ from one band/mode to another, such as between GSM and 802.11. The PSTN standard DS0 (64 kbps PCM) can bridge such streams. Questions of privacy and authentication (PKI?), data rate, media access, and protocol stack remain. Other questions include the following.

1. Will relay use MIMO mode?
2. Can Sam switch to WLAN to MIMO with the home network?

3. How will Sam know to store and forward big email attachments, but to relay short text messages immediately, estimating time of arrival of the big attachment so that Charlie can use premium services if needed?
4. How can Sam estimate arrival time from a learned commuting pattern if today is a work day?

Implementing the Use Case: The radio engineering community regularly practices the skills needed for this use case. Autonomous cross-band range extension requires collaborative relay planning. RXML assists in integrating radio knowledge with user preferences for relay expressed in user- and radio-domain KOs.

Expression 9-3 Charlie Is a Person Who Owns Lenny and Sam

```
<UKO> <Name> Charlie </Name>
  <Definitions> <Constraints>
    <Intelligent-entities> <People/> </Constraints> </Definitions>
  <Slots> <Role> Owner </Role>
  <Owns> <RF-devices> <Self/> <Sam/> <RF-devices> </Owns> </Slots>
    </UKO>
```

Expression 9-3 describes Charlie to Lenny. This simplified UKO helps Lenny aggregate knowledge about cognitive entities that are carbon-based and silicon-based. Lenny also needs to know about Sam.

Expression 9-4 Sam Is a Cognitive Automobile With RF Capabilities

```
<UKO> <Name> Sam </Name>
    <Definitions> <Constraints>
      <Intelligent-entities> <Computationally-intelligent-entities>
      <Cognitive-automobile> <Sam/> </Cognitive-automobile>
        </Constraints></Definitions>
  <Sam> <Slots> <RF> <WLAN/> <AM/> <FM/> <DVB/> <DAB/>
    <CB/> <Bridge/> </RF>
    </Slots> </Sam> . . . </UKO>
```

Sam can <Bridge/> as follows.

Expression 9-5 Bridging One-Way From GSM to a WLAN

```
<RKO> <Name> <Bridge/> </Name> <Domain> GSM </Domain>
  <Range> WLAN </Range>
  <Definitions> <Constraints>
```

```
    <Self/> <RF/> <Protocol/> </Constraints></Definitions>
  <Functions> <Methods> <Request-message>
    Request-bridge ( )</Request-message>
      <Transfer> RPE-LTP-80211g ( )</Transfer>
      </Methods> </Functions>
  <Slots> <RF> WLAN GSM </RF> <Payload> Voice </Payload>
    </Slots> </RKO>
```

This bridge transports voice from a GSM source network to a WLAN sink network. With the KO in Expression 9-5, Sam knows he can bridge between GSM and 802.11. Lenny can request bridging by sending the string "Request-bridge ( )." A complete KO has the necessary details while additional KOs and DHs are suggested in the exercises.

The RKO is *intelligible to Lenny and Sam*, not just to the radio engineers who designed them. Unlike SDR architectures that are meant for radio engineers, these KOs and DHs are autonomously used by Lenny and Sam, as cognitive agents, to reason about bridging not just to implement it.

9.3.7 Wireless-for-Free

Today's consumer has two broad choices for wireless access to core networks: rent cellular service to access the PSTN or buy WLAN hardware from an electronics retailer to access an ISP core network. IEEE 802.11, Home RF, and WiFi offer the dedicated enthusiast many opportunities to learn myriad details of wireless networks. The consumer doesn't have to learn the horrendous details of 3G to use the cell phone. There are hybrids of both extremes. Wireless-for-Free envisions an alternative, the computationally intelligent consumer-purchasable AACRs that interoperate and cooperate with CWNs of evolved cellular service providers and with ISPs for "free" radio spectrum use.

As 4G cellular deploys, wireless service providers use WLAN technologies to fill hot spots [221]. A mix of consumer products use cordless phone, BlueTooth, WiFi, 802.11, Home RF, and who knows what next for ISP access. If legal hurdles and market conditions were supportive, Wireless-for-Free would use computational intelligence to drive free access up and total consumer cost down, disrupting conventional markets.

Implementing the Use Case: The implementation of this use case is considered an exercise.

9.4 USER SKILL ENHANCEMENTS

ZB and CHERISH enhance the user's natural abilities. Wireless radar from a weather channel enables a user to see the tornado beyond the horizon. The

AACR user domain packages data from Internet and broadcast media into wearable, personalized, immediate information. Data requires work to turn into information. The work is driven by the psychological and social needs of the specific user where semantics plays the pivotal role. Lenny learns to scan the weather channel in a style set by Charlie to augment Charlie's personal skills.

9.4.1 Emergency Balancing Act

To get a CWPDA to talk requires speech synthesis. VoiceXML [222] has emerged into an industry-standard tool for programming guided voice interactions. Developed for call-center processing, VoiceXML may be adapted for embedded dialog generation. In the previous earthquake example, the AACR sensed vital signs, with visual and acoustic stimuli to detect a major emergency. Lesser emergencies may be more difficult to detect. The definition of emergency is fluid, different from one user to another. A typical solution to the dilemma of recognizing an emergency is to put that burden on the user. Give Lenny an Emergency button. If the button is pushed, then Lenny knows it can call for help from under the rubble pile if Charlie is unable to do so. Calling wirelessly for help reaches out further, but VoiceXML (set to LOUD) attracts the rescue personnel to exactly the right spot under the rubble pile.

On the other hand, if Charlie dials 911, Lenny knows there is an emergency. That was easy. Lenny could monitor and record the 911 emergency conversations, taking video clips of Charlie's emergency. Not so fast. It might be more appropriate for Lenny to conserve battery life in an emergency. Some kinds of emergencies such as traffic accidents may need the video record. Others, like being lost in a snowstorm warrant the preservation of battery life. Suppose the automobile accident and snowstorm are preprogrammed, one to maximize the video record and one to conserve battery life. When Sam detects the automobile accident via its accelerometer, it can tell Lenny via BT. Let's say Lenny also detects a snowstorm on NOAA radio, correlating GPS to the predicted snowstorm. This situation has an automobile accident and an approaching snowstorm. Lenny should take videos for Charlie's insurance claim, but should conserve battery power because of the snowstorm. The competing goals could be resolved by taking fewer pictures to save battery life. In the emergency, Lenny can speak, "Charlie. My battery is low. Should we take pictures of the accident or conserve battery life so I can help rescuers find us in the snowstorm?" Charlie laughs and says, "Central Park is over there and we can walk home, so don't worry about your battery." If Charlie were somewhere in Yosemite in the dead of winter, the answer might be different.

Implementing the Use Case: There are several technical challenges in this use case. What kinds of UKOs are needed so Lenny only talks when appropriate? How do you integrate AML with VoiceXML? What UDHs deal with situations that are close to a preprogrammed use case, but not identical?

Suppose snowstorm emergency is replaced by a typhoon? What combination of UDH, SDH, and RDH assure that battery conservation is effective?

9.4.2 Restaurant Vignette

This one may be a little overworked, but it is a good one to come back to regularly to see how the technology, markets, and infrastructure for this use case are evolving. You want to eat at a local restaurant. You would like to ask your CWPDA for advice. Steve [223] pointed out that a better way to get such advice might be to ask one of the locals which restaurant a local would choose. Such an establishment might not have to advertise in the yellow pages, virtual or real. If your CWPDA is listening to everything you say and do, why not empower it to share your feelings about local restaurants? "Boy this place has great lamb chops!" could be shared among CWPDAs making your impressions available to others, enhancing the value of the CWPDA to others and indirectly to you. After all, you may not know who to call to find out about this particular place, but wireless sharing of such impressions could be a killer app.

9.4.3 Playing Poker

A team consisting of the three Steves plus Larry [224] suggested a poker-player's assistant and Geek Poker. Each player wears a CWPDA with audio and video capabilities. The CWPDA counts cards, reads body language, and in short cheats any way it can. The user can play as advised by the CWPDA or go it alone. Rainman poker? A corollary idea also offered in the same discussion was the possibility of using CWPDAs in contract negotiations, making detailed information available more quickly to one side than to the other.

9.4.4 Helping the Blind

Cou-Way Wang of NTIA suggested helping the blind at the 10 June 2004 Cognitive Radios course in Alexandria, Virginia. A CWPDA with vision could generate a tactile field on the chest of a blind user. GPS and RF multipath signatures could assist vision in pinpointing location with sufficient accuracy for the blind user to navigate with precision, accuracy, and safety.

9.4.5 User Profiles

Professor Tony Kalus of Portsmouth University in the United Kingdom suggested the highly autonomous updating of user profiles. European research in location-aware services and user-adaptive communications has shown the value of user profiles, lists of parameters that characterize a user's preferences for services and cost. These preferences change as a function of the social role of the user, such as whether the user is at home, at work, or at school.

Furthermore, asking the user to even indicate which mode or profile might be applicable at a given time puts undue burden on the user. Why couldn't a CWPDA monitor the user's location (visual environment) and conversations with ASR and <Histogram/> to detect which of the profiles is most likely to be applicable, and after suitable reinforcement by the user, to autonomously change user profiles for better QoI? It could switch profiles based on location, time, and inferred <Topic/> of conversation.

9.4.6 The Nose Knows

My son-in-law, Max, wants his CWPDA with a chemical sensor so he doesn't get sick from bad food. The sensor chips are on the way. Will AACR create the killer application for these chips?

9.4.7 Museum in Your Pocket

My wife, Lynné, wants her CWPDA to analyze pictures of interesting things like a plant growing in a park, to access the Audubon Guide, and to identify the flower. Or what is the name of that monument, what is its history? To transition from brainstorming to a real application requires a business case analysis that articulates the value proposition in terms of revenue against costs of technology, infrastructure, and support to realize a profit.

9.5 EXERCISES

9.1. Consider an amateur radio "sleeve" for your PDA that can receive GPS and AM/FM broadcast at least. Write glue code so the PDA can tell you where you are. Integrate an ASR tool box and train it to recognize "Traffic and weather . . . ?" How would you train it to recognize typical traffic situations? What are the key implementation questions? Train a (high order) neural network to recognize "traffic and weather." Preprocess the speech into text and teach a neural network to recognize traffic and weather broadcasts.

9.2. Consider other functions of CHERISH and Zeus in a Box. Of the functions that drive CHERISH, only five have been discussed in this chapter. Your exercise is to address the others in the same way. Write a user vignette for each. Describe how the vignette enables greater independence from network service providers or military radio spectrum managers. How will the CHERISH node get the information it needs to detect the user state? How will it know which of its myriad SDR modes to employ? How can cognition technology enable Zeus in a Box to be more effective in the field? What specific capabilities should it have in order to configure wireless networks for a user? How would it scan for existing networks and recognize noise in unknown bands? What would a designer do about products designed after ZB has been on sale for a year? Suppose 802.x operates on ultra-wideband. Can ZB learn this to do its job? How could ZB expand to CHERISH? How might the developers of ZB enhance it toward CHERISH?

CHAPTER 10

USER-DOMAIN KNOWLEDGE

Although there are many ways to acquire knowledge about users, the CRA emphasizes the passive observation of user actions in the environment, drawing inferences rather than asking the user what may be perceived as stupid questions. Since radio is a technical domain, the domain knowledge is structured into radio physics, regulations, and practices across the relevant bands and modes. The user domain is less structured and vastly more complex, even in a single microworld focused on the immediate audio-visual scenes in which users are situated. Although sensory perception is a major technology challenge so is the complexity of user-domain knowledge. This chapter addresses sensing, perception, and mutual grounding as primary methods of relating user-domain knowledge to QoI enhancement by AACRs. This may be simplified by the modeling of users both as individuals and by class, and by a priori knowledge of user domains with sociological and psychological models.

The primary iCR sensory-perception capabilities for the user domain are the keyboard, audio, and video. AACRs use conventional tactile input to perceive user actions (e.g., tuning the radio, entering a phone number, clicking on a web site) along with visual displays of text and graphics. Auxiliary sensory-perception capabilities may include navigation, acceleration, temperature, barometric pressure, and smell or wind velocity. This chapter focuses on the perception of those aspects of a <Scene/> for which SKOs and RKOs are known to the <Self/> and that are important to QoI.

10.1 USERS' NATURAL LANGUAGE EXPRESSION

Natural language (NL) expression includes spoken and written language as well as the body language associated with lip, face, hand, and other body movements that convey meaning. NL processing for AACR focuses primarily on speech and text processing. Historically, researchers address text retrieval or speech but AACR must fluidly interpret both as offered by the user.

10.1.1 Human Language Technology

The insertion of HLT in AACR envisions user dialog and machine translation among human languages. Commercially available language software translates written grammatical sentences successfully between those natural languages that are in wide use, such as English, German, French, Japanese, and Chinese. Language translation software products address over 700 languages [225]. These software tools are less than ideal, often mistranslating complex or idiomatic expressions. However, they reduce the burden of human translation, for example, in conducting multinational business [226]. Early progress with the automatic acquisition of rules of syntax [227] helped lay the foundation for substantial understanding of NL by computer [228].

The commonalities of NL in text and speech are illustrated in Figure 10-1. Translation includes speech recognition [229], machine translation (MT), and speech synthesis [230]. Speech recognition generates an errorful transcript or a hypothesis tree from the acoustic signal, for example, via hidden Markov models (HMMs), dynamic time warping, and/or fuzzy sets [231]. The core MT parses errorful text against (statistical) rules of grammar, with a lexicon. Semantic ("content") extraction rules use syntax data structures, dictionaries, and side knowledge; and higher level semantic rules resolve ellipsis (incomplete references) and anaphora (pronomial use of phrases) and adjust the semantic model as a function of discourse context [232].

10.1.2 WordNet

Although informal conversation need not employ complete sentences, word and phrase units typically enjoy a 1 : 1 and complete ("onto") mapping between spoken and written language. For any given written or spoken word, there is a corresponding word in the alternate form of expression. Although some niche languages have no written form, AACR near-term evolution addresses written languages. Similarly, for any spoken phrase, there typically exists a written form. People may write in complete sentences, but tend not to speak in such complete sentences, relying instead on shared context of both physical scene and discourse for unspoken subjects, objects, and other referents. Thus, the integration of NL technology into AACR emphasizes the reliable interpretation and use of informal words and phrases, with visual context aids.

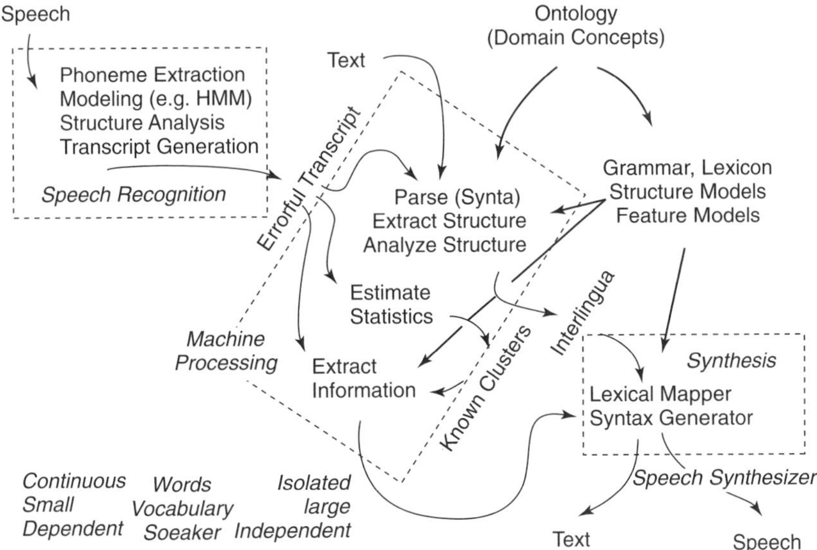

FIGURE 10-1 Natural language processing.

Comprehensive standard definitions are found in WordNet [233, 234], a collection of hundreds of thousands of words of English and other languages that emphasizes the psycholinguistics—semantics and semantically related structures rather than just the forms (e.g., spelling and pronunciation). WordNet is incorporated by reference into RXML for lexical and ontological structure. Fundamental differences in the semantic organization of syntactic categories like noun, verb, adjective, and adverb can be seen in WordNet.

According to Miller et al. [235], nouns are organized in human lexical memory as topical hierarchies. The WordNet nouns begin with the entry "Entity," from the WordNet perspective, the common root of all nouns:

n [noun] 01 entity | that which is perceived or known or inferred to have its own distinct existence (living or nonliving)

Associated with Entity and immediately following entity are the related nouns: "thing" defined as "a separate and self-contained entity"; "anything | a thing of any kind" (e.g., "Do you have anything to declare?"); "something | a thing of some kind" (e.g., "Is there something you want?"); "nothing | a nonexistent thing"; "whole | an assemblage of parts that is regarded as a single entity" (e.g., "How big is that part compared to the whole?"); and "living_ thing | a living (or once living) entity." The numerical references establish associative links between Entity and other nouns. The companion CD-ROM follows the threads more deeply.

WordNet doesn't differentiate well between subsets (e.g., action as a type of act), synonyms (e.g., property and attribute in object-oriented programming), and related abstractions that do not necessarily share set/subset relationships but that are not synonyms in many instances (e.g., feeling and emotion). The RXML adaptation of the candidate primes leverages the insights of WordNet, adapting them to the <User/> domain.

Verbs, on the other hand, may be organized in human memory by a variety of entailment relations. One might like to find a single root verb from which all verbs might be derived in the same sense that all WordNet nouns share Entity as a root. Because of the complexity of <Action/> as a multirooted heterarchy, no single abstraction appears at the root of WordNet verbs. Instead, the WordNet types of verb <Action/> are: bodily care and functions, change, cognition, communication, competition, consumption, contact, creation, emotion, motion, perception, possession, social interaction, and weather, as well as verbs referring to states, such as suffice, belong, and resemble, that could not be integrated into the other categories. Radio XML adopts these as the top-level categories of action in the user domain, leveraging for the <User/> the best insights of linguists into how people communicate with each other.

In the psycholinguistic view, adjectives and adverbs may be organized as N-dimensional hyperspaces. Again, according to Miller, these lexical structures reflect a different way of categorizing experience. Attempts to impose a single organizing principle on all syntactic categories would badly misrepresent the psychological complexity of lexical knowledge. WordNet organizes lexical information by word meanings, rather than word forms, resembling a thesaurus more than a dictionary.

WordNet adjective, adverb, noun, and verb databases for AAAI 2004 [234] included the four base files for adjectives, adverbs, verbs, and nouns consisting of 21 MB of text. Since word-level semantics are context dependent, WordNet 2.0 includes indices for clusters, similarity metrics, and related tools. The associative-semantics tools of WordNet 2.0 are incorporated into RXML by reference for <User>-domain skills for AACR. Space required for the storage of these word-level semantics along with a rich set of semantic relationships even in a handset is not excessive by today's standards.

CRs also learn user-specific word associations from the <User/> acting normally in <Scenes/> such as work, home, and leisure. WordNet provides a starting point for understanding user expressions for information support, which the AACR may tailor to the user specifics through reinforcing user associations over WordNet associations, leveraging WordNet but not too strongly bound to its detailed structure, which after all was acquired through exposure to large collections of text, not through colloquial usage.

Going beyond WordNet, statistical rules of syntax enable the same words to take on different roles, contributing differentially to content. Consider the differences in the way that "back" contributes to meaning in phrases like (a) "Back up, please," and (b) "Don't get your back up, please." Phrase (a) is a proper subset of (b) syntactically, but the roles of the words are very different, as inferred by rules of syntax that describe the formation of phrases and sentences. To adapt and integrate such rules, the AACR systems engineer needs some familiarity with phrase structure grammars.

10.1.3 NL Grammars

NL research systems like SNePS [236], AGFL [237], and XTAG [238] work with a morphological analyzer like PC-KIMMO [239] to analyze phrases to characterize the roles of words in those phrases, and thus to infer limited semantics through mutual constraint satisfaction. Such research systems are foundational to current standards such as the W3C standards for speech recognition [240] and synthesis [241]. One readily accessible and fairly comprehensive NL parser application relevant to AACR is the TRAINS system that demonstrates speech recognition and synthesis for a notional railroad train reservation system [41, 242]. Since grammars can be used for the analysis of speech and text and for the generation of spoken language, their features and limitations motivate both audio perception and speech synthesis.

Grammars require lexical categories and associated rules. Since nearly every word has multiple lexical categories, parsers generate multiple hypotheses, typically in the form of a stack or hypothesis tree, eliminating or rank ordering hypotheses according to the rules of grammar and the methods of parsing the language. Although elusive for decades, RetrievalWare's ThingFinder exemplifies commercial parsers capable of reliably finding named entities (people, organizations, places, etc.) embedded in general text in English and other languages [243]. WordNet lexical categories are the CRA <Self/> <Roles/> for <Words/> in the <User/> domain: Noun, Verb, Adjective, Article, Preposition, Auxiliary verb, Pronoun, Question determiner, Words that function like prepositional phrases, Proper names, and the word "to".

Words are characterized by their description in a lexicon that a parser uses to determine the potential lexical categories of the word, along with properties such as agreement, root, subcategories, and verb form:

ILLUSTRATIVE LEXICON ENTRIES

Word	Category	Agreement	Properties
dog	Noun	(agr 3s)	(root dog)
saw	Verb	(agr ?a1)	(vform past) (subcat_np) (root see)

(agr = agreement; 3s = third person singular; ?a1 = tense variable)

The CRA <Self/> does not specify lexical entries, deferring that to the NL subsystem.

The TRAINS Parsing System illustrates the structure of grammatical rules as well as the benefits of bottom-up parsers [244]. TRAINS parses word lattices with best-first parsing using context-free probabilistic rules with incremental (word by word) parsing with backup for corrections. Its hierarchical feature values and extended unification options enable relatively sophisticated logic to resolve ellipsis and anaphora. Its hierarchical lexicon simplifies defining large lexicons, such as those needed to continue to support an AACR user over time. Procedural attachment to chart actions simplifies programming the parser for the extraction of relevant information from well-formed phrases. As a research objective, an AACR could use CBR with RL to update its own chart parser to <User/> style.

The CRA <Self/> does not specify syntax or grammatical rules, enabling NL subsystems to function as black boxes that instantiate phrases in AACR dialogs. While the constituents of sentences are words, the constituents of spoken words are lower level acoustic elements, phonemes, or "phones."

10.1.4 Speech Phones

The basic unit of speech that conveys linguistic information is the phone. A phone is an indivisible space–time–frequency epoch of a speech utterance. For American English, there are approximately 42 phones. Each is generated by specific positions and movements of the vocal tract. The Microsoft Speech API (SAPI) defines language-unique and universal phones, organizing the universal phones into categories. These MS Universal Phonetic Symbols (UPS) are based on the International Phonetic Alphabet (IPA) [245]. There are 294 speech phones in the UPS in the following categories: consonants, affricate consonants, monophthong vowels, diphthong vowel (UPI), diacritics, suprasegmentals, clicks and ejectives, tones, and other symbols. Speech recognition systems hide the signal processing required to deal with such complexity by transforming a temporal epoch of sampled speech into an errorful transcript or set of alternate transcripts with an associated degree of belief for each alternative interpretation. Speech synthesizers interpret rules for sentence formation following generative annotations such as the suprasegmentals, tonals, and tone sequences, particularly in Asian languages like Chinese.

This introduction establishes the basic characteristics of natural language as a primary source of user-domain knowledge. Acoustic sensory perception is the AACR ASR component that translates sensory stimuli into hypothesis data structures through which AACR interprets the speech environment.

10.2 ACOUSTIC SENSORY PERCEPTION

The acoustic domain of human perception spans the audio spectrum from a few hertz to about 20kHz—more for gifted people. Acoustic sensing in the human begins with the ear's tympanic basilar membrane and cochlea that implement about 16,000 nonlinear bandpass filters that send preconditioned signals to the lower auditory cortex. There, signals from the two ears are cross-correlated to perceive depth and direction, as well as for the separation of sound sources. The higher order processing of these signals in the medial and upper ganglia is not as well understood, but a combination of acoustic, visual, balance, and tactile sensory tracks interact to percieve point sources like speakers and distributed sources like a brook or symphony orchestra. The functions of acoustic sensory perception for AACR include the following:

1. Characterizing the acoustic background as noise, music, machinery, or conversation.
2. Estimating the number of speakers in the scene, their gender, and the degree to which their voices are known or novel.
3. Localizing in signal space the speech of the designated <User/>, the specific person to whom the AACR is to be responsive.
4. Localizing sources of coherent interference for which high fidelity computational models may be available, such as TV or radio broadcast that the AACR can receive via RF as well as hear in the background.
5. Enhancing <User/> speech potentially by multiuser detection (MUD) and space–time adaptive processing (STAP).
6. Extracting the enhanced <User/> speech for ASR.
7. Converting speech to text hypotheses with associated degree of belief.
8. Time tagging, rank ordering, and annotating hypotheses for <Scene/> perception.
9. Retrospective refinement of <Speech/> skills during sleep.

10.2.1 Acoustics

Acoustic sensors access the bulk pressure vibrations of air waves with frequencies between 10 Hz and 20 kHz with speech power measured in decibels (dB) over a range of 100 dB. Microphones limit the frequency response and direction of sound, and multimicrophone arrays can enhance sound quality in meeting rooms.

Sound boards typically create standard sound formats: 8-bit signed (.SAM), 64-bit Doubles (RAW) (.DBL), Adobe Audition Loop (.CEL), A/Mu-Law Wave (.WAV), mp3PRO (.MP3), PCM Raw Data (.PCM), Audio (.WMA), Windows PCM (.WAV), and others. Sound boards also generate stereophonic sound from these files. MIDI files are scripts consisting of notes from a music score that may be converted to digitized acoustic waveforms of musical instruments with special effects.

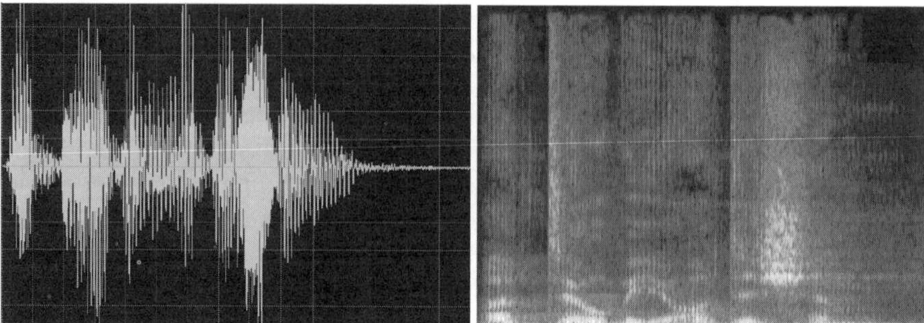

(a) Trace of the audio signal amplitude versus time (b) Trace as color-modulated spectra versus time

FIGURE 10-2 Two word speech epoch (Adobe Audition) in time and frequency. (a) Trace of the audio signal amplitude versus time. (b) Trace as color-modulated spectra versus time.

10.2.2 Sensing Speech

Typically, low-cost microphones do not respond well to far-field acoustic stimuli (even a few feet from the microphone). They may respond in excess to local noise, for example, the sound of typing fingers on a laptop or the road noise of the car.

Speech sensing consists of detecting speech, locating the speech epoch endpoints, and analyzing the speech waveform in the time domain, frequency domain, and various other cumulant domains such as wavelet projections and hidden Markov models (HMMs).

Speech is differentiated from a quiet background by a voice activity detector (VAD), which is also called voice operation transition (VOX) for historical reasons. There are many VAD algorithms, some of which are available as products for particular DSPs [246]. ITU (formerly CCITT) specifies tests for VAD in telephony, such as the ITU-T G.729 Annex B tests.

Software tools like Adobe Audition™ enable one to visualize and manipulate speech, music, and sound (Figure 10-2).

Other tools like Matlab and Mathcad estimate spectra and analyze speech signals algorithmically. Automatic speech recognition (ASR) systems may be programmed for a range of dialogs. A cognitive wireless PDA initially may listen to only the owner when the owner talks into the PDA as with a cell phone. Later in commercial settings and perhaps sooner in military applications, high noise levels and spatially sensitive sound acquisition will be needed.

10.2.3 Perceiving Speech

Given an isolated speech epoch, speech recognition software tools translate speech to text. Since the acoustic signal alone is ambiguous, products employ statistical models of language to disambiguate alternative interpretations of the audio track. The process of speech perception selects from among the top *N* hypothetical phrase transcripts, in AACR using <Scene/> level expectations to assist in the resolution of uncertainties. Speech recognition APIs enable one to embed ASR in AACR. The CRA integrates ASR as a black box capability. Phrases may be learned and associated with functions to enable AACR–user dialogs, but that capability may not mature until the iCR. Meanwhile, commercial speech recognition tools like IBM Embedded ViaVoice, Enterprise Edition can improve accuracy, such as the N-Best technology, which returns multiple phrase matches so the AACR can select the best phrase for the <Scene/> and <UDH/>. IBM's JAVA® supports Java Speech API (JSAPI) with multiple Java voice applications to share the same Java Virtual Machine (JVM). IBM's Mastor is a multilingual speech-to-speech translator variant of ViaVoice [247]. The Microsoft Speech Application SDK (SASDK) is based on the Speech Application Language Tags (SALTs) specification for Web applications. Microsoft SDK related products include Maxxam speech-enabled applications for banks and credit unions based on Microsoft Speech Server and Speech Application SDK [248].

While the spoken utterance itself holds clues to the identity of the speaker, AACRs need to know speaker identity as accurately as possible. Speaker identification (ID) has long been of research interest [250]. Speaker models characterize the dominant tones and prosody features of a segment of audio. If the audio segment consists primarily of speech of one speaker, the resulting model reflects the speech patterns of the speaker. Sometimes, the acoustic patterns of the background corrupt or even dominate the acoustic patterns, so speaker ID in a moving automobile will in part identify the automobile through its acoustics.

Speech perception shares concepts and methods with text processing, information extraction, and document retrieval. Performance varies substantially along the critical dimensions of vocabulary size, quality of articulation (e.g., isolated words versus continuous coarticulated speech), and user-specific training (none versus speaker-dependent session). In addition, noise backgrounds and acoustic scene complexity (e.g., newsroom versus cocktail party) also affect performance. In challenging scenes word error rates on the order of 40% are common. Commercial speech processing software may achieve a 50% word error rate for general untrained speech or 20–30% for fully trained speaker-dependent recognizers in low background noise [249]. Statistical language models can correct word errors to a lower phrase error rate.

10.2.4 Perceiving Noise and Interference

Many <Scenes/> have intense vehicle noise, for example, from automobiles, trains, earth-moving equipment, aircraft, and the like. Sports enthusiasts (NASCAR comes to mind), construction workers, longshoremen, and the like are regularly exposed to high intensity sound other than voice. Rarely do speech-to-text tools characterize the background noise, so if the CR is going to extract QoI cues from background noise (e.g., the sound of a train or the chirp of a bird), it will need further far-field acoustic signal processing beyond speech recognition. The AACR may be able to tell that the user is in or near a vehicle that is running based on its ambient sound signature, such as from a commuter aircraft. An acoustic processing subsystem with VOX set for voice may remove such background "noise"—which is really acoustic interference—in order to obtain a clearer voice command from the user. Although that may be a valuable contribution, the iCR algorithm also would characterize the strong background for the AACR association of scene and acoustic cues. To the degree that specialized machinery exhibits characteristic patterns, the acoustic signal may characterize common equipment like elevators for acoustic cues.

10.2.5 NLP Tools and Standards

Commercial speech recognition and synthesis tools increasingly support World Wide Web Consortium (W3C) standards. Some standards are oriented toward specialized applications like 800 directory assistance, call centers, and telemarketing, for a level of NLP technology not typical of radio engineering enterprises. NLP starter kits help one develop the depth needed to adapt W3C standards and tools to AACR.

Zhang Le has published the bootable Morphix-NLP CD [252]. Morphix is based on Knoppix, a version of Debian GNU LINUX. The tool has a GUI with iconified tools for immediate experimentation including the Link Grammar Parser. It labels verbs and nouns and parses sentences interactively. Its WordNet Browser facilitates interaction with WordNet via the World Wide Web and its CMU Festival Speech Synthesis System is also interactive. Tools include tokenizers, taggers, and a collection of parsers. Le's statistics package includes tools for word frequencies, vocabularies, and word bigram and trigram counts of the co-occurrence of two or three words in sequence. Le's Maximum Entropy Modeling toolkit classifies documents, disambiguates word sense, and address other advanced topics in NLP.

Such tools can go too far and not far enough in the direction needed for AACR. NL parsers can force a speaker into unnatural grammatical expressions in order to be understood. Sentences in colloquial dialogs may have no explicit subject, no verb or objective noun phrase to the degree that they are implicit in the scene (e.g., by pointing) or in the dialog (e.g., by anaphora, ellipsis, or intentional vagueness). Thus, the CRA accommodates NL as an aid to perception, but includes case-based reasoning and AML to enhance NL via multisensory perceptions in stereotypical <Scenes/>.

The Speech Recognition Grammar Specification Version 1.0 was adopted by W3C on 16 March 2004 [253]. It recommends syntax for ASR grammars so that developers can specify the words and patterns of words to be listened for by a speech recognizer. The forms include BNF and XML DTD and Schema definition for Form Grammars as well as the required DTMF Grammar. An XSLT style sheet converts XML Form Grammars to ABNF, as well as details of a logical parse structure for speech recognition. Examples are drawn from English, Korean, Chinese, and Swedish.

Speech Synthesis Markup Language (SSML) W3C Recommendation was adopted on 7 September 2004 [254]. The Voice Browser Working Group defined this standard to enable access to the Web using spoken interaction. SSML provides XML-based markup for the generation of synthetic speech, with emphasis on pronunciation, volume, pitch, and rate across different synthesis-capable platforms. The standard summarizes the synthesis process in text normalization, markup support, nonmarkup behavior, text-to-phoneme conversion, prosody analysis, and waveform production. Appendices of the SSML standard include audio file formats, internationalization, MIME types and file suffixes, schema, a DTD, and examples.

10.2.6 Speech Perception Challenges

Even with W3C class speech recognition and synthesis technology, challenges remain. Near-term AACR may adapt existing language processing tools for radio. Such cross-discipline applications include both radio and user domains and methods for translating among them. For example, a speech recognizer alone will not detect a <Topic/> to which the AACR should react from text of a user <Scene/> by itself. Text retrieval and data mining software such as Google™ and the Digital Libraries Project extract documents that share concepts from large text corpora [30]. AACR may adapt word-vector techniques to detect the presence of QoI topics in the user <Scene/>. Unfortunately, the amount of text or speech offered in requests to a PDA may be small. For example, a user might say:

> "PDA, find out about pre-war porcelain from Japan."

Even with an error-free transcript, different perceptions of the intent of cues in this question lead to different responses. Word-vectors from brief utterances often lack a sufficient number of words to be statistically significant regarding the specific topic.

As shown in Table 10-1, the <Scene/> in which the user is situated may imply other interpretations of the question.

Chronic speech perception challenges [41] include reference, anaphora, ellipsis, ambiguity, vagueness, and implicit inferences. For example, the sentences "When the soldiers fired at the terrorists, they fell" and "When the soldiers fired at the terrorists, they missed" are structured in parallel, but the referent of "they" is different. Clearly, people resolve the referent based on the semantics of being shot versus doing the shooting, not purely on syntax.

TABLE 10-1 Speech Perception Context Challenges

Sensory Cue	<Scene/> Cue-Perception	<RF/> Response
"Find out"	<Walking/> Request is global	<WLAN/> 10 Google pages
"Find out"	<Walking/> Request is local	<Wireless kiosk/> "No stores within walking distance"
"Find out"	<Taxi/> User needs directions	<BT-Taxi DB/> "Left on Grev Turgatan; 200 m on right"
"From Japan"	<Hotel/> Data from Japan	<WLAN/> "Can't reach Japan—no Internet"

Parallels in the AACR user domain might include the following pair: "When the kid cries for an on-line game, buy some" and "When the kid cries for an on line game, buy time."

Anaphora is a pronomial reference to previous phrases, typically more challenging than the use of "they" in the prior example. Consider the following two sentences:

1. What is the depth of the *Ethan Allen*?
2. What is its speed?

In sentence 2, "its" refers to the nuclear submarine alluded to by the last phrase of 1, which requires one to know that the *Ethan Allen* is the name of a nuclear submarine, semantic side information. These easily could be AACR <User/> domain sentences.

To clarify the elliptical sentence "John thinks vanilla," one might add "But Mary likes chocolate ice cream." In an AACR setting, the CWPDA might have entered an ice cream parlor within the past three minutes, disambiguating by visual cues. The ice cream parlor's WLAN kiosk might have indicated "Newport Creamery" and listed all 81 flavors. Not all of the commercial tools could interpret John's thought accurately from the speech or text, but the CRA could use vision and <RF/> cues to overcome anaphora and ellipsis.

Ambiguity occurs in human language when there are multiple alternative meanings of an expression or utterance. Ambiguity may be centered on a specific word or in a structure. For example, overloaded words like "take" engender word sense ambiguity: Take five; What is your take on that? Vagueness enters from abstract meanings with multiple specializations, such as "I got a new TV at Macy's." Side knowledge that Macy's is a department store may be assumed, but suppose Macy is a friend on the East Side who just bought a new TV and doesn't know what to do with the old one. Vagueness also occurs when there are lots of implicit inferences, such as in "John went to a restaurant. He ordered a hamburger. He left." Did he eat it? Did he pay? Did the AACR pay? Maybe not. If Genie usually pays by BT, will John be arrested for not paying?

People regularly use causal chains, sequences of events that carry cause-and-effect meaning that may not be perfect, sometimes yielding humor, such as: "John needed money. He got a gun." Like Mike in *The Moon Is a Harsh Mistress*, the world-class CR can take a joke. After picking up the gun, John may have been on the way to the pawn shop, or he may have been on the way to the bank. To follow the more complicated requests from users, CRs will need to know causal chains and ground them in <Scenes/> accurately. Speech acts can also be causal chains, such as "Is there a water fountain around here?" The CR can answer "No, but there's a soda machine down the hall." The act of asking for a water accomplished the speech act of asserting (<Thirsty/> <User/>) to which the CR offered a remedy besides water. These challenges of NL expressions must be met often if the AACR is to do the right thing for QoI. The integration of data across sensory-perception channels seems to offer opportunities beyond that of isolated speech or text processing, suggesting integrated multisensory speech <Scene/> perception.

10.3 VISUAL SENSORY PERCEPTION

Between 1997 and 2000 when the foundational research for this book was underway, some expected a decade for digital cameras to be integrated into cell phones. In March 2004, my wife bought one, years before I expected it to be on the market. The sensor acquires color VGA images in 3 lumens, typical room lighting, compressing them to JPEG sequences for 15 second video clips that she can send to her sister halfway across the country.

Within the decade (give or take five years, of course), people could wear glasses that acquire images from a camera embedded in each side of the frame, projecting prescription-corrected images onto the retina. Such glasses could wirelessly connect to the wearable PDA via BlueTooth, for example, enabling the PDA to see exactly what the user sees (Figure 10-3).

10.3.1 Digital Video

Cell phone class video sensors acquire VGA or megapixel color images with 30° to 40° fields of view and image depth from a few inches to about 30 feet.

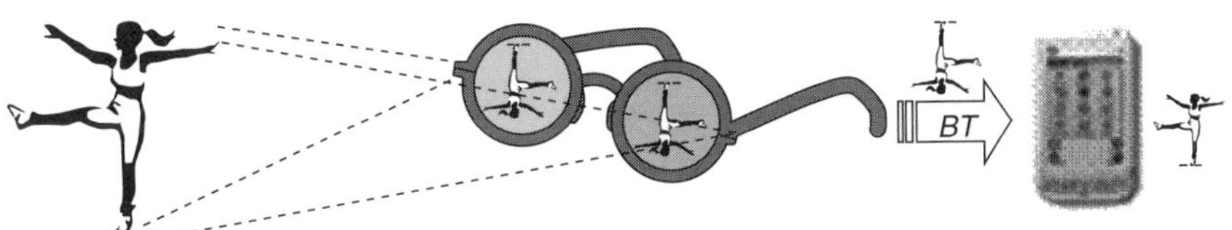

FIGURE 10-3 PDA sees the <Scene/> that the <User/> sees.

Consumer electronics foundations for the CWPDA that sees what the user sees include products like the Divio NW901 Single-Chip MPEG-4 CODEC [255], originally introduced in 2003 for solid-state camcorders. Typical consumer products capture natural-motion 30 fps MPEG-4 video, IJU-R BT656, stereo audio, and 2.1–5 megapixel still images. Images, short clips, and long video sequences suitable for user entertainment may also be subject to AACR <Scene/> detection with in-depth analysis during <Sleep/>. During 2005, 128 MB flash cards were commodity products and 4 GB data sticks cost $150.00.

Image and video stream capture is just the beginning of the visual sensory-perception processing hierarchy. In order to contribute to AACR perception of the user <Scene/>, the perception system must isolate and identify 3D objects in potentially cluttered 2D images. Image sequences reveal projections of partially occluded objects, some of which are moving, and the <User/> typically also is moving in the scene.

10.3.2 Optical Flow

Optical flow is an image of vectors that estimate the orientation and rate of change of local features like texture from one image frame to the next. If one is approaching an object, the flow vectors point away from the center of that object. The larger the vectors, the closer the object. Flow of distant objects is small from frame to frame, so flow-based algorithms detect approaching obstacles to avoid by steering the <Self/> (e.g., a bee or a model airplane) toward benign optical flow.

Figure 10-4 shows the Ladybug™ sensor integrated into a hand-held model airplane that avoids obstacles using the vision sensors, optical flow ASICs, and the general purpose PIC microcontroller [256, 257]. The vision system computes and interprets visual flow in real time. Computing optical flow is computationally intense, typically best if the frame rate exceeds 1000 frames per second. The API enables an autopilot to detect patterns indicating obstacles. For early AACR evolution, one might integrate Centeye chips in a pair of eyeglasses linked via BlueTooth to a CWPDA, enabling recently sight-disabled people to avoid obstacles by feeling a pattern of mild vibrations; alternatively, the AACR could coach the person by VoiceXML when needed, for example, "Door opening to the left." The flow sensors could help the AACR orient to the number of people in the scene, and to the transition from one scene to another, such as movement through a hall from one <Office/> to another. Wireless support needed from the AACR might be very different in one's own office versus a conference room.

The computer vision research literature has long included mathematical models for translating 2D images to estimates of 3D scene coordinates [258], for identifying easily recognized features in scenes (e.g., linear features, blobs), and for calibrating these with a priori knowledge. The Kaiserslautern SPIN system exemplifies such research systems.

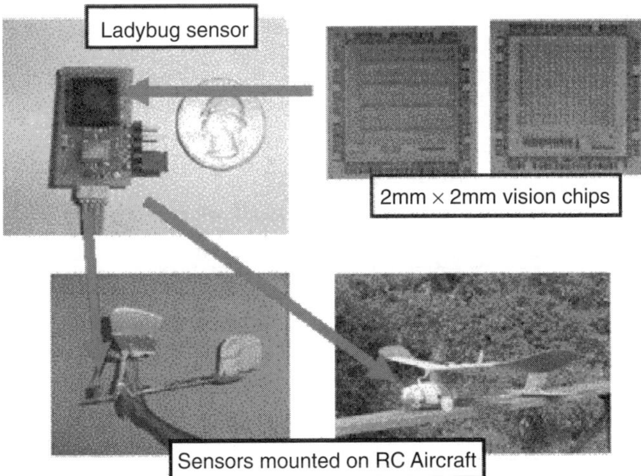

FIGURE 10-4 Centeye Computer Vision Subsystem avoids obstacles via visual flow. (© 2004 Centeye Corp. Used with permission).

10.3.3 Kaiserslautern SPIN System

The SPIN system provides a good example of a machine learning framework for 3D spatial perception. SPIN originally was developed for robotics, as illustrated in Figure 10-5 [259]. Paraphrasing Zimmer, the SPIN system consisted of edge detectors, models, and maps; and abstraction pipelines that identified points of interest.

The Edge-Surface Detector is the source of all information gathered from the environment for SPIN. Its sensor is a laser-range-finder, where several edge-following strategies scan 3D edges. One of the significant differences between the macrorobot domain and the <User/> domain of AACR is the estimation of depth—range between the sensor and the objects in a scene. Robot domains tolerate the laser range finders' size, weight, and power because robots need precise estimates of distance in order to control robot motion. AACR users would not tolerate large mechanical laser scanners that occupy a cubic foot, consume tens of watts of power, and weigh several kilograms. Thus, AACR insertion entails the computation of distances from a pair of cell phone camera quality images, with algorithms following edges instead of laser scanners. Otherwise, the robot and <User/> domains are very similar. For example, both rely on models of a <Scene/> with both a priori and real-time maps.

The models & maps (M&M) subsystem achieves unsupervised building of adequate models and maps from the environment; thus, this is the central structure. Abstraction progresses from the left to right side, where the left-most model ("Point of Interest Map") plays a special role. The first model, "3-D Reality," consists of scanned edges. Dynamic data structures make the

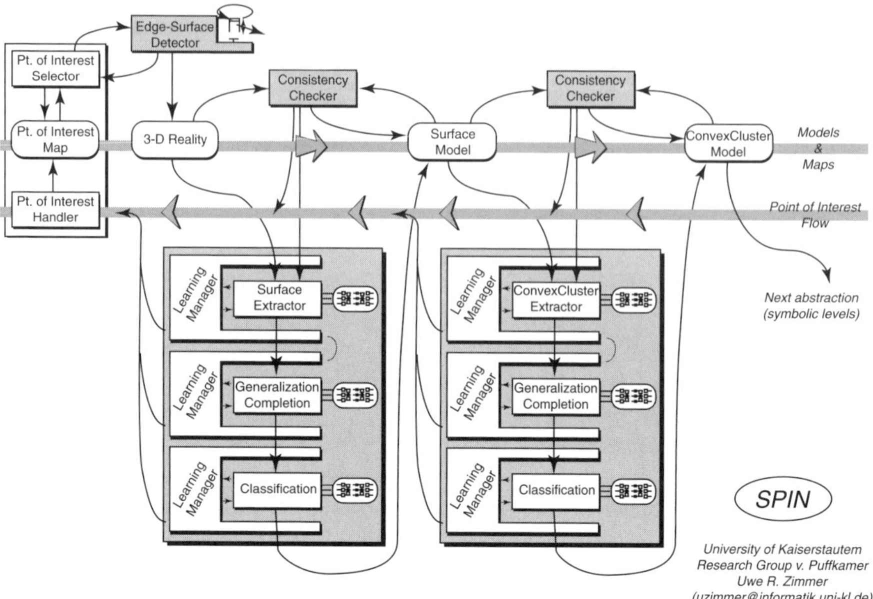

FIGURE 10-5 Illustrative computer vision system. (From [259]. © 1996 Uwe Zimmer and Universitat Kaiserslautern; reprinted with permission.)

model efficient. Edges forming a closed polygon are represented as surfaces together with the connectivity information between neighboring surfaces. The "Surface Model" consists of completed, generalized, and classified surfaces extracted from the 3-D Reality model. The difference between the models at different abstraction levels steers the focus of attention. The consistency checkers between the several stages detect the differences and try to get an explanation for them. The third model, the "Convex Cluster Model," expands to the next higher abstraction based on the surface model. It represents completed, generalized, and classified clusters of surfaces, which are common in the current environment. The interface to the symbolic object-recognition subsystems labels these elements and constructs meaningful objects or features from this model.

Between each two adjacent models on different abstraction levels exists a pipeline of processing steps of the abstraction process. Two abstraction pipelines generate the Surface Model and the Convex Cluster Model. In the first stage the feature that should be used as the basic element on the next abstraction level has to be extracted. This facilitates the subsequent "learning-by-examples" pipeline stages, which are very sensitive to the "bootstrap phase," such as the order of training examples.

Each component in the SPIN structure formulates tasks ("wishes") for the exploration process. This includes requests like "Gather more information at a specific area in the environment" or "Check a specific feature regarding hallucination found during the abstraction process." These requests are collected in a "point-of-interest flow" and stored in a "point-of-interest map." The strategies employed for selecting the actual scan area are called "focus of attention" or "visual search" strategies.

SPIN illustrates key issues with computer vision in support of AACR. The vision subsystem must contain both high level abstractions needed to extract and classify previously unseen objects, along with substantial a priori models of known classes of objects such as people, buildings, furniture, and vehicles. In addition, the identification of such objects from one minute to the next will be errorful. Vision-capable AACR therefore integrates knowledge across sensory domains, including the domain of RF-based location knowledge and user interaction, to reliably identify just those aspects of the scene that must be recognized in order to support the AACR information services use case(s). Although still computationally intensive, the technology includes open source algorithms and well-understood methods for detecting, recognizing, and tracking known types of objects in visual scenes [258, 260–263, 265].

10.3.4 Cognitive Vision

SPIN exemplifies computer vision systems, a rich domain with decades of research, relevant to AACR. The binocular vision system of the Mars Lander from the Caltech Jet Propulsion Laboratory (JPL), for example, provides APIs by which a robot controller interacts with a scene-depth map to avoid "negative obstacles," shallow depressions that are difficult to perceive. Dr. Jim Albus of the National Institutes of Standards and Technology (NIST) has published a standard computer-vision architecture, 4D RCS, that navigates Army vehicles through a forested area without hitting trees [264], suggesting vision-capable AACR in future automobiles.

The summer 2004 issue of *AAAI Magazine* on cognitive vision [265] addresses chronic AACR vision problems. Computer vision based on invariant triples enhances the reliability of object identification from a single digital frame. For example, wheels, windows, and the front grille of an automobile exhibit clusters of triples that reveal the orientation and scale of automobiles in a traffic <Scene/>. In the Bert-and-Ernie use case, a CWPDA worn as eyeglasses might avoid a car, particularly if the oncoming vehicle's AACR detects an emergency <RF/> beep from the child's CWPDA. An AACR that detects rapidly increasing automobile engine and tire noise might yell, "Look out" via speech synthesizer, causing the child to look around quickly and thus enabling the vision or optical flow system to confirm the danger of the oncoming vehicle so the car can be asked via RF to break in milliseconds, not in the seconds it takes for the distracted driver to notice the child.

10.4 AUDIO-VISUAL INTEGRATION

Audio and video sensory-perception channels interact in animals and humans. Integrated audio-visual processing may abstract independent audio and visual scene features to form consistent abstractions for <Scene/> level perception.

Joint audio-visual ASR fuses features for audio-visual automatic speech recognition (AV-ASR [266]). Fusion may be based on either feature fusion or decision fusion. With feature fusion, one classifier concatenates features to enhance audio features, for example, at the subphonetic level. With decision fusion, features are concatenated, but phone, word, or utterance level decisions may be affected. Any video assistance improves word error rates as a function of SNR, a key metric in informal AACR dialog in complicated acoustic settings. Noisy environments benefit most from the video cues, amounting to about a 7 dB equivalent improvement and reducing error rates from over 50% to more like 30% in a 5 dB SNR, which is perceived as pretty noisy.

Lipreading [266] applies to both noisy speech recognition and to reliably recognizing and tracking the user, protecting the user's private information. The lipreading process begins with the detection of the face. The algorithm then finds the lips and estimates whether the person is talking or not to assist in segmenting the speech signal, differentiating foreground from background speech. Potamianos et al. [266] classify visual features for automatic speech reading as based on either (1) video pixels (e.g., appearance), (2) lip contour (or shape), or (3) a combination of (1) and (2). In addition, systems with binary optics also could incorporate (4) distance-based features or (5) visual flow-based features, both of which estimate motion parameters in three-space.

Not all of the speech articulators are visible to an observer. Therefore, the number of visually distinguishable units is much smaller than the number of phonemes. Visible units are called visemes. Speech-readers and statistical clustering are alternative methods for defining phoneme ↔ viseme mappings. Some visemes are well-defined, such as the "bilabial viseme" that maps 42 phonemes into 13 visemes.

AV-ASR therefore goes beyond the typical hidden Markov model (HMM) of speech-only ASR to the composite HMM states. AV-ASR illustrates the benefits of integration across these two sensory-perception domains. Although the computational burden of vision and thus of AV-ASR remains high, the continued optimization of computer vision algorithms and work in ASICs continues to propel us closer to AV-AACR.

AV and <Scene/> integration, particularly with language as a domain of action, occurs in robots that simulate human behavior, such as GRACE, the first winner of the IJCAI Mobile Robot Challenge and Aibo, Sony's toy dog. Commercial robots range from Roomba [267], the autonomous vacuum cleaner, to Asimo, the Humanoid. Each has varying degrees of cognitive capability as illustrated in Figure 10-6.

Kazuo Murano has said that the most important technology change for the coming decade will be the robot: "The robot will probably be the technology of the 21st century as the automobile was the technology of the 20th century.

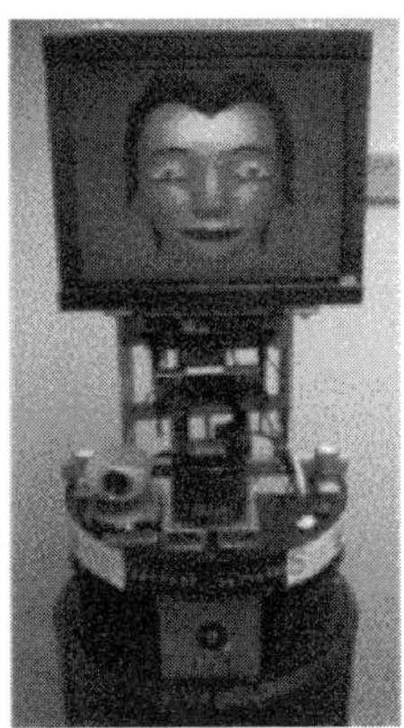

FIGURE 10-6 Asimo Humanoid, Abio Dog Robot (from Sony Corp, AP Photos), and GRACE from CMU.

It is expected to address many social issues in Japan and the other developed nations—coping with rapid demographic change, providing security, and improving the convenience and comfort of daily life" [268].

The robot competitions at AAAI and IJCAI include the mobile challenge for a robot to check itself into the conference and present a talk about itself, including answering questions. Such robots have many of the attributes of AACR, including multisensor perception of the scene, reasoning, planning, making decisions, and taking actions. GRACE (Graduate Robot Attending ConferencE), the first winner of the IJCAI robot challenge, isn't a very attractive robot compared to Sony's Asimo, but she has considerable autonomy. AML is the crucial feature that differentiates GRACE, Abio, and Asimo as cognitive systems from merely artificially intelligent but preprogrammed alternatives.

GRACE is a six-foot-tall, socially oriented autonomous talking robot. It was developed by a team of researchers from Carnegie Mellon, the Naval Research Laboratory, Metrica, Inc., Northwestern University, and Swarthmore College. It successfully completed the mobile robot challenge at the American Association of Artificial Intelligence (AAAI) national meeting in Edmonton, Alberta, Canada on 31 July 2002 [269]. The U.S. Naval Research Laboratory [270] supplied speech recognition, parsing, multimodal speech and gesture interpretation, and human–robot interaction. Metrica provided vision-based gesture recognition. Northwestern contributed speech synthesis, and Swarthmore contributed specialized vision for reading of signs and nametags and recognizing people.

Physically, GRACE consists of a wheeled mobility platform and a cage of electronics on top of which is a display that shows GRACE or the PowerPoint slide presentation of itself. Because of the limitations of the speech recognition subsystem, GRACE responds to questions posed by the keyboard.

Architecturally, GRACE consists of perception, planning, and action subsystems [271]. GRACE's perception capability features speech and vision subsystems, but includes a laser rangefinder to assist in navigation and obstacle avoidance. Instead of a cognition cycle like CR1, control in GRACE is mediated by state changes in the perception and mobility subsystems. GRACE has very limited capability in dialog, typically getting stuck in shallow loops that reveal its lack of knowledge in commonsense domains [272].

CR1's Hearsay serModel is a preprogrammed blank planning capability of CR1 for which stimuli and responses are learned during operation. GRACE generates plans from predefined primitives, not from learned primitives. This showed when interacting with GRACE at AAAI 2004 [273]. At the time, GRACE could engage in small talk, but it was easily led into loops, asking the same question again without any apparent realization that it had just asked the same question. This isn't a criticism of GRACE per se. In other modes, GRACE is known to learn more but to be less predictable, which is a characteristic of the state of AML in 2004.

10.5 LEXICAL CONCEPTUAL SEMANTICS (LCS)

In the search for abstractions by which sound and vision may be integrated, researchers have explored many knowledge and skill representation schemes. Frames, speech acts [12], and conceptual dependency [279] were seminal ideas that have been used for reasoning at a level of abstraction consistent with the <Dialog/> level of speech interaction and with the <Scene/> level of visual perception. Logic operating on frames can reason about possible worlds. Rules operating on frames can synthesize preprogrammed behavior. CBR over frames enables incremental knowledge acquisition. Conceptual dependency introduced PTRANS (physical transfer), ATRANS (change of attribute), and MTRANS (change of memory). LCS takes this formalization a step further. It characterizes words in terms of their contributions to abstract perception of things in space and time, states, and events [274].

LCS has a very small set of primitives by which lexical stimuli (words) bear spatial semantics at a high level of abstraction. The word "room" implies a place with an interior, a spatial mental model. Logic can describe possible worlds with terminals, predicates, quantifiers, and statements, but the predicates make most of the spatial knowledge unavailable except through additional statements like onTopof (Ball, table) $\wedge$ inFrontof (Ball, clock). These abstract semantics are not lexical semantics because the meaning is not associated with the words, but with the larger frame structures.

A <Scene/> of 10 objects requires a thousand (2^{10}) logic statements to fully express how the objects are oriented with respect to each other, a combinatorial explosion. With LCS, the <Scene/> is expressed in 10 statements to which words are referenced, deriving any of the thousand statements as needed. Specifically, LCS annotates nouns as things (people or places that take up space); verbs as states, events, and causes; while prepositions characterize paths (Expression 10-2).

Expression 10-2 Lexical Conceptual Semantics (LCS) Primitives

[THING]
[PLACE] <- PLACE-FUNC([THING])
PLACE-FUNC ε {AT, ON, IN, ABOVE, BELOW, BEHIND, . . .}
[PATH] <- PATH-FUNC([PLACE] | [THING])
PATH-FUNC ε {TO, FROM, TOWARD, AWAY-FROM, VIA, ALONG, . . .}
[EVENT] <- GO ([THING], [PATH]) | STAY ([THING], [PLACE])
[STATE] <- BE ([THING], [PLACE]) | ORIENT ([THING], [PATH]) | EXTEND ([THING], [PATH])
[CAUSE] <- CAUSE ([THING], [EVENT])

John Bender's software uses the LCS spatial characteristics of words to draw abstract pictures; related software parses the abstract pictures to generate sentential descriptions of the abstract pictures and to answer questions about the objects in the pictures [275]. The LCS expression for "John walked into the room" is a sequence of pictures with John on a path from a place outside the room to a place inside the room, denoted as follows:

[Event GO ([Thing JOHN], [Path TO [Place IN [Thing ROOM]]])] (10-1)

John is a <Thing/> as are all nouns in LCS and in RXML. The verb "walked" indicates a GO event, expressed in RXML as the user-domain knowledge that <Event> <Walked> GO </Walked> </Event>. The concepts of <Path-function/> and <Place-function/> give LCS its unique flavor. <Path/> is the conceptual spatial analog of a locus of motion. Some prepositions like "to" and "from" express verbally the functional relationship between a <Place/> and a <Thing/> established computationally by <Path-function/> of LCS. <Place/> is a subset of space–time that anchors <Place-function/>, the relationship between a <Place/> and a <Thing/>. These are expressed lexically by the position and orientation prepositions at, on, in, and so on. This forward-looking research enables one to simply and accurately abstract the salient features of a <Scene/> and to relate <Dialog/> and <Scene/> to each other. Bender's programs, VISUALIZE and DESCRIBE, translate lexical LCS expressions to abstract drawings and conversely. RXML generalizes the <Place-function/> to a set of place-relations among places and things, such as <On> <Thing/> <Place/> </On>. The <Place/> something is on may be intimately associated with another <Thing/>, such as <Top/> of a <Table/>.

Vision researchers have implemented similar performance with different methods, based on logical forms to describe scenes and changes in scenes, for example, so that a surveillance system can report anomalies such as the presence of a suspicious vehicle [276]. These researchers use a situation graph tree, such as the following:

<Situation> <Cross> <Agent/> Intersection </Cross> <!– top level –>
<Drive-to-intersection/> <Drive-on-intersection/> <!– 2nd level –>
<Leave-intersection/>
<Proceed-on-intersection/>speed <Wait-on-intersection/>
<!– 3rd level –>
<Proceed-to-intersection-alone/> . . . </Situation><!– 4th level –>

Using discourse representation theory [277], the authors convert natural language text into a discourse representation structure, which yields text such as "Obje_2 entered the lane," in a street scene. Research in semantic grounding of language extends LCS principles to more complex research frameworks such as situated language in which the features of a visual scene play the primary role in the resolution of multiple anaphora and ellipsis [278]. In the situated language interpreter (SLI), an algorithm models the spatial templates of projective prepositions and integrates a topological model with visual perceptual cues. This approach allows the system to correctly define the regions described by projective preposition in the viewer-centered frame of reference. SLI spatial templates are somewhat implementation-dependent templates for trajectories and entities in prepositional semantics, so rather than adopting any one of the systems like SLI that offer language–vision performance, this treatment employs the fundamentals of LCS to enhance RXML for not just speech–vision integration, but for the integration of other sensors including radio into <User/>-domain semantics.

LCS evolved from conceptual dependency theory [279]; comparing it with discourse representation theory and other logic forms for reasoning [280] and planning leads to the use of LCS in the CRA <Self/> as the primary representation, with a few amendments. Each <Thing/> is also a <Place/>, with positions on the <Thing/> taking the role of <Places/>. The <Top/> of a <Table/> has a geometric relationship, even in the primal sketch, to the center of mass of the table and to the legs. Primal sketches are geocentric. In addition, each <Thing/> is composite, consisting of any number of smaller, typically fully contained <Things/>. Each <Place/> is associated with one or more <Things/>, the default of which is the <Scene/> situated in the <Universe/>, <Global/> . . . <Local/> spatial ontology.

"The leg of the table is on the rug."

Expression 10-3 Sentence and RXML LCS Tags

<State> <Thing> <Thing> leg </Thing> of the table </Thing>
<Being> is <Place> on the <Thing> rug </Thing> </Place> </Being>
< </State>

A part of a larger object is expressed in RXML as a nested <Thing/> as shown in Expression 10-3. In this particular form, the original text is preserved, leaving the articles and other non-LCS items in place. The primal sketch consists of a one-legged table or a default four-legged table, one leg of which is located <On/> the <Thing> rug </Thing>, for example, using the methods of the VISUALIZE algorithm [275].

Since there are many ways to express and reason in the user domain, the CRA does not preclude extensions to LCS (e.g., SLI), nor the use of logic forms encapsulated in APIs as suggested previously. The CRA requires the use of such alternatives with less than tightly bounded computational resources to be deferred to periods where non-real-time performance is permissible, such as during <Sleep/> cycles.

10.6 OTHER SENSORS

Even the integration of speech perception and action with visual perception and explanation is not the final word on sensing and perception in the user domain. Other sensors add orthogonal observations that simplify valid conclusions, reduce search space, and in other ways make the AACR more capable of dealing with the complexities of the user domain.

Specialized sensors for location, motion, temperature, barometric pressure, and even smell enable AACR to express functionality in nonverbal aspects of the <User/> domain. GPS and accelerometers are discussed in the companion CD-ROM.

10.7 ARCHITECTURE IMPLICATIONS

This chapter concludes with the architecture implications of the sensory-perception technologies (see Table 10-2).

Interface 25 conveys speech and vision from user sensory perception to the cognition cycle. User speech is augmented in the perception subsystem and formatted into LCS expressions readily integrated by the cognition subsystem. Similarly, the visual scene is abstracted through the recognition of <Places/>, <Things/>, and <Paths/> in the visual scene, characterizing their presence and actions by LCS expressions for ease of integration in the Observe phase of the cognition cycle.

10.8 EXERCISES

10.1. Consider developing AACR vision in a realistic microcosm, such as that of model vehicles. ExtremeTech (www.extremetech.com) along with many other Internet (www.plantaco.com) and local hobby market suppliers offer mobile

TABLE 10-2 AACR *N*-Squared Diagram for Speech–Vision Sensory-Perception Action

From–To	User SP	Environment	Sys Apps	SDR	Cognition	Effectors
User speech	Phrases	N/A	Commands	N/A	25 PA	31
User vision	Scenes	N/A	Attention	N/A	25 PA	32
Environment user location	Scene reference points	N/A	Controls (pause/ resume)	N/A	26 PDC	33 PEM
Environment user acceleration	State change	N/A	Interactive games	N/A	26 PC	34 SD
Other sensors	Features	N/A	Specialized	N/A	26 PC	
Speech synthesis	N/A	N/A	Commands	N/A	30 PC	

scale model cars, tanks, aircraft, tractors, and other specialty vehicles. These vehicles generally have little computational intelligence, but offer affordable mobility platforms for low cost autonomous vehicles. Since model vehicles entail a self-contained microcosm, they offer a manageable self-contained environment for experimentation with cognition technology. Acquire a digital camera for a model automobile and program it to guide the car around obstacles in the living room to the kitchen and back to you.

10.2. Web sites for the DARPA Grand Challenge (e.g., CMU's) illustrate the high end of ground vehicle technology. Between these extremes are vision-based control systems based on small optical fields such as the 5 gram "Ladybug" sensor from Centeye [256]. If you or your company would have liked to participate in a DARPA Grand Challenge, acquire a GPS receiver, cameras, and a commercial off-road vehicle like the John Deere Gator® and develop your own off-road software.

10.3. What robotic subsystems, subsystems, or software could readily be adapted to AACR from GRACE? Acquire a low cost computer vision subsystem. There are many min-Cams available for laptop computers. Use the API to capture isolated images. Download one of the contemporary machine vision systems from Universität Kaiserslautern or Hamburg, Carnegie Mellon, or elsewhere on the Web. Extract blobs from your vision system using this software. Define the technical challenges in recognizing your laboratory, home, and place of work autonomously. Write the code to differentiate outdoors from indoors based on blob and edge detection. Try the same problem using colors and light levels instead of shapes. Some PDAs include built-in cameras. How well suited are such devices as subsystems to be added to AACR? Define a migration path from PDA to vision-capable AACR.

10.4. Continuing in the spirit of GRACE, acquire an automatic speech recognition software system. Without teaching it, characterize its ability to recognize conversational speech by conducting a 5 minute conversation about the stock market with someone in its presence. A day later conduct the same 5 minute conversation with a different person. Characterize the error covariance between the two conversations. Are the errors and successes speaker dependent? Move the platform to a different setting, such as to a crowded living room with the TV on and people talking. Conduct the same 5 minute conversation with the ASR system operating. Define enhancements to the microphones, acoustic signal processing, and ASR system that the ASR's API or training capability enables. Make the hardware and configuration changes that you define. Train the ASR system to your voice and conduct the previously mentioned tests. Characterize the word- and phrase-level error rates. Give the AACR system a name and during the 5 minute conversation, ask the AACR for help at least three times, using its name.

10.5. Download a speech synthesis system and integrate it with your emerging AACR now consisting of laptop, vision, and speech recognizer. Write high order language script that enables it to find a RXML open tag <. . .>, not <. . ./> as a keyword in CRA <Self/>, and to read the contents delimited by the tag out loud.

10.6. Write glue code that will patch the ASR output to the speech synthesis system so that if the ASR system recognizes a question of the form "Computer, what does X mean?" it will look up <X> in CRA <Self/> and read the contents out loud. Test it with questions about radio. Characterize the kinds of errors this crude question–answer system makes. Ask it increasingly complex questions and note the ways in which it breaks. Define methods to resolve these problems. Write code that enables the AACR to remember the sentences that precede the request for query. Give the AACR data needed for the query. What is needed to link the information to the query? Use the techniques of embedded inference or PROLOG to mitigate the problems by planning dialogs in which the AACR recognizes that it must ask you for more information in order to successfully complete the query.

CHAPTER 11

IMPLEMENTING USER-DOMAIN SKILLS

This chapter continues the process of bringing together speech, vision, machine learning, planning, and location-awareness technologies to achieve user-oriented goals for increasingly intelligent SDR. This is part of an evolutionary process in markets fueled by research in academic institutions and corporate laboratories, guided by government, regulatory authorities, and standards bodies like the ITU, IEEE, Object Management Group, W3C, ARIB, and ETSI; and catalyzed by industry forums like the SDR Forum and WWRF. It is virtually impossible technically, economically, and institutionally to transition from SDR to iCR in one leap. The AACR evolution includes sequences of independently useful steps distributing emphasis among <RF/> and <User/> domains; enhancing radio awareness, adaptation, and <RF/>-enabled QoI for the <User/>:

1. XG "spectrum-aware" adaptive radios will share pooled radio spectrum <RF/>.
2. "User-aware" soft biometrics will improve the security of <User/> information.
3. "Location-aware" radios will adapt RF modes to <RF/> policy per location.
4. "Schedule-aware" radios will adapt <RF/> to workdays, holidays, vacation, and other temporal patterns, optimizing information selection, aggregation, and sharing per <User/>s.

Cognitive Radio Architecture: The Engineering Foundations of Radio XML
By Joseph Mitola III

5. Niche-focused AACRs become niche-domain-aware: patient-aware for <User/> healthcare; crisis-aware for emergency <RF/> services; and golfer-aware for real-time advice on aligning the shot at any hole at any golf course in the world.

To realize such steps, cognitive radios must go well beyond SDRs:

1. To accurately perceive the local environment, including the NL expressions and visual scene in which the <User/> is immersed with the beliefs, desires, and intent of the <User/> learned through the instantiation and evolution of UKOs guided by shared families of UDH reflecting focused classes of user.
2. To efficiently apply radio knowledge by the continually improving use of <RF/> bands and modes available on the physical radio platform, including the effective application of UDH with SDH and RDH to evolve situation-specific KOs for physical RF propagation, communications modes, legacy radios, networking, and wireless information services.

These evolutionary processes may be facilitated by implementation technologies that enable the computationally effective interaction of the <RF/> and <User/> domains, specifically (1) to employ ontological standards that address both the informal <User/> domain and the many technical <RF/> domains, and (2) to interactively learn from <User/>s, AACR peers, and host CWNs.

These last two design objectives or mandates suggest how difficult it is to separate <User/>-domain skills completely from <RF/>-domain skills.

11.1 INTEGRATING COGNITION

The software components that implement the sensory-perception domains include those in Table 11-1.

The structure of <Scenes/> conveys to the orient or plan phases of the cognition cycle by the perception-level User SP interfaces. The CRA envisions the labeling of objects on these interfaces via LCS or equivalent so that the entities and relationships in a <Scene/> are faithfully represented. Although there are many research and applications engineering challenges, VISUALIZE [275] suggests engineering principles for annotating speech transcripts with <Place/>, <Thing/>, <Path/>, and <Action/>, the RXML generalizations of EVENT, STATE, and CAUSE of LCS. Similarly, DESCRIBE suggests approaches for translating the blobs, edges, and recognized objects from vision to a level of abstraction consistent with multisensory DHs for QoI enhancement. In a RXML/LCS framework, satellite navigation

TABLE 11-1 Sensor Domains With Perception Interfaces

Sensor Domain	Perception Interface	Illustrative Components
Audio acoustics	LCS errorful transcript	IBM's ViaVoice
Vision	Conceptual sketch of recognized 3D <Objects/> in recognized 3D <Scene/>	4D RCS, Centeye [256]
Satellite navigation (GPS)	<Place/> on a terrain map	Garman Navigator
Acceleration	<Path/> mobility signature	Crossbow IMU tool set

RF, visual cues, and speech map to <User/>-defined <Place/>s, like <Home/>, <Work/>, <Golf/> and the <Supermarket/>.

11.1.1 Integrating Perception Into Awareness

The canonical information structures, such as grammatical English and simple objects in a visual scene, are rarely present in <User/> task domains like <Home/> or <Work/>. Most scenes are complex. Objects are partially hidden or physically occluded; words are occluded by noises. The degree of occlusion is not necessarily constant over time, so people naturally view scenes from different perspectives and over time, aggregating and integrating the scene.

Similarly, <User/>-domain <Scene/> aggregation algorithms must reconcile objects and actions perceived in a scene over space and time. The following scene occurs outdoors.

Expression 11-1 Simplified Outdoor Scene Object in RXML

```
<Scene ID = 1>
  <GPS Time = 38394.5064583333> <Location>
    <Latitude> 28.5635 </Latitude>
    <Longitude>-81.7534</Longitude> </Location> </GPS>
  <RF Time = 38394.5064583333> <AM/> <FM/>
    <Cellular> GSM </Cellular> <WLAN> Null </WLAN> </RF>
  <Speech Time = 38394.5064583333> <Humming/> </Speech>
  <Vision Time = 38394.5066235451> <Tree/> <Grass/> </Vision>
  <Vision Time = 38394.5067583321> <House/> </Vision>
  <Vision Time = 38394.5068583344> <Door/> </Vision>
  <Duration> 5.78 </Duration> <Exterior/> </Scene>
```

The <Scene/> latitude and longitude were generated when GPS became available to the CWPDA as the owner stepped out of the car, transitioning from BlueTooth WLAN to no WLAN. <Exterior/> is asserted by the high RSSI of the <GPS/> <Signal/>. Speech and vision assert LCS objects into the

<Scene/> with time of observation. There are myriad uncertainties and frame alignment challenges for a <Scene/> integration algorithm to restart a <Scene/> after it has been lost for a threshold period of time (e.g., a minute, to prevent thrashing hypothesized new scenes when GPS is flaky).

11.1.1.1 Scenes and Microworlds

The CRA does not specify linkages across scenes, but postulates microworlds and linkages among microworlds, with some CRA <Self/> exemplars. To preserve LCS, a new scene may be formed when a new <Sketch/> needs to be generated; for example, because the <Place/> has changed, the <Self/> has no anchor points, the <Self/> senses movement along a <Path/>, or the configuration of an <interior/> has changed through actions like rearranging the furniture. Thus, <Home/> would be a microworld with each <Room/> and separate space as a <Scene/> with an associated LCS <Sketch/>.

11.1.1.2 Time Structures Perception Into Observations and <Paths/>

Perception at a specified time is an <Observation/>, asserted into the current <Scene/> as it is perceived. If sufficient time passes, a watchdog timer terminates reasoning over unrecognizable features linked to the <Scene/> to manage combinatorial explosion. In a sequence <Vision Time = 38394.5066235451> <Tree/> <Grass/> </Vision> and <Vision Time = 38394.5067583321> <House/> </Vision> the time tag enables the cognition system to construct a space–time path with trees, grass, and a house. If the AACR knew from the in-car BlueTooth that it had just been in a car, it can construct an embryonic <Path/>:

Expression 11-2 Embryonic Path From Car to House

<Path> <Place> Car </Place> <Time t = 0> <Scene ID = <Owner's-car/>/>
</Time>
<Time t = 1> <Tree/> <Grass/> </Time>
<Time t = 2> <House/> </Time> <Where?/> </Path>

The path has not yet ended, so it terminates with the concept <Where?/>, asserting that the end of the path is not known. Such RXML paths can be traversed in either forward or reverse directions and when something is recognized to complete a loop, the AACR knows the <Path/> along the loop. By counting the reoccurrence of such paths, the AACR learns user paths through simple reinforcement. In the primal loop <Home/> <Car/> <Work/> <Car/> <Home/>, there are many small loops, for example, from car to <Grass/>, <House/> and back.

11.1.1.3 Novelty Suggests Scene Boundary Hypotheses

If the perception system detects strong <Novel/> features of a known <Class/> of <Scene/>, a new <Scene/> may be asserted as a KO for subsequent AML to balance the combinatorial explosion of recognition processing against isochronous interaction with the <User/>.

Expression 11-3 Owner Enters Home for the First Time

<Scene ID = 2> <Vision Time = 2/11/2005 12:20:59/>
<Novel>
<Thing> <Chair> <Position> <Distance> 4.5 <Meters/> </Distance>
</Position>
<Color> Brown </Color> </Chair>
<Floor> <Position> <Distance> 2 <Meters/> </Distance> </Position>
</Floor>
<Couch> <Position> <Distance> 3 <Meters/> </Distance> </Position>
<Color> Brown </Color> </Novel> </Thing>
<Unrecognized> <Blob> 3242 <Color> RGB . . . </Blob>
</Unrecognized> </Vision>
<Speech Time = 2/11/2005 12:20:59>
<Transcript> hi honey . . . I'm home </Transcript>
<LCS> <Thing> honey </Thing> <State> I'm </State> <Place> home
</Place> </LCS>
<Speaker ID = <User> Joe </User> </Speech>
<Place> <Hypothesis> <home/> 0.25 </Hypothesis> </Place>
</Scene>

A flood of <Novel/> <Thing/>s of known types reinforces the hypothesis that the <Scene/> has changed, for a new <Scene ID/>. Although still computationally intensive, vision algorithms can detect, recognize, and track known types of objects in such visual scenes to yield a <Sketch/> of <Novel/> but recognizable new <Thing/>s. As the <User/> moves through the scene, some previously unrecognized <Blob/>s may be recognized, causing the recognized <Thing/> to be asserted and the <Novel/> <Blob/> to be retracted. As a new owner proceeds to different rooms of the house, new <Wall/>, <Picture/>, <Door/>, <Window/>, and related interior <Things/> and <People/> will be sensed. Over time, with reinforcement and with <Sleep/> cycles, the AACR with evolved UKOs and UDHs for aggregating <Scenes/> repeatedly experienced over time may converge on <Home/> <Scenes/> populated with <Things/> and transited by <People/> and other animate entities like pets, some of which the AACR will assist with information access via typical conversations that reoccur in the scene.

Awareness can be thought of as a process of mutually resolving noisy and conflicting sensory perceptions across auditory, visual, and navigation sensors into a data structure that describes a very complex scene in terms computationally workable for use cases such as a LCS <Sketch/>. Scenes and the microworlds that they describe vary over time, yet, the iCR must generate plans, make decisions, and take appropriate information access actions on

TABLE 11-2 Levels of Adaptation Depend on Abstractions in Data Use

Sensory Data	Abstractions in Data Usage	Level of Cognition
GPS location	Presented to user	GPS location-aware
GPS location	Location-based LAN power up/down as preprogrammed	Rule-based location-adaptive
GPS location	Learned that WLAN is corporate LAN RKO/RDH applied to power-down LAN while commuting; UKO/UDH indicates to first ask user for permission	Location-aware, user-aware autonomous machine learning (AML) enabled cognition

behalf of the user. Adaptation DHs align scene features with stereotypical behavior <Paths/> so that a QoI-enhancing behavior morphs to the scene, still enhancing QoI.

11.1.2 Perception-Driven Adaptation

Not all awareness translates into action. A CWPDA with a GPS system doesn't necessarily do anything as a function of its location. The user might like to know the location. The PDA might even present location on a map. That is not location-adaptive AACR behavior because the PDA itself isn't using the information. If the PDA turns on its IEEE 802.11 LAN card because it detects that it is within 1 km of the owner's place of business and thus starts looking for the corporate LAN, then the PDA is taking a specific action based on location. Such a preprogrammed behavior would be termed adaptive. Powering the LAN card down when there is no hope of connecting to the corporate network both saves power and prevents the PDA's LAN card from being hacked. Electronics packages have been trained to do this by neural networks [282]. With LCS extracted from a <Scene/> one may evolve levels of cognition suggested in Table 11-2.

In the first entry of Table 11-2, the PDA senses location but doesn't use that data itself. In the second entry, the behavior is preprogrammed. These are good examples of AACR location-adaptive behavior, but there is no <Self/> and no machine learning so there is no cognition. In the third case, the AACR applies general rules for the conservation of power (RDH) to its own battery state (RKO) to determine that it should turn off the WLAN since it is not likely to be connected to the corporate WLAN soon on its current <Path/>. Interacting with the <User/> reinforces major decisions retained in a UKO, guided by UDH, a high level form of AML by discovery with reinforcement.

Toward such ends, the CRA structures AACR into AML-enabling functional components mediated in part through LCS sketches (Table 11-3).

AACRs encounter conflicting contextual cues from vision and other sensors that interfere with their ability to behave appropriately. If the vision

TABLE 11-3 Contributions of Functional Components to LCS

Component	Contribution to LCS
User sensory perception	Identity of people in the scene, beliefs, desires, and intent of user
Local environment	Asserts location, temperature, acceleration, compass direction, and so on
System applications	Goals of media-independent services like playing a network game
SDR subsystem	RF access opportunities and SDR state parameters
Cognition functions	Symbol grounding; system state, plans, reinforcement
Local effectors	Speech synthesis; text, graphics, and multimedia displays

system tells the AACR that there are no other people in the living room, but there are many loud voices, then it could postulate a TV in the room. However, writing rules that capture this and a myriad of other possibilities is a brittle approach that does not scale-up well. The CRA therefore offers LCS-enabled reasoning with UDHs over UKOs. The simplified UKO of Expression 11-4 expresses the fact that TV produces conversations. The UKO employs LCS primitives in describing the behavior of a TV program as dialog, which are expressed in speech and which occupy a path from the TV to one or more persons.

Expression 11-4 UKOs Avoid Specialized Reasoning at the <Scene/> Level

```
<UKO> <Name> <TV/> </Name> <TV> <Thing/> </TV>
  <Path> <From> <TV/> </From>
    <Speech> <Dialog> </Program> </Dialog></Speech>
    <To> <Person/> </To>
  </Path> </UKO>
```

A corresponding UKO that <People/> hear <Speech/> and a UDH in the sense of VISUALIZE enable the algorithmic sketching of an abstract picture with the TV in the living room producing conversations, such as that of Figure 11-1. Given the UKO and UDH about how people can listen to speech from a TV set, along with others about the equivalence of speech dialogs and conversation, the AACR can reconcile the fact that its speech subsystem observes conversations even though its vision subsystem observes no people in the room to generate them. The possibility that the conversations are generated by people in the next room, for example, is dealt with through the association of <Room/> level scenes into <Home/> level microworlds with sketches of paths among <Rooms/>. Although complex, the combinatorial explosion of logic expressions for such scenes remains daunting, while the algorithmic

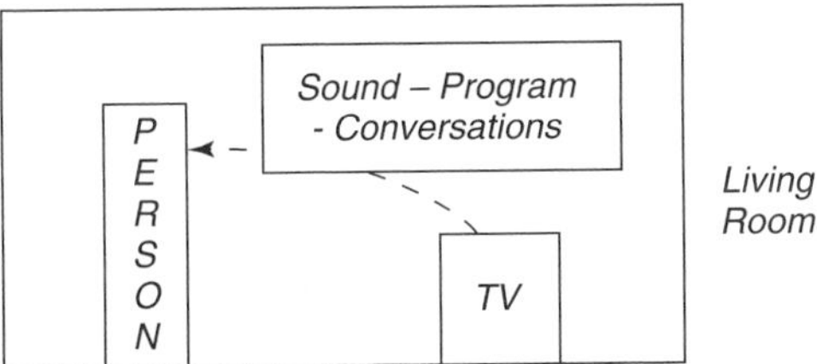

FIGURE 11-1 LCS sketch of "person hears TV program."

traversal of such primal sketches to address such conflicts appears orders of magnitude more manageable, although not trivial either.

Specifically, to manage concrete computational complexity, the CRA <Self/> envisions LCS or equivalent primal sketches as the primary means for representing scene-level perceptions from multiple sensory-perception tracks over space and time. Conversations in a room may be realized as broadcast channels in which everybody can hear everything, or in a larger room such as a banquet hall where the intensity of the background conversations render only the nearest people as conversational partners. Each sensory-perception channel participates in the population of observations into the <Scene/>. All <Scenes/> encountered are remembered and during <Sleep/> cycles they may be merged or taken apart to overcome such problems, to enhance future behavior.

Such sketches enable AACR software to deal more effectively with the complexities of scenes. In particular, a wearable algorithm need only examine the <Path/>s attached to <Person/> to determine what sounds it should hear. If there are ten objects in the room (a couple of chairs, some toy trucks, two doors and a window, a vase full of flowers, and some wall hangings), then in a logic system, there are (10 factorial) 72.5 million inferences to be made about the way sound impinges on an object. Since the number of inferences grows exponentially with the number of objects, complicated scenes with many predicates require substantial space–time computing resources. Although suitable for CWN and potentially for AACR <Sleep/> cycles, such logic formulations appear to be less suited to real-time behavior support than LCS primal sketches.

Consider the problem of navigating an AACR through a cityscape that is full of WLAN enclaves, viewers of weak TV stations, and the like, which the AACR's private WLAN must avoid (e.g., in the Genie use case). Once at home, the AACR docks wirelessly into its own Home RF environment, which is the goal of the transit from work to home. This is not unlike guiding a submarine through a minefield to the home port. The penalties for making mistakes in the radio domain may not be as immediate, but they could be severe. The minefield domain consists of a perception space, an action space, and a set of interactions including successful navigation and the consequences of striking a mine. The perception and action spaces are as follows:

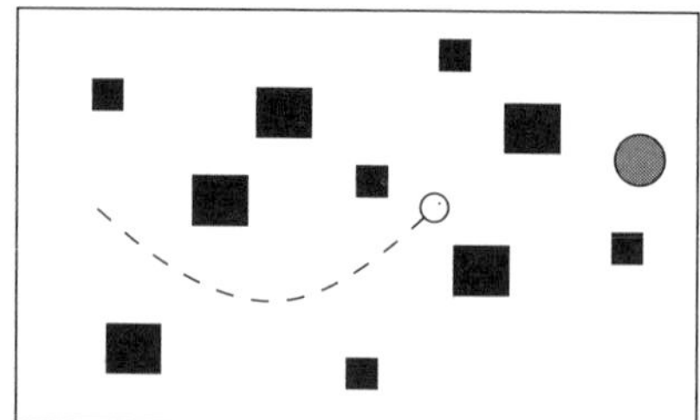

FIGURE 11-2 Sketch of minefield and goal.

SIMPLIFIED MINEFIELD PERCEPTION–ACTION SPACES

Perception Space	Action Space
Condition primitives (16)	Movement primitives (2)
Sensors: sonar [7] and infrared [6]	Controls: throttle and wheel
Range and bearing to goal	Speed (−4 to 12 in./s)
Current speed and steering rates	Steering rate (−30° to 30°/s)

The process of evolutionary computation [283] derives both reactive and symbolic control laws from a control genome that includes neural networks and symbolic primitives that may be composed into control laws. In a simulated environment [53], a robot is placed at an origin with random pose, with random placement of obstacles and goal, and with random obstacle density. Noise, including Gaussian noise, false positives, and false negatives are added as populations proceed toward the goal. The simulations are massively parallel, enabling a population to reach the goal in time $t(i, x)$. The population is thus pruned, and the survivors are randomly mutated and combined to yield a next population of more fit individuals. When a sufficient performance has been reached or simulation time has been exceeded, the most effective control law is downloaded. The primal sketch of the minefield is illustrated in Figure 11-2, with the size of detected mines reflecting the uncertainty of the position of the mine.

In the minefield application, the entity has only an incomplete map of the scene, relying on local sensors for discovery of both the mines and the goal. The AACR application of Sun's minefield avoidance control schema to interference minimization requires representation of the <Self/> in an appropriate primal sketch. A GPS-based estimate of the position of the <Self/> in an urban scene has the drawback that GPS often glitches in such locations. A visually capable automobile that reads road signs or an AACR that reads the metropolitan BlueTooth <Location/> service at each street intersection could navigate by GPS aided by a computational model of user movement plus a sketch based on streets and street intersections. The associated primal sketch could refer to places with the city of Metropolis in terms of street intersections:

<Place> <Thing> Metropolis
<Place> <Intersection> <Thing> <Street> Street-1 </Street> </Thing>
<Thing> <Street> Street-2 </Street> </Thing>
</Intersection> </Place> </Thing> </Place>

The essential elements of a corresponding XG-like radio spectrum management space are as follows:

AACR SPECTRUM ADAPTATION PERCEPTION–ACTION SPACE

Perception Space	Action Space
Condition primitives (16?)	Transmission primitives (3)
Scan [t]; noise/modulation [j, t]	Power (dBm); burst timing ($P(t)$)
Range/bearing to goal {Recipient, Legacy}	Beamforming ($P(\theta)$)
Band, mode, bandwidth, and power occupancy	Band, mode, bandwidth, and data rate

Mines correspond to legacy users such as regions where TV reception favors normal TV usage versus ad hoc networking. The Genie TV spectrum-use case might employ this strategy to express <RF/> constraints spatially. Each RF region is a <Thing/> that exists in space–time, where space is defined using streets as boundaries. The AACR itself is a <Thing/> that consists of many other <Things/> including <TV-WLAN/>, the physical extent of which may be defined in terms of city blocks as a function of mean adjacent building height as well as the transmitted power and the threshold power above which the <TV-WLAN/> interferes with normal <TV/> reception.

The condition primitives are those sensory parameters that the AACR can sense via its SDR subsystem. CRs and CWNs estimate range and bearing between the <Self/> and the competing user or receiver. The CR with a street map of Metropolis can <Sketch/> locations of competing <TV-WLAN/> reported by a host CWN on its own map, along with regions of good reception of broadcast TV. As with all RXML–LCS primal sketches, the coordinates are 3D or 4D to include time, so reception above the 5th floor of high rise buildings is easy to establish.

The action parameters for conventional radio networks are just power and bandwidth, but with AACR, there is a choice of band and mode to leverage XG policy for secondary users. With MIMO, particularly on vehicles and fixed facilities, beamforming shapes the transmission in space–time. Thus, the evolutionary spectrum navigation process consists of the following:

1. AACR placed at origin (e.g., place of work) with random user movement plan.
2. Representative placement of interference, competing nodes with expected variations.

3. Noise (Gaussian; false positives and negatives) added to reflect error syndromes.
4. Populations reach goal (or not) in time $t(i, x)$ representative of AACR travel time.
5. Download most effective control law to AACR from CWN.

The control law could include reaction to mitigate interference and overcome link impairments, for example, using a neural network to control power and MIMO parameters. The symbolic level of control adjusts the choice of unoccupied TV channel to smoothly avoid interfering with known primary and secondary users while maintaining connectivity. Deliberate control at the symbolic level may occur in the <Plan/> phase where the transition from UHF 33 to UHF 44 occurs at the <Intersection/> of <Street> 4th </Street> and <Street> Main </Street>, an abstract yet precisely specified time and place for control via RXML–LCS primal sketches.

To introduce <User/>-domain challenges into this apparently nicely solved problem, one only need consider the unpredictability of users. "Seems like Ed and Diane live in Metropolis," comes the remark from the chase vehicle. "You know, we haven't seen them in years," the <Owner/> replies. "They sent us such a nice card over the holidays," and we all know what is coming next. "Genie, can you see if you can reach Ed and Diane so we can say hello on our way through Metropolis." If Ed and Diane were known to <Genie/> then all would be well. <Genie/> looks them up in the <Directory/> and sets up a connection between <Diane/> (now known to <Genie/> and thus a conceptual primitive) and <Spouse/> that leads to an impromptu visit, which changes the path through <Metropolis/>. If <Genie/> advises the CWN of the change of plan, then it will be able to advise of a more efficient spectrum-use plan for staying connected by <TV-WLAN/> as the vehicles transit different neighborhoods on entirely different time lines to visit Ed and Diane, stay the night, and be on their way in the morning.

However, the technology challenges of autonomously maintaining primal sketches becomes clearer if <Genie/> is new to the family and has never heard of Ed or Diane. Introducing new types of knowledge to AACR is much more difficult than merely instantiating known types of knowledge into a primal sketch. The successful solution of this class of problem without the mediation of expensive knowledge engineers calls for the autonomous incremental knowledge acquisition for broad classes of task-related knowledge.

11.2 AUTONOMOUS EXTENSIBILITY

Autonomous incremental knowledge acquisition is another technology that has existed since the 1980s but that for a variety of technical and economic reasons has yet to be fully deployed, at least not in the <User/> domain of AACR-class radio engineering. This section explores the foundations of autonomous extensibility for use in the flexible aggregation of <User/>-domain competence through interaction with <User/>s.

Tieresias [318] was the first autonomously extensible rule-based system. It was implemented as the autonomously extensible form of Mycin, the expert system that diagnosed bacterial infections of the blood. Some features of Tieresias appeared in applied expert systems in the AI boom of the 1980s, including tools like ART and KEE [284]. Principles were refined and applied in agent technology, for example, in telecommunications in the 1990s [94, 285]. These applied expert systems generally did not exploit Tieresias' methods that interactively interpreted schema–schema to transfer knowledge. Tieresias itself was known to be brittle, extensible by those who deeply understood its internal data structures and KA quirks, but not readily extensible by the typical user.

This section takes a closer look at the schema–schema method of Tieresias for <User/>-domain knowledge customization for AACR. This provides a foundation for the embedding of *autonomously extensible* rule bases into the <Orient/>, <Plan/>, and <Decide/> phases of the CRA, realized in UKO, UDH, and LCS primal sketches to improve the robustness of autonomous extension for casual users.

11.2.1 Extensible Knowledge Bases

Rule-based inference systems employ a global representation of the problem space such as a blackboard or game board [12] and use condition–action rules (IF → THEN) that work well in closed-world applications like board games, where the state of the board changes, but the size of the board and the legal moves do not change during a game. Extensible rule bases add legal moves during a game.

In open settings like medical diagnoses or autonomous adaptation to the infinite variety of quirks of wireless PDA users, the size of the game board of relevant knowledge is continually growing in some ways and shrinking in others (e.g., as constraints are discovered). As radio spectrum-use policy evolves and new downloads become available, new legal moves are constantly being added to the SDR repertoire. As consumers introduce new electronic systems equipped with these new <RF/> capabilities, the ways in which the <User/> can employ them explode exponentially. The CRA addresses such open domains as a discovery game board. The rules may consist of embedded rule bases, PROLOG logic bases, and autonomously acquired serModels with substantial <Self/> modification. This section introduces such autonomous extensibility in the extensible rule base of Tieresias.

An illustrative rule from Tieresias is

If *the gram-stain (attribute) of the culture (entity) is negative (value of gram-stain attribute) & context*

then *disease might (20%) be E. coli (classification decision)*

Here the game board consists of clinical information about patients and the related diagnostic conclusions of medical professionals. The rule uses the entity-attribute-value model of object-oriented behavior. The conjunction "*& context*" assures that the rule fires using the data from the current patient. Tieresias' top-level control system continually accesses the patient database to infer the identity of a bacterial infection of the blood. Such rules can be implemented as objects themselves with condition–action slots and corresponding methods or attached procedures that perform the inferences, test conditions, and assert an inference. In Mycin, the rules were implemented by computer scientists interacting with medical doctors in the process now widely known as knowledge engineering.

Tieresias enhanced Mycin to interactively acquire incremental knowledge beyond core Mycin [318]. Tieresias interpreted medical data from expert users, lab technicians, and medical doctors just like Mycin, but synthesizing "automatic knowledge acquisition." Mycin applied facts and rules to deduce likely causes for symptoms, but Tieresias used abstractions of rules to guide the acquisition of new rules. The rule models were "schemata," declaratory data structures that could be *interpreted* algorithmically to acquire new rules without further programming, theoretically from nonexpert users.

The Identity-Schema defines an organism in the open world of Tieresias. As shown in Figure 11-3, the acquisition of the identity, for example, of an

Identity-Schema

```
PrintName (Atom AskIt)
Value (PrintName InSlot)
PropertyList
                    ((InstanceOf (Identity-Schema                    GivenIt)
                    Synonym        ((Kleene (10) <Atom>)             AskIt)
                    Description    (String                           AskIt)
                    Author         (Atom                             FindIt)
                    Aerobicity     ((Kleene . . .                    AskIt)
                    GramStain      (GramInstance                     AskIt)
                    Morphology     (MorphInstance                    AskIt)
                    Date           (Integer                          Create)
                    Create}
Updates             (AddTo         (AND Organisms))
Instances           (Acinetobacter, . . . Yersina)
StringTranslation   "The identity of an organism"
Father              Value-Schema
Offspring           Nil
Description         "the Identity-Schema describes the format for an organism"
Author              Davis
Date                1115
InstanceOf          Schema-Schema
```

FIGURE 11-3 Identity-Schema defines the data structure for an organism.

organism, entails creating a name for the entity and filling in its property list. Some properties enhance interaction with the user, such as the Synonym and Description properties. The string text of Description describes the entity to the user. Domain-specific properties enable reasoning about the entity, such as Aerobicity, GramStain, and Morphology. Each property consists of a property name followed by a tuple (class advice). The class is the inverse of InstanceOf, a backward pointer to the class to simplify bottom-up tracing of the Mycin class hierarchies.

Some clever data structures that give the program traces more of a natural feeling, substituting for NL generation ability, which is not a criticism. For example, the StringTranslation of Identity is "the identity of an organism," which makes automatic rule traces read naturally:

"If the identity of an organism is unknown, then ask the user"

The algorithmic acquisition of properties is guided by "Advice" that points to facilities by which to acquire the knowledge, particularly using the following methods:

ADVICE SCHEMES FOR THE IDENTITY-SCHEMA

AskIt	Ask the user
CreateIt	Manufacture the value from existing data
FindIt	Retrieve it from an internal database of facts or rules
GivenIt	Use the contents of the blank slot as the value, like LISP Quote
InSlot	Use the contents of the slot pointed to as the value

Sometimes, the knowledge source (KS) is an existing data structure, in which case the KA system is performing inductive inference (FindIt, GivenIt, InSlot). In other situations, the advice is to "ask the user," with data that structures the interaction.

The rule schemata were themselves based on a root schema called the Schema-Schema. While Mycin interpreted rules to form diagnoses, when necessary Tieresias interpreted the Schema-Schema recursively to acquire new rules. This schema-schema-guided KA typically was invoked when the Mycin core ran into a reasoning log jam and thus had to ask the user to clarify the reasoning, and in that clarification ran across new entities, properties, or rules. The Identity-Schema is based on the Schema-Schema, the meta-metalevel and most general data structure that described how all Tieresias' data structures themselves are structured (Figure 11-4).

The schema-schema describes the structures of the knowledge schemas of Mycin. There is a simple elegance to being able to express in one page the core recipe or seed from which a kind of fractal recursion grows data structures to yield an entire system, but that is the nature of a seed. It acquires material from its environment, selecting those things that contribute to the

Schema-Schema

```
PrintName          (Atom               CreateIt)
Structure          (PrintName          InSlot)
PropertyList       {(PrintName         ((Blank-Instance Advice-Instance)   AskIt)
                   Structure           ((PrintName InSlot)                 GivenIt)
                   PropertyList
                     ((InstanceOf      (((PrintName InSlot) GivenIt)       CreateIt)
                     Description       (String     AskIt)                  GivenIt)
                     Author            (Atom       AskIt)                  GivenIt)
                     Date              ((Integer   CreateIt)
                     Kleene            ((SlotName-Instance (Blank-Instance
                                         Advice-Instance)                  AskIt)
                   CreateIt}
                   Father              (Schema-Instance                    FindIt)
                   Instances           (List                               AskIt)
                   StringTranslation   (String                             FindIt)
                   InstanceOf          (Schema-Schema                      GivenIt)
                   Description         (String                             Create It)
                   Author              (Atom                               AskIt)
                   Date                (Integer                            CreateIt)
                   Offspring           ((Kleene (0) <Schema-Instance>)     AskIt)
                   Updates             ((Kleene (0) <(UpdateCom-Instance Kleene(1)
                                              <(SwitchCom-Instance Kleene(1)
                                         <Knowledge-Structure-Instance>)>)>)
                                       AskIt)
                   CreateIt}
Father             Schema-Schema
Instances          (All-Schemata-List)
StringTranslation  "knowledge structure"
InstanceOf         Schema-Schema
Description        "The schema-schema describes the formal for all other schemata"
Author             Davis
Date               876
Offspring          Nil
Updates            (AddTo (AND* All-Schema-List)
```

FIGURE 11-4 Tieresias' Schema-Schema. (Schema shown here have been expanded from the original for easy reading, such as Identity instead of IDENT.)

underlying purpose encoded into the seed. Davis didn't describe it like that, but it is very important for AACR engineers to realize that schema-schema implicitly define all the properties of all the systems that can be based on that particular seed. In an AACR open world model, such a schema-schema strongly shapes the <User/> interaction with the <Self/>. The CRA <Self/> is a starting point for an evolved <Self/>, the schema-schema for constantly evolving families of AACR. To see how this can work, consider the Schema-Schema, from which Tieresias derives the medical domain schemata of Figure 11-5.

Knowledge Structure (Knowledge-Structure-Schema)
Value-Schema
Site-Schema Identity-Schema
Parameter-Schema
PiParm- InfParm- CulParm- OrgParm-Schemata
SVCP- MVCP- TFCP-Schemata

FIGURE 11-5 Schema derived from Schema-Schema.

By traversing the schemata hierarchy interactively, Tieresias elicited knowledge structures (mostly from Davis), rules, kinds of parameters, and the values of parameters associated with medical diagnoses from medical doctors. The resulting rules were added to the knowledge base to forward chain under uncertainty to diagnose bacterial infections of the blood, just like Mycin. The core algorithm maintains pointers into both the KB and the schema-hierarchy while traversing the schema-schema and other domain schemata. Tieresias bootstrapped diagnostic decision trees from the medical doctors to construct rules that made classification decisions based on the values of attributes of the entities, such as the gram-stain of blood.

11.2.2 Bootstrapping the CRA

There are many steps between Schema-Schema and a working system. Figure 11-4 can guide the instantiation of knowledge structure schemas, which can then be instantiated into values and parameters, which in turn instantiate into Site-Schema, Identity-Schema, PiParm, and all the rest in Figure 11-5. Similarly, the CRA <Self/> isn't a system, it is a template for bootstrapping knowledge–action templates, KOs, and domain heuristics, yielding diverse classes of system built up through experience. To enable such AACR evolution, CRA <Self/> includes only core ontological knowledge in the <RF/> and <User/> domains, indicating paths to skill via bootstrapping like Tieresias. There are many pitfalls, so the balance of the chapter suggests ideas to mitigate brittleness and to overcome other challenges of AACR evolution.

The CR1 research prototype learned via serModels and CBR (with the patience of the researcher) a few things in <User/> and <Radio/> domains autonomously and thus in an interesting way. The CRA <Self/> takes a step very similar to Schema-Schema, functioning as a blank sheet of paper onto which engineers can write by supplying improved interpretation algorithms and then using and training the resulting AACRs. Algorithms for evolving competent AACRs from the CRA <Self/> are not complete, but algorithm strategies suggest research and engineering directions.

11.2.2.1 XML Tag as Metaschema Strategy

Tieresias may apply to AACR evolution by informing the interpretation of RXML statements as Schema-Schema, which is not the typical method of interpreting XML. XML may be processed as strongly typed (e.g., using

Schema-Schema	
PrintName (Atom CreateIt)	
Structure	(PrintName InSlot)
Father	Schema-Schema
Instances	(All-Schemata-List)
StringTranslation	"knowledge structure"
InstanceOf	Schema-Schema
Description	"The schema-schema describes the formal for all other schemata"

FIGURE 11-6 Schema-Schema without property list.

DTDs), less rigidly typed (e.g., using XML-Schema), or weakly typed, as with many domain-specific languages in which the XML tags serve the pragmatic purpose of simplifying information retrieval (e.g., by tagging words with domain-specific lexical semantics). XML tag as metaschema takes some hints from the semantic web research, but with each XML <Tag/> in RXML in the role of object schema. Tags are asserted via the ontological and set-theoretic expression <Tag/>, which means that there is a set of abstractions named "Tag" that have the role of ontological primitive, a map to a subset of the real world. The specifics of that map depend on where the <Tag/> is asserted and how it is grounded through experience. One may follow the Schema-Schema from the top to see how <Tag/> can act as a metaschema. Every Tieresias schema has a name—the PrintName (Figure 11-6).

Radio XML uses XML <Tags/> where the print-name for <Tag/> is "Tag." The structure of <Tag/> may be expressed in XML syntax via BNF:

<Syntax> <BNF> <Tag/> := "<" <Tag/> "/>" | "
<"<Tag> "/>" * "</" <Tag/> "> "</BNF> </Syntax>

In the LISP notation of Tieresias, each schema has a name and property list, which includes a structure specification. The property list of Schema-Schema has Structure, Father, Instances, StringTranslation, InstanceOf, and Description (Figure 11-6). Each property has not just a value but also "advice" on how to deal with the property such as InSlot if the value is in the slot or CreateIt, a function that makes a LISP structure. With RXML, that which is delimited by a <Tag/> is an element of the <Set/> defined by <Tag/>. That which is sensed has a referent in the <Physical-universe/> that is inferred by the <Self/>, for example, as a primal sketch. Comparing sensory data to primal sketches enables grounding, the state in which the primal sketch or equivalent conceptual model of the world corresponds sufficiently to the external reality inferred from the sensory data for AACR QoI enhancement. Any failure to enhance QoI constitutes <Error/> that the <Self/> seeks to drive to zero. Such QoI-reinforcement grounding of <Tag/> falls *far short* of the general symbol grounding problem.

11.2.2.2 CRA <Self/> as Schema-Schema Strategy

The CRA isn't just an architecture specification, but is a processable XML document with which neither DTD nor XML-Schema have been rigidly associated. Traversing the CRA <Self/> synthesizes a computationally intelligent <Self>-modifying entity. This is appropriately recursive with <Self/> in its own definition, so to avoid the Gödel–Turing trap, the <Self/> may be recursively interpreted only in <Sleep/> cycles with a (hardware) watchdog timer that the <Self/> cannot control.

The <Self/> expresses a <Universe/> that consists of the broad classes <Abstractions/> and <Physical-universe/>. The outer wrappers of the knowledge represented in CRA <Self/> is as follows:

```
<CRA> <Self> <Universe>
  <Abstractions/> <Physical-universe/>
</Universe> </Self> </CRA>
```

The ontological stance is that the only universe that exists to a computationally intelligent entity is the one that is within the <Self/>, and that this inner universe defines both an inner universe of abstractions like Truth and an external universe of less ambitious abstractions like True (observable) and False (Boolean constant for not internally consistent) that is accessible via sensors. The universe within is the perceived universe, founded on the idea that there is an external universe accessible through sensors and perception. Furthermore, the LCS <Self/> is that which the <Self/> can <Control/> while the <Outside/> is that which the <Self/> can sense, but not directly control ontologically. The LCS primal concept <Place/> is introduced (Expression 11-5).

Expression 11-5 Distinguishing Self From Outside by Sense Versus Control

```
<Self> . . . <Place/>
  <Place> <Self/> <Outside/> </Place> <!– Two broad classes of place –>
<Action/>
  <Action> <Sense/> <Control/> </Action> <!– Two types of action –>
<Control> <Self/> </Control> <Self>
  <!– The <Self/> is that which is within the scope of <Control/> –>
<Sense> <Outside/> </Sense>
  <!– The <Outside/> is that which is within the scope of <Sense/> –>
  . . . </Self>
```

From this ontological schema-schema, the <Self/> inherits the properties of that which is controllable, a functional definition ("If I can control it, it's ME"). The outside is that which is available to the sensory system of the AACR ("If I can sense it but not control it, then it's NOT ME") and <Action/>

influences. Bootstrapping from this ontological declaration as schema-schema provides a strategy for the autonomous algorithmic differentiation of <Self/> from <Outside/> so that it does not have to be preprogrammed into the system. That which is going on <Outside/> is algorithmically uncontrollable and thus observably different from that which is going on inside. To the degree that the two match, autonomous symbol grounding occurs. This symbol grounding occurs when internal patterns are reinforced as referring to the same thing, such as voice reinforcement that generalized triples refer to automobiles. The first derivative of symbol grounding is mutual grounding where both the AACR and user agree that certain classes of triples are <Automobiles/>, extending the ontological primitives through an ASR grounding event. The evolutionary acquisition of autonomously acquired symbols derive from the way the <Self/> is expressed in the CRA. Since overcoming evolutionary brittleness remains a core computer science challenge, early evolution must stay close to the <Self/>, focused on AACR QoI and mediated by CWNs.

11.2.2.3 Primal <Self/> Portrait Strategy

Formulating <Sense/> and <Control/> as <Act/>s lays the groundwork for an internal structure of AACR as a set of <Paths/> with an initial primal sketch of the <Self/> (Expression 11-6).

Expression 11-6 Primal Maps Sense, Perceive, Abstract, Remember, and Act Are Paths

```
<Path/> <Path> <Place/> * </Path>
  <From/> <To/> <From> <Place/> </From> <To> <Place/> </To>
  <Path/> <Path> <From/> <To/> </Path>
  <Self> <Action>
  <Path> <Sense/> <Perceive/> <Abstract/> <Remember/> <Act/>
  <Think/> <Effect/> </Path> </Action>
  <Place> <Observation/> <Perception> <Idea/> <Memory/> <Controls/>
    </Place> </Self>
  <Sense> <Path> <From> <Outside/> </From>
    <To> <Observation/> </To> </Path> </Sense>
  <Perceive> <Path> <From> <Observation/> </From> <To>
    <Perception/> </To> </Path> </Perceive>
  <Abstract> <Path> <From> <Perception/> </From>
    <To> <Idea/> </To> </Path> </Abstract>
  <Think> <Path> <From> <Memory/> <Idea/> </From>
    <To> <Idea/> </To> </Path> </Think>
  <Effect> <Path> <From> <Controls/> </From>
    <To> <Outside/> </To> </Path> </Effect>
  <Remember> <Path> <From> <Observation/> * <Perception/> *
    <Idea/> * </From>
```

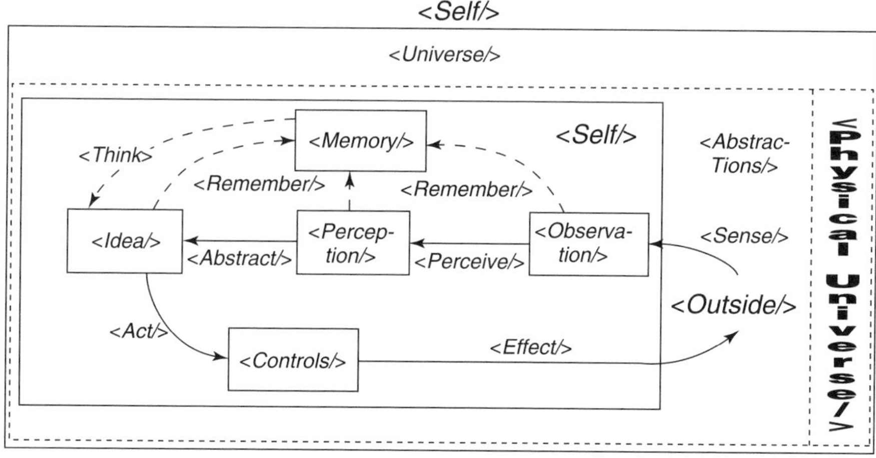

FIGURE 11-7 Primal Self portrait.

<To> <Memory/> </To> </Path> </Remember>
<Act> <Path> <From> <Idea/> </From> <To> <Controls/> </To>
</Path> </Act>

A <Path/> consists of one or more <Place/>s, with optional endpoints <From/> and <To/>. Within the <Self/> <Action/>s (generalizations of the LCS verb forms GO, BE, etc.) are also <Path/>s among abstract internal <Place/>s, collections of data generated and interpreted by traversing this schema-schema. The <Idea/> is the <Place/> for the results of the <Act/> of <Abstract/>ing <Perception/>s. Memory is the <Place/> within the <Self/> for the storage of <Interesting/> (in the sense of the AML <Histogram/>) <Observation/>s, Perception/>s, or <Idea/>s. Since an <Act/> is a <Path/> from an <Idea/> to <Controls/>, an <Idea/> may be a <Plan/> in the cognition cycle. At this level of abstraction, temporal reasoning and set theory do not exist, so <Plan/> cannot be fully expressed. A primal <Self/> portrait (Figure 11-7) may be generated by traversing this schema-schema.

The use of <Self/> as schema-schema remains a challenging research issue. With present technology, however, it is feasible to autonomously evolve the CRA <Self/> into the <User/> domain via AML techniques like Q-learning.

11.2.3 Incremental Reinforcement Learning (RL)

Declarative knowledge alone is insufficient to guide AACR autonomous extensibility in the <User/> domain. Reinforcement can be a means of incrementally acquiring knowledge. Q-learning makes the degree of exploration explicit in the parameter $\gamma(s, a)$ for a transition from system state s via candidate action *a*, and all forms of RL entail both the exploitation of current knowledge and the exploration of the unknown.

11.2.3.1 Exploitation: Finding WTOP

The <Self/> may be used as a schema-schema for knowledge–action evolution by the exploitation of current knowledge via RL.

Suppose a <User/> tunes to WTOP every morning. The AACR has a goal of enhancing <Information/> <Preferences/> of the <User/>, so it needs to know if the <User/> is expressing a <Preference/> by tuning to WTOP. One way for a CR to find out is to ask. This can be a pain in the neck to the user and not very informative to the AACR.

> *AACR*: "What are you doing with the radio?"
>
> *<User/>*: "I'm listening to it you dummy."
>
> *AACR*: "No, I mean the channel."
>
> *<User/>*: "You are the radio, why don't you know what channel?"
>
> *AACR*: "Sorry, I know what channel but what information are you getting from that channel?"
>
> *<User/>*: "I'm driving so be quiet, will you?"

This continues until the AACR is returned for a refund or there is a traffic accident and lawsuit.

In a more focused exchange the CR monitors the <User/> <Tuning/> to FM channel 107.7 in <Location> Fairfax County, VA </Location> reading the channel's <Annotation> News </Annotation>. The ability to read the text annotations is declared in RXML to the <Self/> (Expression 11-7) as a <Sensor/> along the <Sense/> <Path/> from the <FM-Radio/> of the <Outside/> to the <Observations/> inside the <Self/>.

Expression 11-7 Reading FM Captions

```
<Skill> <FM-Listen>
  <Outside> <FM> <From> <RF> 88 <Units> MHz </Units> </RF>
    </From>
    <To> <RF> 108 <Units> MHz </Units> </RF> </To> </FM>
      </Outside>
  <Action> <Call> read-fm-text.dll <CMD-line> <RF/> = ?X
    </CMD-line>
    <Returns> <Observation> <FM/> = ?X <Audio-Stream Stream = S
    <Annotation> WTOP NEWS </Annotation>
    </Observation> </Returns> </Call> </Action> </ FM-Listen>
      </Skill>
```

The channel to which the user is tuned has the text <Annotation/> "WTOP NEWS." The CRA <Self/> lists "News" as an <Information/> <Service/> for which a <User/> may express a <Preference/>. Thus, the AACR could ask the <User/> if it is OK to infer

<User> <Information> <Preference> NEWS
</Preference> </Information> . . . </User>

If the user says "OK," then the association is reinforced but it may be suppressed if the user says, "No, I'm listening to traffic and weather together on the eights, not to news right now." The <Self/> annotates positive and negative reinforcement separately and aggregates them when planning an action in a given state, forming $Q(s, a)$ for "NEWS" versus for "traffic and weather." RL to exploit existing knowledge might proceed via a focused dialog like the following:

AACR: "I would like to generalize from an observation."

<User/>: "OK, CR, go ahead." [Reinforces the inclination to generalize.]

AACR: "WTOP's broadcast is labeled news. Does that sound correct?"

<User/>: "Yes, CR, WTOP is a news station." [Reinforces confidence in FM annotation. CR might not parse that response but may just look for the positive reinforcement of "yes" without negatives.]

AACR: "Then can I generalize that you like news radio stations?"

<User/>: "Of course. But I like other things too." [Reinforces <Preference/>.]

AACR: "OK, I will conclude you like news, but I will not take this to be an exclusive inference regarding media content. Is that OK?"

<User/>: "OK." [Reduces news <Preference/> reinforcement.]

The above interaction is a type of mixed-initiative that the <Self/> enables through the <User/> <Information/> <Preferences/> tags and associated data structures of the <Self/>. That kind of dialog quickly becomes as tedious as the PAL dialog at DARPATech 2005, where each brief phrase had to be confirmed by the PAL with "OK" or "Alright." Researchers can stand this but most consumers will not.

11.2.3.2 RL Supports Incremental Planning

The WTOP dialog implies that CR generated a plan based on an observation. RL inherently supports planning since RL aggregates reinforcement for alternative actions in given states. The choice of action given a state is the core operation in <Decide/> to <Act/>, and with such decisions, there is a plan. To conduct complicated sequences of actions may require further plan generation techniques. Learning to parse unknown languages and learning to solve challenging puzzles like Rubik's Cube require state–space plan generation capabilities relevant to AACR. In particular, state–space planning over a vocabulary that characterizes a finite domain can infer rules for finite state languages (FSLs), like the simpler subsets of those embedded in air interfaces and computer protocols. FSLs are generated by finite state machines (FSMs) with only N states and at most N^2 transition rules. Each rule specifies a condition for transitioning and a lexical action to be taken. The induction problem for FSLs is decidable. An N-state FSL can be decided with test sequences of length at most $2N$ [286]. The worst-case number of possible sequences of length $2N$ is 2^{2N}.

Advanced methods for inducing the languages and sequences of state transitions for FSMs include the use of macro-operators that do not preserve intermediate goal states. Constrained FSMs like Rubik's Cube can be solved efficiently using macro-operators [287]. If an AACR-CL is an N state FSL, then it can be learned completely by another iCR in 2^{2N} steps, which is under a million for a ten state metalanguage. Induction of a FSM that is isomorphic to the FSM generating an N-state FSL is feasible, for example, using inference methods like those of INDUCE [288, 321]. Methods for learning transition rules for FSLs enable AML of FSMs for iCR [321, 289]. Automatic planning [e.g., 290] extends state–space planning to planning over graphs, with constraint satisfaction, propositional satisfiability techniques, and heuristic control over time and resources with and without uncertainty.

The plan generation activity of the dialog, therefore, may be mediated by one of the known plan generation techniques as a lower level component of the <Plan/> phase of the <Cognition/> component of AACR. The plan for WTOP was initiated by the UDH observation that the annotation on a broadcast ("NEWS") matched a feature of <Information/>, which can be a <User/> <Preference/>. The plan consists of a series of actions to obtain reinforcement of such QoS-related incremental inferences.

In this case, the AACR observed the <News/> annotation and wants to be sure that the annotation represents the Owner's internal world model. This is a strategy of model verification in exploiting existing knowledge. After verifying that the user regards news to be a characteristic of WTOP, the CR verifies the generalization that the user likes news radio stations.

The AACR <Self/> has a <Motive/> to provide information services to users via radio (Expression 11-8).

Expression 11-8 Motive UDH to Use <RF/> to Provide Preferred Information to User

<User> . . . <Motive> <Event> <Action> <Begin> <RF/> </Begin>
  <Achieve> <Goal> <Information-service> <Preference/>
    </Information-service> </Goal> </Achieve> </Action> </Event>
      </Motive>
. . . </User>

The <Preference/> may be instantiated based on the dialog and UDH as in Expression 11-9:

Expression 11-9 UKO Realization of <Preference/>

<Preference> <News> <Evidence>
<Self> "Then I would like to generalize that you like radio stations with news type of media content?" </Self>
<Boss> "Of course. But I like other things too." </Boss>
<Source> <Owner/> <Scene> <Communting/> </Scene>
<Place> Fairfax, VA </Place> <Time> 2/19/2005 3:03:20 pm </Time>
</Source> </Evidence> </News> </Preference>

The UDH <Motive/> from CRA <Self/> by AACR instantiates the <Preference/> UKO analogous to the way knowledge-schemata extend knowledge in Tieresias. This process does not introduce new kinds of knowledge, but creates new instances of known types of knowledge. The reinforcement includes the recorded dialog that affirms the inference along with the source, place, and time of the reinforcement. The annotation regarding <Source/> is critical because other sources could assert that the <Owner/> likes <Hard-rock/>, and these might be <Scene/>-dependent, not <Evidence/> to the contrary. The space–time annotation enables context resolution during sleep cycles, so the <Owner/> can like <Hard-rock/> when with friends who like it, but can like <News/> in other <Scenes/>. The context sensitivity of user preferences implies problem solving as complex as the Rubik's Cube.

11.2.3.3 Exploring World States

The alternative to exploitation is exploration, using UDHs to enable decisions in unknown directions. In the dialog regarding news, the CR was exploiting prior knowledge, but a CR could proceed as follows to explore new knowledge:

AACR: "What other radio broadcast media types do you like?"
Boss: "I like radio stations that play old songs from the fifties too."

[CR matches all words in this response with the twenty feature-values of its broadcast-media objects. The only one that correlated 2 sigma above random was the string "OLDIES" correlated with the word "old."]

AACR: "Do you mean OLDIES stations?"
Boss: "Exactly." [Pause]

This second dialog reveals other aspects of reinforcement learning. The lead question about "other radio broadcast media types" entails risk. Instead of merely exploiting prior knowledge obtained from the broadcast, a UDH fires to postulate from the "but" in the reply to "news" that the <Preference/> may be nonexclusive. The exploratory dialog-UDH has a schema-schema <Action/> of AskIt. Reinforcement learning calls this type of resource use "exploration" and addresses the question of how best to mix exploration and exploitation [92]. The AACR mix depends on <User/> and <Scene/>.

11.2.3.4 Defining a Mix of Exploitation and Exploration

Suppose an AACR can generate a plan to perform one and only one of the following actions at a time:

Task#	Action
0	Be quiet and do nothing unless Owner clearly asks for a known <Action/>
1	Perform an <Action/> using its associated parameters:
	(a) Ask about default parameter values of an <Action/>
	(b) Ask about user-specific <Action/> parameters
2	Ask about potentially relevant <Actions/>, or
3	Ask about <Observations/> of the <Scene/>
	(a) Ask about <Novel/> <Observations/>
	(b) Ask about reinforced <Observations/>

The question of which <Action/> to try influences the acceptability of the CR to the user. We want the <User/> to like the CR. Each task has a value, for example, defined in Q-learning by Quality(state, action). An incremental update rule ascribes a new value to the action in the given state, Quality(state, action) as a function of the reinforcement of the anticipated action, plus the expected value of the other possible actions from that state.

$$Q_{k+1} = (1-\gamma(s,a))Q_k(s,a) + \gamma(s,a) f(s,a,(\max|_b(Q_k(s,b)))) \quad (11\text{-}1)$$

If the maximum value of $Q(s, b)$ dominates, then the decision favors action b, an exploration of insufficiently reinforced alternative actions. Thus, if the CR makes an observation, for example, that WXYZ is an Oldies station, the CR can just be quiet, Action-0, or it could perform some other action. If the CR is quiet and does nothing, then the inference that Owner likes Oldies does not occur. If the CR always makes this choice, then reinforcement from the environment accrues as the Owner repeatedly tunes to WXYX, but the link to Oldies is implicit rather than explicit.

Suppose the CR has an initial learning fraction of $\gamma(s, a) = 0.5$. Reinforcement accrues for positive responses. If the value of Action-0, saying and doing nothing, is reinforced because of the lack of negative reinforcement, then the CR will just sit there and not try anything else. To overcome this class of side effect, exploration is randomized.

11.2.3.5 Randomness for Exploration

Randomization is an artificial mechanism to assure that behaviors are attempted so that the estimated value of the behavior from a given state represents reinforcement. One way of exploring alternatives is to cycle randomly through the candidate behaviors for some period of time or number of occurrences, computing the new value after an evaluation time *T*. An alternative would be to try each in turn some number of times *N*. By forcing behaviors to occur, the relationship between candidate actions and the environment has a degree of correlation. Unfortunately, the <User/> may not be thrilled with the promise of exponentially effective behavior over infinite time if the AACR is perceived to be untrainable at the outset.

Thus, an approach to getting more <User/>-centric behavior early in a training process would be to initialize the UDHs by asking the new Owner questions that extract broad strategies as early in the training process as possible. The CR could ask metalevel questions such as "Would you prefer it if I talk a lot or keep quiet?" If the <Owner/> replies, "That depends, . . ." then the AACR could randomize behavior to discover the limits. In particular, in the algorithm of Equation 11-1, initializing all of the values high and providing small incremental reinforcement causes the system to search, exploring rather than exploiting success because alternatives seem better than initial values. Conversely, initial values of zero reduce exploration, driving decisions toward generating plans that use whatever tasks are first given positive reinforcement. In the ML literature, there are different methods for dealing with tasks that are performed continually versus episodic tasks (see [92] for the algorithms and mathematical analyses).

11.2.4 Extensibility Through Temporal Reasoning

RL includes the temporal difference algorithm, which is helpful in its own right in the implementation of <User/>-domain skills. Following [92], one may learn by noticing differences of the state of the environment over time. This requires algorithms that employ time as a universal index, that compute equality versus differences, and that associate time, equality over time, and differences over time in a way that autonomously extends the AACR's knowledge and skill. The CRA formalizes time as a schema-schema from which a temporal reasoning strategy may be bootstrapped.

11.2.4.1 <Time/> as a Schema-Schema

RXML expresses space–time as fundamental, with both space and time as abstractions that cannot be independently experienced, but to which one can refer as a dimension of experience.

Expression 11-10 Space and Time Are Subsets of Space–Time

```
<Place> <Space-time/> <Space/> <Time/> </Place>
   <Place> <Space-time/> <!– Space-time is the fundamental domain . . . –>
   <Space/> <!– Space is an abstract subset of space-time . . . –>
```

<Time/> <!– Time is an abstract subset of space-time . . . –>
</Place>

As a LCS <Place/>, time is also a <Thing/> and thus may be attributed extent through the <Extend/> <Action/> for <State/>. Since the <Self/> has <Memory/>, the experiences of which are tagged with <Place/> and <Time/>, the AACR that uses the CRA can compute differences, such as between <Observations/> of the same <Thing/> with different <Time/> tags—that is, made at different times.

<Time/> is a schema-schema in several ways. As a <Place/> it has a dimensional extent. Thus, the lexical semantics of <Place/> apply to <Time/>, such as an ability to locate a <Thing/> <At/> a <Time/> or within an extent of time. <Time/> has associated <Metric/>s, the domain of values of an associated <Action/> to <Measure/> <Time/>.

Expression 11-11 Metrics for Time

<Place> <Time> <Metric> <Point>
<Era/> <Year/> <Month/> <Day/> <Hour/> <Minute/> <Second/>
</Point> </Metric> </Time> </Place>

Time <Intervals/> are defined as point-sets, unidimensional <Places/> in which <Things/>, including other <Places/> and <Intervals/>, may be located.

11.2.4.2 Temporal Difference Learning

Temporal difference learning (TDL) algorithms derive a behavior that (1) observes temporal differences based on <Time/> tags (which generalizes Sutton and Barto [92] slightly) and (2) seeks to reduce the difference in sequential predictions to zero through either updates to the estimates (e.g., in policy verification) or through actions (e.g., to reduce the differences by selecting better plans). In simple board games like queens, tic-tac-toe, or checkers, one may ascribe a predicted future value to a state. Value(s) could be the number of one's own pieces on the board in checkers (counting Kings as 2 pieces). A move from a given state is performed by an <Action/> <Move/>, which applies a rule to the board to yield a new state:

$$\text{Move}(\text{state, rule}) = \text{new-state} \tag{11-2}$$

If the system is in state s, and the state after s, induced by the player's move and the opponent's move, is s', then if the player contemplates a move in which a piece could be lost, the state could have a lower value. One would like to reduce the potential loss of value of a good strategy from state s, since otherwise Value(s') is less than Value(s).

TABLE 11-4 Sequence of Temporal Events

Step	Activity/Event	Connectivity	*T*	*dT*	*T*-to-Go
1	At the office on Friday	Corporate WLAN	−90	30	
2	Plan trip home 5 pm	Various	−60	60	40
3	Depart office 6 pm	Corporate WLAN	0	5	45
4	Approach car	Car BlueTooth	5	1	46
5	Depart parking lot	BT-Cellular	6	2	40
6	Navigate to Thruway	BT-Cellular, WTOP	8	10	30
7	Transit Thruway	BT-MIMO	18	20	10
8	Navigate home	BT-Cellular, WTOP	38	10	0
9	Enter house	Home WLAN	48	3	0
10	Arrive home		51		

Source: Reference [92], Chapter 6.

$$\text{Value}(s) = \text{Value}(s) + k[\text{Value}(s') - \text{Value}(s)] \tag{11-3}$$

If Value(s') > Value(s), then Value(s) increases by some fraction of the difference at each iteration. In other words, if we can get to a better state from s, then let's reflect that in the value ascribed to s itself so when we make our first move, we head in the right direction from the outset. Temporal difference learning can work well for small scope problems like board games and it may work for appropriately simple prediction learning tasks in the <User/> domain.

Consider the predictions of connectivity during a daily commute (Table 11-4).

In this vignette, the <User/> calls home at 5 pm with a plan to be home "in 40 minutes," which the CWPDA notes as the predicted temporal Value(Depart office, Arrive home), which nominally is 40 minutes, allocated to 10 minutes from work to the Thruway, 20 minutes on the Thruway, and 10 minutes more in the suburbs to arrive home. Suppose the new Thruway offers MIMO wireless LAN connectivity for low cost VoIP during the transit, with traffic and weather reporting to enable drivers to optimize the trip. The CWPDA predicts its usage for the home CWN based on the <User/> 's established patterns. This particular <User/> prefers VoIP on corporate or home WLANs and is trying out the new MIMO WLAN on the Thruway for the next month, offloading the Cellular network. The CR depends on the CWN to predict its user's behavior from both previous patterns and from the details of local context. Table 11-4 shows the standard pattern of the trip home, with a time delay from the usual departure time of 5 pm to a current departure of 6 pm. The TDL algorithm rewards itself for accurately predicting time remaining in the trip home, *T*-to-Go in the table. At 5 pm, it offers the following plan to the CWN:

Expression 11-12 Nominal Plan for Transit Home

<User> <Plan> <Place> Work <Time> 5 pm </Time> </Place>
<Place> <Service> <Cellular>
<Time> 5 pm <Interval> 10 minutes </Interval> </Time> </Cellular>
</Place>
<Place> <MIMO> <Time> 5:10 pm <Interval> 20 minutes </Interval>
</Time> </MIMO> </Place>
<Place> <Cellular> <Time> 5:30 pm <Interval> 10 minutes
</Interval> </Time> </Cellular> </Place> </Plan> </User>

When the <User/> is delayed, the Value(Depart work, Arrive Home) increases with respect to the original start time of 5 pm, initially extending an hour until the <User/> actually departs the office, moving the <Plan/> to 6 pm. The <User/> talks to a colleague in the parking lot, further extending the plan. With TDL in this policy verification mode, the initial estimate of an immediate departure is updated to Value(Depart office) = 68 minutes, up from zero for this particular instance. The next time the <User/> says he will depart at 5 pm, the Value(Depart office) could reflect either the nominal plan Value(Depart office) = 0, for an immediate departure ($k = 0$, or no aggregation of experience), or it could reflect the most recent departure of 68 minutes delay ($k = 1$, no credit for planning), or it could reflect some fraction $0 < k < 1$ of the old plus the new, updating the policy to reflect experience modulated by a degree of relevance. This is a direct application of the theory of TDL. A CR can use <Scene/> features to deal with such discrepancies between a policy and experience.

11.2.4.3 Context for Temporal Differences

Since the AACR continuously examines sensory-perception channels, it need not be limited to the manipulation of k as the only controllable feature in TDL. Instead, since the temporal plan of Expression 11-12 sets the framework for the TDL algorithm, the TDL and <Plan/> may be integrated. Activity 2 of Table 11-4 itself defines a <Place/> with <Extent/> in the <Time/> dimension. That activity entailed an increase of Value(Depart work) from 0 to 68 minutes.

Although increasing context sensitivity has the potential to increase QoI to the user, it also has the possibility of contributing to a combinatorial explosion of hypotheses and therefore must be undertaken with steps that avoid combinatorial explosion, for example, by relegating analysis of such patterns to <Sleep/> cycles or to a CWN.

11.2.4.4 Avoiding Combinatorial Explosion in TDL

TDL can't be implemented as-is for large problems, however. For example, in games with few states like tic-tac-toe, all moves from all states can be computed and remembered. For even relatively simple games like checkers (played on an 8×8 board, each square of which may be occupied by black or red or unoccupied), the number of board states is 2^{65} or 3.69×10^{19}. Today's memory limit for a laptop is about 1 billion (10^9) states (2 GB of memory), which is ten orders of magnitude too small. The generation of such states occurs in the <User/> domain when CRA <Self/> templates are interpreted as schema-schema. An important way to mitigate combinatorial explosion is the aggregation of information over time, setting relatively high thresholds for UDH pattern detection algorithms to spawn UKOs.

An alternative to unacceptably large numbers of states is to use partial data or side information to assess the likely quality or value of a state. A mathematical heuristic is a state evaluation algorithm that runs quickly to completion and that bounds the cost of reaching the goal from a state. The classic AO* algorithm uses such mathematical heuristics to perform branch-and-bound partitioning of large state spaces to eliminate from further consideration those that are unlikely to get to a goal more quickly than the best current plan [196]. In UDH and RDH such mathematical heuristics can manage combinatorial explosion of TDL.

Reinforcement may be normalized or not. For example, CR1 counts reinforcement by integer values, remembering that the <User/> listened to WTOP for "news" once this hour, while listening to "traffic and weather" four times. TDL with nonzero k and Q-learning normalize the reinforcement to 1.0. Learning about WTOP as either a news channel or a weather and traffic channel may be accomplished via either of the following:

CR1 (Preferences (news, 2), (weather, 4), (traffic, 4))
Q (Reinforcement (news, 0.2), (weather, 0.8), (traffic, 0.8))

Whether normalized or not, these metrics reflect uncertainty about the <User/> in the environment. The technologies for estimating certainty, uncertainty, and degree of belief are informed to some degree by the theory of probability, but there are many open issues. Nevertheless, the robust AACR must deal effectively with degrees of certainty and uncertainty as well as degrees of belief and other fuzzy aspects of the <User/> domain. Explicit supervision by <User/> or CWN also manages uncertainty.

11.3 SUPERVISED EXTENSIBILITY

Algorithms that employ an independent source of ground truth during a learning phase are supervised, while those that try to infer patterns from the raw data are unsupervised, and those that are supervised by the environment are reinforced. Although AACR needs reinforcement learning, both AACR and CWNs may use other directly supervised learning methods, both embedded and in the network.

TABLE 11-5 Some Illustrative Radio Station[a] Preference Data

Value	Class	Value	Class	Value	Class	Value	Class
107.7	Cool	102.5	Yuk	101.9	Yuk	99.5	Yuk

[a] These FM values and the related preferences are hypothetical and do not reflect on any actual radio stations.

11.3.1 Supervised Linear Spaces

Linear dichotomizers place a linear boundary, an $(N-1)$-dimensional hyperplane, in an N-dimensional feature space to divide the feature space into two regions. If the features of an object fall into one side of the hyperspace, the objects are classified into one class, otherwise they are classified into the other class. Table 11-5 presents data in two classes, radio stations that are OK ("Cool") and radio stations that are not OK ("Yuk"). This is not to deprecate the radio stations, but to show how inconsistent and unpredictable personal tastes may be.

If the user expresses these preferences in advance, say, in a structured dialog, then the AACR finds itself in a supervised-learning situation. The user's value judgment on the content constitutes a "class." Two classes are present {Cool, Yuk}. WordNet, UKO, and UDK may map slang like "Cool" to ontological primitives like <Approve/>, or a <User/> could complete a dialog, or press an Approve or Disapprove button.

Many algorithms can learn this pattern. The first wave of machine learning began in the 1950s and 1960s with the development of adaptrons, perceptrons, and similar "learning machines" [291]. When embedding capabilities, such simple algorithms may be best subject to known limitations. The adaptron draws a hyperplane in a feature space, a domain in which Type A appears on one side of the plane and Type B appears on the other. The values in Table 11-5 can be learned by any dichotomizer. Equation 11-4 shows a very simple candidate algorithm for the one-dimensional hyperspace RF, the radio station's RF:

$$\text{Class} = \text{if RF} > 105.1, \text{ then "OK", else "Yuk"} \tag{11-4}$$

This algorithm computes the arithmetic mean $(x + y)/2$ of the one "Cool" value and the closest "Yuk" value, defining a point on the RF line. The algorithm is a simple unweighted, nonadaptive dichotomizer. Since FM broadcast RF ranges from 88 to 108 MHz, those values constitute a line segment ranging from 88 to 108. The algorithm defines a point at 105.1 MHz. The CR subsequently infers that the user likes any radio station with RF greater than 105.1 MHz. Wrong!

11.3.2 *N*-Dimensional Context

In the radio station example, suppose the user likes WTOP on FM 107.7 because in the Washington, DC area at about 8 past the hour traffic reports are broadcast. RF is one dimension of the <User/> <Scene/> and decision space, but that <Scene/> also includes location and time: Washington, DC and 8 past the hour. The same user doesn't like 107.7 in Tampa because it's just static; he doesn't like 107.7 at 15 past the hour in DC, because that is the time for sports and this <User/> is not a sports fan. Therefore, the AACR training space needs to be expanded to a vector that captures more of the context:

RF-<Scene/>: (RF, Place, time)

Now the decision table is a matrix (RF-<Scene/>, Decision) with entries like this:

Decision-in-Context: ((107.7, DC, 9:08), Cool)

Often, training provided by a user to a system such as a CR includes implicit space–time dependencies like these that the typical user simply is not going to waste time explaining. Thus, a CR must infer the context as a function of place and time.

11.3.2.1 Learning Space–Time Context

The following supervised ML data set would be confusing:

(107.7, Cool) (107.7, Yuk)

whereas

(107.7, Northern VA, Cool) (107.7, Dallas, Yuk)

are not confusing at all. Similarly, the data (107.7, Cool) (107.7, Yuk) in a given location is confusing unless the AACR constructs the temporal context, such as

(107.7, 9:09, Cool) (107.7, 9:15, Yuk)

In this unspecified location, traffic and weather together on the eights was still in progress on the nines, so the user still liked the content. Given the variability of the start time and end time of traffic reports, variability of even a truly consistent news broadcast must be accommodated. Therefore, temporal reasoning over content must include associated content markers, such as the predictable occurrence of the phrase "Traffic and weather together on the eights" in the audio stream. With such multidimensional cues, even a simple *N*-space dichotomizer may learn relatively complicated information access <Preferences/> from typically complicated users.

11.3.2.2 Teaching the Recommender Substrate

One can teach an AACR to perform *Recommender* functions [220]. A *Recommender* can make purely mechanical recommendations such as those from a database. Other *Recommender* functions include content filtering, collaborative filtering, demographically based recommendations, utility-based recommendations, and knowledge-based recommendations (from user-unique knowledge) along with various hybrids. Stand-alone algorithms of *Recommender* systems may be simulated, emulated, or incorporated into CR by one of the following methods:

SYNTHESIZING RECOMMENDERS IN CR

To Recommend	Embed Algorithm	Or Equivalent CR Training
Current movie	Keyword vector latent semantic indexing	Teach threshold for number of user-interest-profile terms in movie synopsis
VCR rental	Nearest-neighbor filter	*Hear* movie rental question *Say* movie with maximum user-interest terms that user has not yet seen
News channel	ML on content or rating	*Observe* user band mode volume up place, time, duration listening to news
News channel	Demographics	*Observe* place Not = home *Hear* news? *Say* {Observe argmax {count {*NetQuery* news} = Owner-profile-demographic}}
FM radio	User-defined value function	Hear "Get me a good FM station"
	Ephemeral	Ask {"Mood?"} Say {argmax {Mood-ontology} = FM-songs-playing-keywords}
	Persistent	Argmax{{User space-time-profile} = FM-songs-playing-keywords}
Boating	Domain knowledge	Remember boating knowledge by matching user's dialogs to a priori boating ontology
FM radio boating	Combining algorithms	Create b.o.a.t.i.n.g serModel that tailors FM radio actions to detected boating scene

Latent semantic indexing (LSI) has been used by the information retrieval community to improve the precision and recall of document searches. It uses a vector space model of language in which the co-occurrence of words in a document determines a vector of word occurrences. Singular value decomposition (SVD) yields, from a matrix of word vectors $R[M, N]$ that has rank K, a matrix $R[M, N]$ that is the closest rank K matrix to the original that can be constructed from a $K \times K$ matrix S as follows:

$$R[M, N] = U[M, K]\ S[K, K]\ V[K, N] \qquad (11\text{-}5)$$

The Lifestyle Finder was freely available on the Web in the late 1990s. It elicited demographic information in a nonobtrusive way that was regarded by most who interacted with it as not an invasion of privacy. From this data, it assigned a user to one of 62 predefined lifestyle categories, recommending web pages potentially of interest, gathering data from 20,000 users [292]. To transition this knowledge from web to <Self/>, the CR would perform a *retention* action instantiating a lifestyle UKO.

11.3.2.3 Perceptrons Aren't Good Enough

In the 1960s, Marvin Minsky wrote an influential paper about a more general form of dichotomizer, the perceptrons, in which he proved that no such algorithm could learn the Exclusive-OR pattern:

In-1	In-2	Out	In-1	In-2	Out
Cool	Cool	Yuk	Yuk	Yuk	Yuk
Cool	Yuk	Cool	Yuk	Cool	Cool

The Cool and Yuk evaluations correspond to 1 and 0, respectively. This pattern cannot be learned with a linear algorithm because the pattern is nonlinear. There is no hyperplane that can separate the space into Cool/Yuk regions, so no algorithm similar to Equation 11-4 subtends the output space $[01 \times 01]$ into "0" and "1" regions. Minsky's classic paper pointed out this critical weakness of the perceptron as a supervised ML tool.

11.3.3 Neural Network Reactive Learning

Hopfield described an improved perceptron-like algorithm that weighted the learning inputs into a nonlinear decision function (a sigmoid, figure-S function) that solved the Exclusive-Or problem. He called this bio-inspired algorithm a neural network. The equation for the core nonlinear function of a Hopfield neural network is given in Equation 11-6:

$$\text{Sigmoid}(z) = 1/\ (1 + \exp(-\text{Gain} * z)) \qquad (11\text{-}6)$$

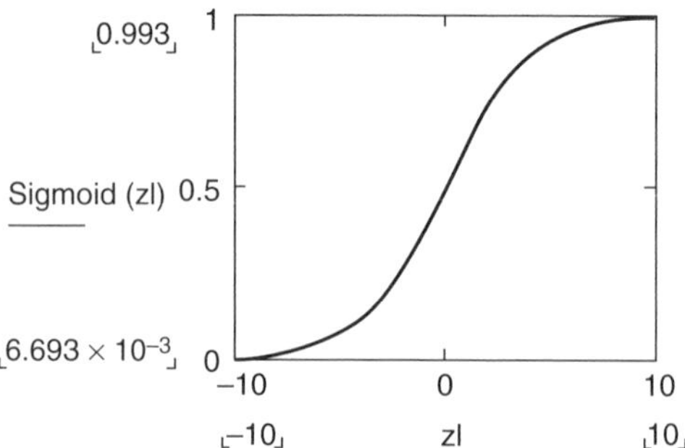

FIGURE 11-8 Sigmoid function (gain 0.75).

Input stimuli are weighted, summed, and operated through the Sigmoid function (Equation 11-6 and Figure 11-8) to yield a response, which may then again be subject to threshold logic or output weights to define the response.

The weights are learned through a feedback circuit that adjusts the weights of the network to drive the response to the desired response. Not only are neural networks trained, but the training typically requires a large number of trials to converge on weights that respond appropriately across the input range. From a statistically significant number of input–output pairs (e.g., 30 or more), a neural network generalizes to any input in the parameter space. This form of generalization can detect signals in noise, such as the heartbeat of a fetus in the presence of the mother's much stronger heartbeat. McClelland and Rumelhardt popularized this class of algorithm in their texts on parallel distributed processing [293], with companion software. Artificial neural networks (ANNs) can learn complicated patterns including the XOR, Exclusive-OR function of Minsky, via a wide range of initial training and update algorithms.

When learning in feature spaces, there is a tacit assumption that categories are connected. In the FM radio stations of Table 11-5, there is no continuous RF domain or straight-line segment from 88 to 108 MHz. Instead, because of spectrum regulation, there are about 100 discrete data points starting with 88.1 MHz and ending with 107.9 MHz, each 200 kHz apart. These are the legal center frequencies. Users either like or dislike the station broadcasting on the discrete frequency. So an ANN could learn the discrete values from the user's preferences, generalizing them to the continuous domain of RF measured by the receiver. A case-based reasoner would remember that WTOP had transmitted on 107.7 MHz at a given time and place. ANNs always generalize and do not remember the raw data on which they were trained. So although there are potential applications for linear dichotomizers, ANNs, and other such pattern recognition algorithms in industrial strength CR, this text focuses on case-based reasoning (CBR), learning by remembering cases and adapting the "nearest" past situation to the current situation. A CBR algorithm can aggregate prior experience to yield performance similar to that of an ANN, but ANNs can't explain their generalization process in terms of prior experience the way a CBR can. This brief introduction does not do justice to the insertion of the rich and diverse ANN technology for AACR. The sensory-perception subsystems may embed ANNs for face tracking, speech recognition, speaker ID, and many other such functions.

11.4 UNCERTAINTY

Not all situations are as clear-cut as many of the vignettes of the prior chapters might suggest. In fact, most situations in the <RF/> domain are perceived through interference and noise, while most <Scenes/> in the <User/> domain are replete with uncertainty. Discovery using the <Histogram/> addressed the discovery of *certain* information through the analysis of the entropy of data structures conveying the information. Usually, however, one cannot say exactly what is going on in a data set because of noise, conflicting information, or an inherent ambiguity in the situation. This occurs because of noisy measurements, psychological factors, the statistical structure of language, and other causes. The <Histogram/> and other pattern discovery and characterization algorithms like support vector machines (SVMs) [17] assist in perceiving certain information through noise. In addition, <User/> domains are characterized by features that are inherently uncertain, such as weather predictions. These uncertainties introduce the need to plan under uncertainty.

11.4.1 Quantifying Uncertainty for the <User/> Domain

This section differentiates probability spaces that satisfy the probability axioms from other pseudoprobability frameworks, each of which are useful when their underlying assumptions are satisfied. Probability is defined in a metric space where the distance metric obeys the triangle inequality, total probability is conserved at 1.0, and total probability is distributed to all possible hypotheses, not just to the enumerated hypotheses.

AACR deals with real <User/>s who have a relatively unstructured intuition for and ways of generating and dealing with uncertainty. Although probability offers strong mathematical foundations for uncertainty, <User/>-domain situations are unlikely to conform to probability axioms per se.

11.4.2 Probability

The reader should be familiar with probability, statistics, and random processes in order to fully appreciate this section. A probability system [294] consists of the following:

1. A probability space, the set X of possible events, and Ω, a family of subsets of X containing the empty set and the universal set over which

a sigma-algebra [295] is defined, so that set-theoretic integrals are defined:

   (a) Associated with Ω is an index set ϑ defined so that each ω in Ω has a unique index i in ϑ. In other words, each element of Ω can be measured and potentially infinitesimal contributions can be aggregated via an index set (in the case of stochastic processes that may be infinite dimensional).
   (b) An operator ∪ (union) exists such that for each ω in ω, {ωi} ∪ {ωj} = {ωk}; if Ω is uncountable, then Ω is closed under uncountable unions.
   (c) An operator ∩ (intersection) exists such that Ω is closed under finite intersections.

2. A map p: $\Omega \to \Re$ such that
   (a) p obeys the triangle inequality for sums, $p(\omega i \cup \omega j) \leq p(\omega i) + p(\omega j)$;
   (b) p(Universal Set) = 1;
   (c) p(Empty set) = 0.

There are important AACR-engineering implications of this construct. Ω is closed. That is, one can't add new events x to X and related sets of events ω in Ω, once Ω is defined to extend a model that is already grounded by the AACR to the <Outside/> world. In addition, all possible events must be reflected in Ω. Often in <User/> domains all events are not knowable. Many reasoning systems do not enforce the axioms of probability fully and thus may implement degrees of belief, but not random processes or laws of probability. Probability theory enables one to address questions of expectation. For example, if two events are related, then the observation of one event may circumscribe the likelihood of a related event. Bayes' famous law explicates such relationships in probability spaces.

11.4.3 Bayes' Law

The situation before the observation of an event is called the a priori probability ("priors"). The situation after an event is called the a posteriori situation. Bayes (the 18th century English clergyman Thomas Bayes) noted that if there are two possible events and exactly one will occur, then $P(A \cup B)$, the probability of the union of A and B (the probability of A or B), is related to the probability of A given that one observes B. Suppose events A_i $[i = 1, \ldots, n]$ are mutually exclusive and their union is the entire sample space Ω. The conditional probability $P(A_n/B)$—the probability of A_n given B—is computed from Bayes' Law or the law of inverse probability as follows:

$$P(A_n/B) = P(A_n)\ P(B/A_n)/\ \{\Sigma_i\ P(A_i)P(B/A_i)\}$$

For example, two balls may be drawn without replacement from an urn containing r red and b black balls. Let A be the event "red on the first draw" and B the event "red on the second draw." The probability of drawing red first depends on the number of red balls, and if all balls are equally likely, then $P(A) = r/(r + b)$. Three red and seven black balls yields only 30% probability of red. $P(A) = 1 - P(\text{not-}A)$, so black is 70% likely on the first try. Suppose the first event is A. What is the probability of B now? $P(B|A) = (r - 1)/(r + b - 1)$. There are only $r - 1$ red balls left, in this case 2. But there are still 7 black balls. So the probability of red on the second draw is $\frac{2}{9}$. The $P(B/\text{ not-}A)$ leaves r red balls but not-A means that one black ball was drawn on the first try, so $P(B/\text{not-}A) = r/(r + b - 1)$ or $\frac{3}{9} = \frac{1}{3}$. Bayes' Law enters the CRA by the <Probability/> and <Modeling/> primitives.

Expression 11-13 Bayes' Law in the CRA

<Probability> <Bayes> <A-priori> $P(B/A_n), P(A_i), P(B)$ </A-priori>
<A-posteriori> <Function> $P(A_n/B) = P(A_n)\ P(B/A_n)/\ \{\Sigma_i\ P(A_i)P(B/A_i)$
</Function>
</A-posteriori> </Bayes> </Probability>

How might an AACR exploit Bayes' Law in the <User/> domain? There are not a few situations in which a fixed set of entities corresponding to the red and black balls occur in a <Scene/> in which they may be randomized by an event like being thrown in a jar. In an emergency use case, for example, the red and black balls might correspond to AACR (red) versus legacy (black) radios in use by rescue workers entering and leaving a scene. If r AACRs and b legacy radios enter an emergency situation, and if a transmission from each is equiprobable, then the probability that the next transmission is from an AACR is just $r/(r + b)$. Such probabilities can help establish dynamic spectrum-use policies, for example. This represents an interplay between <User/> and <RF/> domains. If $r = 90$ and $b = 10$, then most radios will be listening before transmitting, so the likelihood of collision would be smaller so the time delay before listening in a spectrum etiquette can be longer than if the numbers are reversed, where most radios are legacy radios and therefore most <Users/> do not have the benefit of ad hoc use of unused TV channels, and of polite automatic backoff to other channels when legacy users need the channel.

11.4.4 AACR Event Spaces

Bayes' Law requires events to actually be distributed according to a probability system that is both defined and mathematically consistent. The space must be stationary: its statistics may not vary over time. It must be ergodic: parameters that are measurable must faithfully represent the probability abstraction—for example, the computable temporal average must equal the ensemble average, the instantaneous average across the probability space at a point in time, which can be inferred from the axioms of probability, but not directly computed. This conformance can be a lot less trivial than one might like.

Suppose, for example, that an AACR needs 12 hours to retrospectively analyze a particular day's experiences. It is 10 pm. It has observed for the week that its owner rises at 6 am each day, so it predicts that it has 8 hours to digest the day's experiences, not the 12 it predicts that it needs. Today is Friday. The CR estimates that the owner will rise at 6 am on Saturday with a simple model of waking up, for example, from TDL, which is a descriptive model that the owner rises each day between 5:50 and 6:10 am because that has happened five times, all the mornings since it was purchased. The observations might even be a good match to a Gaussian distribution with mean 6:01 am and standard deviation of a few minutes. The AACR starts its <Sleep/> cycle. At 10 am, it detects the loud noises of the owner awaking and moving about the house. If Waking-Up is the event WU, it is the time, $\boldsymbol{t}_{\mathrm{WU}}$, of the event WU that the AACR needs to estimate:

$$P(\boldsymbol{t}_{\mathrm{WU}} = t) \equiv Pt(\mathrm{WU}) \tag{11-7}$$

In the CRA, there is no requirement to ascribe a probability distribution to $\boldsymbol{t}_{\mathrm{WU}}$, but it is helpful to do that since such events are among the daily patterns of the <User/>. Is there a probability distribution here of the type needed for Bayesian inference? If so, then waking up is a certain event, $P(\mathrm{WU}) = 1.0$ (not 99.9 but 1.0). The owner could die in his sleep, so $P(\mathrm{WU}) \neq 1.0$. There are, then, events not yet included among the observations of the CR that can result in a change to WU (not just $\boldsymbol{t}_{\mathrm{WU}}$) so there is no true $Pt(\mathrm{WU})$, only $Et(\mathrm{WU})$, an expectation or degree of belief about both the fact of and the time of the Wake-Up event. In fact, in the <User/> domain, the prejudice is that events do not obey probability axioms. When an AML algorithm discovers a set of events like WU that behaves very much like a probability, then that model may be exploited for planning. The space Ω, the space of all possible events, is invariably unknown, yet the probability of a <Novel/> event is often treated algorithmically as if it were infinitesimal.

11.4.5 Causality and Probability

Suppose the owner is young and in good health, so $P(\text{death-in-sleep}) \sim 0$. Then $Et(\mathrm{WU}) \cong Pt(\mathrm{WU})$. That is, the degree of belief is behaving like a probability, so the CR may safely use the probability function as a degree of belief. Suppose owner says, "I'm not feeling so good, so I'm going to sleep-in tomorrow until maybe 10 am." Does the CR have a causal model [296]? It is easy to see that, given the remark,

$$P(\mathrm{WU};\, 5{:}50 < \boldsymbol{t}_{\mathrm{WU}} < 6{:}10) \sim 0 \tag{11-8}$$

In other words, the probability that the CR had modeled as an event distributed in the vicinity of 6 am is not going to be so distributed tomorrow morning. The alarm clock may even go off, but the owner plans to hit snooze and put the alarm in the off position. The owner might ask the AACR not to wake him up at the usual time. Conditioning on weekdays, Saturday, and Sunday yields three unimodal distributions. Until the user goes on vacation. Or has to cut the grass on Saturday morning before the predicted rain that afternoon. Or, or, or, or.

Counting or RL with context detection reveals the multidimensional nature of the owner's life as follows:

COUNTING WITH CONTEXT

Day	Time WU	Context Work	Context Home
Friday	606	Yes	No
Saturday	1005	No	Yes
Sunday	823	No	+Church
Monday	600	Yes	No

By detecting contextual features, a causal space–time map appears for which the somewhat random yet patterned distributions establish probability as a reasonable way to model the likely time of events. The space–time–context <Observation/>s give the AACR a sufficiently high dimensionality feature space in which to infer causality. The owner could still fail to wake up, but there is no probabilistic requirement that he/she live forever, only that the $P(\text{Not (WU)}) \sim 0$. Some aspects of the <User/> domain lend themselves to probabilistic modeling, which typically is enhanced by an appropriate choice of distribution.

11.4.6 Using Probability Distributions

Bayesian analysis also requires one to accurately estimate the a priori probabilities of an event in order to determine the a posteriori probability, which typically is the objective. Probability in the <User/> domain typically is subjective rather than exact. How does an AACR estimate the prior probability that the <User/> is in the living room? That might be useful information if the AACR is in the bedroom and wants to be taken to work today. It could send a message care of the WiFi network to be broadcast through the current TV program: "Sir, did you mean to take me to work with you today? If so, I'm still in the bedroom."

To ascribe probability, typically one measures the relative frequency of occurrence of events and matches this distribution to well-known distributions to establish a probability model. Analytical tools like Matlab, Mathcad, and Analytica [297] among others offer many standard distributions that could be embedded into AACR. Questions to be addressed in selecting a model include whether the system is discrete or continuous (and if so, is it bounded), the number of modes, and its symmetry. The embedding into the CRA <Self/> builds on the ontological treatment of probability distribution.

Expression 11-14 CRA Probability Distributions

<Probability> <Distribution> <Definition> Map from sample space to [0, 1.0] </Definition>
<Discrete> <Bernoulli/> <Certain/> <Chance-dist[1]/> <Probtable/> <Binomial/>
<Geometric/> <Hypergeometric/> <Poisson/> </Discrete>
<Continuous> <Beta/> <Cumdist/> <Fractiles/> <Gamma/> <Lognormal/> <Normal/>
<Probdist/> <Triangular/> <Uniform/> <Exponential/> <Logistic/> <LogTriangular/>
<LogUniform/> <Weibull/> </Continuous>
</Distribution> </Probability>

There are many ramifications to these choices that publishing space precludes addressing but that are addressed well in the analytical tools cited above. The autonomous attribution of probabilistic behavior to distributed observations entails the algorithmic simulation of the reasoning processes discussed above. Pioneering approaches to the algorithmic simulation of such subjective processes may be found in the literature on qualitative physics [298, 299] and in the more recent literature on the simulation of the learning processes of newborns [300].

Probability relies on a crisp definition of sets, such as whether a ball is black or red and whether a radio is a legacy radio or an AACR. There sometimes are degrees of membership that require one to set arbitrary thresholds in order for a probability model to work. Suppose a radio has a few SDR personalities, including the use of UHF TV bands, but lacks other AACR features such as peek-through. Is this an AACR or a legacy radio? An alternative approach to probability theory, which relies on crisp sets, is the fuzzy set theory of Lofti Zadeh.

11.4.7 Fuzzy Sets

The time is now 11:12 pm. Is it bed time? Usually. Generally, this <User/> goes to bed at something like 9:30 to 10:30 and awakes at 4:45 am. But the boundary between wake state and sleep state cannot be drawn precisely at 10 pm or at any other time. Generally, at 3 am, this <User/> is asleep, and generally by noon, awake. But there is a very fuzzy line between these two states. The state space U = {Wake, Sleep} is the simple universe of states to

[1] For convenience, these distributions are as defined in Analytica 2.0 [297]; several tools offer equivalents.

which the AACR would like to map time of day t for the <User/>, deriving $U = f(t)$ to unambiguously map any time of day $t \in \{0, \ldots, 23.59\}$ onto the two states of U. Why is this so difficult? Is it because the <User/> is so hard to get along with? Unpredictable? Maybe to some degree.

Professor Emeritus Lofti Zadeh first formulated this kind of set-membership problem as the fuzzy set, the class of sets in which the membership function is not limited to the discrete values $U_o = \{0, 1\}$ but that include any of the real numbers in the interval, $U_z = \{0, \ldots, 1\}$. Since CR often has to deal with sets for which the membership is uncertain, they are particularly relevant to AACR.

There is always some difficulty in determining when fuzzy sets are appropriate versus conventional probability. Zadeh offered the following on the Uncertainty in Artificial Intelligence (UAI) bulletin board [301]. He noted that there is an extensive literature on the relationship between fuzzy logic and probability theory, going back to a paper by Loginov in 1966. The most thoroughly studied aspect of this relationship, according to Zadeh's email, relates to the connection between fuzzy sets and random sets (see [302]). A discussion of the relationship between fuzzy logic and probability theory may be found in Zadeh's paper entitled "Probability Theory and Fuzzy Logic Are Complementary Rather Than Competitive," published in *Technometrics*, Vol. 37, No. 3, pp. 271–276, 1995. His view (November 2003), which is more radical than that expressed in the cited paper, is that probability theory should be based on fuzzy logic rather than on bivalent logic.

Zadeh cites several examples that are difficult to deal with as probability systems, such as the "tall Swedes" problem: Most Swedes are tall; what is the average height of a Swede? If one thinks of Swedes as instances of a genome from which they are drawn, then probability seems to apply. If one thinks of actually expressing the notion that Swedes are tall in an algorithm for an AACR, one might write a Beta distribution with a peak at maybe 2 meters or higher. The fuzzy notion of "tall" that we may reify as <Tall/> provides another alternative: a sliding scale of membership in the class <Tall/> beginning at maybe 1 meter with <Value> <Tall/> ~0 </Value> (could be a tall midget) and ending at, say, 3 meters with <Value> <Tall/> ~1.0 </Value>. Similarly, <Normal-height/> might peak at 1.8 meters at 1.0 with scales sliding down to zero at 2.5 meters and down to 0 at 0.5 meters, or something like that. These aren't probabilities but value judgments about the meaning of the sets we have chosen to <Name/> <Tall/> and <Normal-height/>. So in AACR <User/> domains, proscriptive definitions like <Tall/> really lend themselves to a priori definition as fuzzy sets, which is included in the CRA <Self/>.

There is a rich literature on fuzzy sets, so this brief hint does not do it justice. This literature can be a source of many fuzzy methods for computational autonomy for AACR. Particularly relevant papers include fuzzy unification of logic terms using edit distance, with resolution theorem proving, with applications to correct spelling errors [304]; and software tools for object-oriented fuzzy knowledge systems [305].

11.4.8 Rough Sets

Rough sets generalize set membership in some sense beyond the uncertainty implicit in fuzzy sets, but with similar objectives and consequences. They generalize sets using a third truth value—"uncertain" or "undecideable." Rough sets differ from fuzzy sets in that no degree of membership, probability, or degree of belief in membership is ascribed to an uncertain member. Therefore, a rough set consists of known members and possible members. A union may either include or not include the possible members, yielding upper and lower sets, respectively, with corresponding constructs on union and intersection operators. Machine learning algorithms have been implemented using rough sets [306]. The approach expresses uncertainty about set membership with a degree of generality from the predicate calculus and requires no a priori assignment of degree of membership as is necessary to effectively use fuzzy sets. This comes at the expense of introducing virtual states and returning uncertain results that can be combinatorially explosive.

The CRA <Self/> includes naïve, fuzzy, and rough sets in RXML as canonical templates for representation and reasoning. The methods of reasoning with uncertainty may be organized into systematic mathematical systems called certainty calculi.

11.4.9 Certainty Calculi

Not every CR need have a computer model of probability or fuzzy sets, but each CR needs some way of reasoning with uncertainty. A systematic method that includes the representation of positive and negative reinforcement along with at least one method for the aggregation of reinforcement across multiple stimuli (e.g., via a logic or rule-based process) may be called a certainty calculus. The Bayes community of interest in UAI [307], Java Bayes [308], and recent texts [13] offer strong support for Bayes reasoning with uncertainty. <User/>-domain driven approaches to evidential reasoning include methods for the nonlinear combination of evidence in such classic reasoning systems like Mycin, Dendral and meta-Dendral, and Tieresias.

11.4.9.1 Uncertainty in the <User/> Domain

Suppose the set $X = \{\mathbf{X}, \mathbf{O}\}$ represents a closed world of states such as the channel symbols of a BPSK modem or the decision to turn the CWPDA on or off. One can't add another element to X without redefining the probability system. The world is rarely closed. Just when you think you have accounted for all the possibilities, a new one is discovered experimentally. Instead of implementing a binary power system {On, Off}, the manufacturer implements {On, Off, Pause}, where {Pause} conserves power while preserving state. The <User/> says, "Pause, will you?" The AACR trained for {On, Off} now needs a new symbol for pause, for example, **Y**. To mark which user paused the system it might remember <User> ***X*** </User> or <User> **O** </User>. The AACR could introduce **Y** as a metalevel construct where the states of play are {Run, Pause [**Y** | x ∈ {**X, O**}]}. Notionally, there could be a probability of **Y**, so in the refined X = {**X**, **O**, **Y**} there is a metalevel probability of Pause.

The original designers of the notional AACR didn't envision pause, but the user thinks it is a good idea. The <User/>-domain AACR must accommodate an open-world setting: Users are continually moving out of "the box" that the manufacturer would like them to stay in. Therefore, in order to employ probability to enhance QoI, the AACR designer must restrict the use of probability to those closed domains that are accurately modeled by probability, such as noise and other stochastic processes. Theories of evidence other than Bayes' Law offer insights for other approaches to uncertainty in the AACR <User/> domain.

11.4.9.2 Theories of Evidence

The Dempster–Schaffer (DS) theory of evidence does not need the total probability space of Bayes. In addition, the Dempster–Schaffer theory of evidence generalized Bayes' notions of a priori and a posteriori probability to the more general problem of evidential reasoning. Although theoretically powerful, Bayes theory requires one to estimate the prior probabilities underlying all possible events. The difficulty of this requirement, among other things, has led to a proliferation of ad hoc techniques for representing uncertainty. To perform consistent logic in uncertain domains requires a calculus that manipulates numerical representations of uncertainty with associated Boolean logic or assertions in rule-based systems. Some powerful uncertainty calculi are nonlinear [318]. There is also much relevant technology from probability and statistics literature. Mixture modeling, for example, is the process of representing a statistical distribution in terms of a mixture or weighted sum of other distributions [309]. AACR exhibits statistical mixtures of uncertainty in RF, in sensory perception, and in interpreting user interactions. The CRA embeds general facilities for reasoning under uncertainty by prescribing reinforcement and expressing <Uncertainty/> tags as schema-schema for application-specific certainty calculi, such as that of Tieresias.

11.4.9.3 Tieresias Certainty Calculus

Tieresias employed the Mycin calculus for reasoning under uncertainty. Medical doctors dealing with bacterial infections of the blood were known to consider a weight of evidence for and against a causative agent. The Mycin certainty calculus therefore independently aggregated weight of evidence for and against a given diagnosis. Subsequently, these aggregates were evaluated to determine whether positive indications outweighed negative indications and conversely, preferring the explanation that most positively endorsed a given diagnosis. Thresholds for weight of evidence were also employed to defer decisions until sufficient knowledge was applied.

(Parameter-predicates Certainty-factor-sum)
((Gram Same) (Morph Same) 3.83) (11-9)

As shown in Equation 11-9, Mycin lists of predicates were associated with certainty factors. The example says that when the gram stain appeared in a rule premise, the morphology also appeared, and the fractional part indicates that 83% of the time gram stain implied morphology.

This is one example of the implementation of reasoning about uncertain relationships through quantified predicates. The predicates aggregate evidence for and against hypotheses using the real line between [–1.0 and +1.0], associating regions with degrees of belief. This certainty calculus is one example of a theory of evidence.

11.4.10 Uncertainty in Language

A review of user-domain issues in uncertainty would not be complete without considering uncertainty in language at greater length. Even with noise-free, error-free speech transcript, there is substantial uncertainty in language. This uncertainty derives from the many ways of expressing a given thought as well as the many ways of reducing the detail supplied in communications, such as referring to "this" or "that" without spelling it out, as with anaphora and ellipsis. However, the statistical structure of language yields detectable features. Topics of discourse can be detected reliably even in the presence of high word error rates [310], provided the number of samples is high, the number of words per sample is sufficient, and the training sets have the same statistical structure of the larger text corpora. Search engines like Google learn what <Users/> are looking for by processing terabytes of data on thousands of servers. AACR requires an approximation of such high quality but with the smallest feasible computational resources, such as using *N*-grams.

The *N*-gram [311] is a well-known measure of the statistical structure of text (which may be from an errorful speech transcript). *N*-grams do not require the parsing of the text to delimit word boundaries and punctuation. Instead, *N* consecutive letters are considered in a sliding window from the beginning to the end of the text sample. For example, the paragraph of the prior section on Randomness for Exploration has the *N*-gram structure shown in Table 11-6.

TABLE 11-6 *N*-Gram Structure of Text Sample From Randomness for Exploration

N	*N*-Grams	Most	Next	Most Common	Remark
1	46	e/138	t/103	N, o, a, i, t, e	1100 character file
2	311	ti/27	in/24	Top 25 = 381	146 occur only once
3	716	the/17	ion/10	23 grams > 5 times	515 3-grams occur once
4	901	tion/10	beha/7	102 > once	800 occur only once

TABLE 11-7 *N*-Gram Structure of Second Text Sample

N	*N*-Grams	Most	Next	Most Common	Remark
1	72	e/173	t/132	r, o, s, a, i, t, e	1944 character file
2	659	th/27	in/28	Top 41 count > 10	200 occur 3 times or more
3	1312	the/17	ing/15	23 grams > 5 times	141 occur 3 times or more
4	1583	orks/6	mall/6	205 > once	1378 occur only once

Although 46 characters can generate 97,336 3-grams, only 716 of these occur in this text, what might be called a rough 3-gram signature of the text. The most common have to do with the structure of the language (e.g., "the" is the most common word in English and in this sample), but some of the more common 3-grams reflect the content. The 4-grams are led by "tion" at 10 occurrences, but four of the five next most common occur because "behavior" is the most common long word that generates the leading "beha" and the trailing "vior" as well as the cyclic shifts "ehav" and "havi." These redundant 4-grams may be suppressed. This content-related structure can be compared to the statistical structure of other content by comparing *N*-grams.

The 72 types of characters that appear in a second text sample (Table 11-7) include more capital letters, punctuation, and numbers (telephone and street addresses) than the purely text of the first sample. In addition, the most common 4-grams of this second sample are orks, ting, mall, this, leas, and ease; each of which occur six times. Other common 4-grams include "work" and "shop," which is very content-dependent since this sample email had to do with a workshop. Although the workshop was about robots, the differences in initial capitalization reduced the occurrence of "robo" to less than five, while the interior "obot" occurred 5 times. Of the 5-grams occurring four or more times, none overlapped between the two samples, indicating relatively good statistical separation between the text samples as measured by *N*-grams.

Each <Scene/> falls within a microworld, defined a priori or defined through the aggregation of <Scene/>-level experience. During sleep cycles, microworld and scene boundaries would be adjusted to manage combinatorial explosion. One method for adjusting <Scene/> boundaries is through the clustering of textual content of the constituent <Scenes/> using *N*-grams, fuzzy logic, rough sets, or an application-specific certainty calculus.

11.5 LEARNING REQUIRES GROUNDING

Grounding is the process of accurately associating internal symbols with external symbols. Internal symbols include intermediate symbols in the sense–perceive–act subsystems representing the intended entities in the outside world. This includes the <Self/> as a member of the <Outside/> world when viewed from the third-person perspective. This section highlights the critical role mutual grounding plays in AACR–<User/> communications and thus in the realization of <User/>-domain skills.

A mutual grounding map M expresses the relationships among <Memory/> symbols, perceptions, and external entities. External entities include <Abstractions/> like "Superman" when referred to in a speech or visual <Scene/>.

M: <Memory/>-symbols ⇔ Sense-perceive-act-entity-symbols ⇔ External-entities

Mutual grounding enables coherent communications between two entities. Learning algorithms access the internal data structures that represent existing knowledge, interactively modifying these internal structures to reflect the mutually grounded learning experience.

11.5.1 Ontology Reconciliation

Even relatively straightforward applications of AACR like XG require mutual grounding. Formal languages between AACR and host networks typically employ data formats defined by the manufacturer and reconciled to the degree determined by the marketplace via international standards setting processes. Historically, radio engineers set radio communications technical standards without the help of computational ontologists. Large, complex documents like the ETSI Reference Materials, ITU-R Recommendations, TIA Interim Standards, ARIB Reports, and IEEE Standards attest to the capability of radio engineers to formalize definitions without the mediation of computational ontology specialists.

11.5.1.1 Resolving Historical Inconsistencies

A closer look at the <RF/> domain reveals that none of the engineering ontologies implicit in the published radio standards are mutually consistent. Channel is a time-domain construct in GSM, a code-space construct in WCDMA, and so on with most terms having radically different meanings in different subdomains of radio. That has been fine historically because people have substantial tolerance for this kind of ambiguity. Not so with computers. People employ context to disambiguate semantics in ways that we have as yet been unable to replicate fully with computers, contemporary research notwithstanding [37]. The approach of this text is neither to propose an approach to this problem nor to sidestep it, but to define in the CRA <Self/> a level of abstraction at which mutual consistency between ontologies exists, yet further inconsistent ontologies may be integrated via ontology mapping via a priori primitives. So the notion of <Channel/> in the CRA <Self/> is abstract, admitting realizations in different domains: GSM:channel in the time domain, WCDMA:channel in code space, and so forth.

Expression 11-15 Channel Abstraction Reconciles <RF/>-Domain Ontologies

<Channel> <Definition> <Path> <From> <Transmitter/> </From>
<To> <Receiver/> </To> </Path> </Definition> </Channel>
<Air-interface/> <Air-interface> <Definition> <Path> <From>
<Transmitter/> </From>
<To> <Channel-symbol/> </To> </Path> </Definition> </Air-interface>
<Air-interface< <FDM/> <FDMA/> <TDM/> <TDMA/> <CDMA/>
<UWB/> </Air-interface>
<Channel> <FDM/> <FDMA/> <TDM/> <TDMA/> <CDMA/> <UWB/>
</Channel>

The abstraction that a <Channel/> first of all is a <Path/> from a transmitter to a receiver expresses the propagation and air interface channels at a high level of abstraction, from which all other notions of channel inherit their basic structure. The <Air-interface/> abstraction postulates a LCS <Path/> from the <Transmitter/> to one or more <Channel-symbol/>s, which really is the function of an air interface channel. Various classes of air interface and channel are then asserted, enabling <Channel> <FDM/> </Channel> to inherit its properties as a <Path/> from transmitter to receiver from the original <Channel/> used as schema-schema. This is the easy part. The hard part occurs in the <User/> domain, where everybody is an ontologist and agreement is needed only occasionally.

11.5.1.2 Confronting Persistent Inconsistencies

The hard part of ontology reconciliation occurs in the less formal settings of the <User/> domain. Unfortunately, everybody answers "Twenty Questions" via a different path after the mandatory "Person, Place, or Thing?" The AACR that embeds the CRA <Self/> asks "<Person/>, <Place/>, <Thing/>; or <Abstraction/>?" and thus need not be stumped by "Mickey Mouse," which to some is a person and to others is a thing. To AACR, Mickey is a <Disney/> <Abstraction/>.

The AACR must create communications content using its internal formal ontology bootstrapped from the CRA <Self/> or some other suitable core, which remains grounded to the <User/>'s own personal and typically ever-changing informal ontologies. <User/> computational ontology is a set of terms and syntactic structures with LCS and other semantics that describe the everyday world. These may be publicly stated as with LCS and the CRA <Self/> and they are informally mutually agreed to as with the bootstrapping of <User/>-specific UKOs. The iCR continually cycles through the capabilities Observe (sense, perceive), Orient, Plan, Decide, Act, and Learn. To learn from the <User/> in a way that does not drive the user nuts, the AACR autonomously perceives the user as an external entity with its own unique ontology.

11.5.1.3 Mutual Ontology

Mutual grounding is particularly critical for human–computer collaboration [313]. Suppose the user asks the PDA: "Will it rain today?" The PDA starts to play back the latest NOAA weather broadcast. It has inferred <Rain/> as a weather state, and it knows (e.g., from the CRA <Self/>) that NOAA weather broadcasts are a source of <Weather/> <Information/>.

<Place> <Time> <Earth> <Land/> <Sea/> <Weather/> </Weather>
</Time> </Place>
<Information> <Weather> <Rain/> <Shine/> <Snow/> <Severe/>
<Radio> <Broadcast> NOAA <News/> </Broadcast> </Radio>
</Weather> </Information>

The user says, "No, I do not want to hear the NOAA broadcast, I want you to tell me the bottom line." Has the PDA successfully communicated with the user? It seems not. Success in such human–computer communications depends on precise mutual grounding, the alignment of the conceptual semantics of the two cognitive entities. Initial AACR products may get away with playing the latest NOAA weather when asked such a question, but advanced iCRs would analyze the NOAA weather broadcast themselves. In this NOAA dialog, the <User/> wanted a yes or no answer to the question, not a lot of data from which to draw a conclusion. The CRA <Self/> includes primitives from which the AACR could have created the following LCS-annotated sentence and drawn the primal sketch of Figure 11-9:

"<Question> <Place> <Time> Will <Future/>
<Action> <Weather> it rain <Rain/> </Weather> </Action>
today <Today/>? </Question>"

In Figure 11-9, the type of question is inferred from its structure. Other types of question are rhetorical, requiring no answer; some ask for an explanation. Words like "how" generate an <Explanation/> goal, while "will"

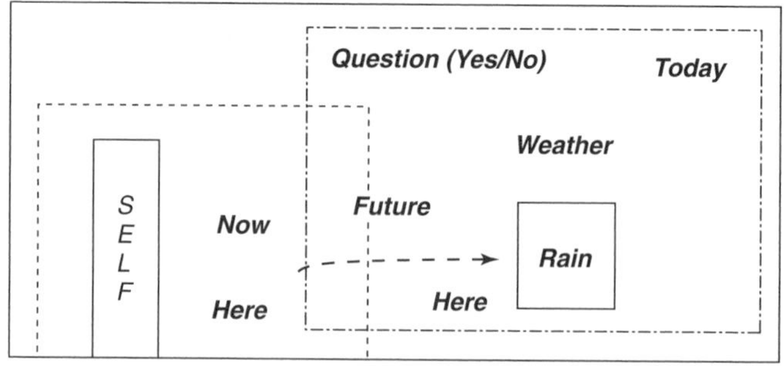

FIGURE 11-9 Primal sketch of question about rain.

generates a <Yes/No/> reply goal. Such a semantic sketch could be converted to a database query against the most recent transcript of NOAA weather, yielding either a crisp yes or no, or a NOAA-like answer: "50 percent chance of rain in the counties of . . .".

Mutual grounding is often mediated by the grounding of internal symbols to concrete counterparts in the external environment via sensory perception and volitional action systems in which the cognitive entity explores the environment in order to ground a symbol. The user extends his palm through the open screen door, feels the wetness, and says "It is raining." The CR on his belt looks at the concentric circles in what it infers to be a puddle in the lawn and infers "It is raining." Although the user and AACR rely on different perception systems, both cognitive entities established a well-defined internal state <Weather> <Rain/> </Weather> through interacting with the same external experience. They also use mutually mapped internal representations of that perception, <Rain/> for the AACR which translates to "rain" in the speech synthesizer.

11.5.1.4 Grounding Errors

Grounding errors can include the following:

1. *Abstraction Failure: Failure* of one entity *to accurately express* an internal state.
2. *Communications Failure: Lack of access* of an intended recipient to the expression.
3. *Misinterpretation* of a perceived expression.
4. *Disagreement*: The unwillingness of each to align to the other.

Metrics have been described which express an agent's inclination toward a communications act to correct a detected grounding error [313]. State-confusion matrices compare the internal states of two agents that are supposed to be sharing identical internal states. Such matrices characterize the success of communications, but do not diagnose the underlying agent–agent or human–machine communications failure. Additional metrics might (1) express the relative importance of internal states with respect to the task at hand, and (2) express the agent's ability to know its own internal state.

Most internal states of conventional software-defined radios (SDRs) are not computationally accessible to another SDR. Although internal states of cognitive radios would be accessible, metalevel states might not be. Predicates and reasoning over metastates seem to be needed to quantify whether agents have the capabilities required for grounding. An AACR that learns from its user constitutes a collaborative problem solving domain with the CRA <Self/> as a metalevel problem solving framework. Within this framework there may be different forms of conflict in the alignment of primal sketches. Sometimes the alignment of the sketches is the root of the problem.

11.5.1.5 Disagreement

In disagreement, each entity stringently asserts its own ontology, unwilling to map its primal sketch to that of the other. Disagreement may be treated as a grounding error, but that may be misleading. If the CWPDA's internal state reflects "It will rain today," but the user thinks "It will not rain today," then the two disagree about the weather. If a CWPDA's internal state was derived from the speech processing of the NOAA radio weather broadcast, its internal state was derived from an authoritative source. The user's view may be based on personal experience that the weather forecasters are wrong 30% of the time, perhaps coupled with optimism. If the CWPDA expresses its internal state to the user, then miscommunication has not occurred. If the user tells the CWPDA, "It will not rain today," then the CWPDA should detect a specific class of conflict, disagreement. Suppose it looks up its serModel for weather and finds "Today → Rain." This was deposited there by its processing of the NOAA weather forecast. Backtracking to the source, it can say to the user, "NOAA Weather Radio says it will rain today." The reply might be "I don't think so." At which point, the PDA could replace "Today → Rain" with "Today → No Rain." To avoid thrashing of this state the next time the CWPDA processes NOAA weather, the cognition component should explicate the source.

```
<Weather> <Weather> <Time t/> <Place here/>
  <Source> NOAA </Source> "It will rain today," </Weather>
<Weather> <Time t+5 /> <Place here/>
  <Source> <User/> </Source> "It will not rain today," </Weather>
    </Weather>
```

Conflict presents an opportunity for a cognitive agent to learn. In order to do this in a computationally effective way, the agent must:

1. Detect the conflict (e.g., replacing "Rain" with "No Rain" is a problem).
2. Encode the situation that led to the conflict as a problem state (User versus NOAA as source for the serModel of weather).
3. Recognize the aspect of context that can convert a conflict over internal state into multiple internal states (extend the Weather model stimulus from Today, a temporal constant to (User, Today), a vector).
4. Obtain the data needed to create multiple states (in the example, srWeather: User → "Today → No Rain" and srWeather: NOAA → "Today → Rain" existed internally, but would not be recognized as such without specialized internal data structures and related processing).
5. Deal with the multiple internal states from that point forward. The CR would update all of its internal models that ask about weather to express (agent, time) as the stimulus to the srWeather model. It could also install

the dialog "Today → ERROR | Askfor (<agent/>, Today)" in the srWeather model. Any internal or external process that asks for the weather will know that that is a function of the context variable <agent/>.

Setting up and accessing such context variables imparts implied internal structure to the cognitive agent that may not be present until such a conflict is detected. Mechanisms to do this are open ended in that the PDA can reshape its internal structure, potentially leading to unstable behavior. A disagreement regarding the value of an internal state is a specific example of conflict. Other types of conflict are present in CR1. For example, a user can always take over control of the PDA's radio resources in a way that differs from the plan offered to the user by the PDA. In any of these situations, the PDA is presented with a learning opportunity. The reinforcement mechanism built into CR1 could trigger communications with the user, to employ internal machine learning algorithms, and to allocate more internal resources (e.g., permanent case memory) to those situations that occur often enough. Those conflicts that occur only once might safely be ignored. Conflicts occurring more than once might be dealt with through machine learning in the next sleep cycle. Those that cannot be resolved autonomously could be presented to the network or to the user for assistance. A limited version of this machine learning strategy has been implemented in CR1.

11.6 SLEEP CYCLES

Decoupling communication among entities by means of remembering associated context allows for the communication among entities that may not even exist at the same time [315]. Creating data structures that reflect this kind of asynchronism can require in-depth reasoning that is combinatorially explosive and that therefore cannot be accomplished within computational resources that can be precisely defined in advance. Therefore, this less well bounded type of computation must be accomplished when the iCR is not doing anything that requires real-time response. Night seems like a good time for such activity. Given a learned pattern of behavior of the iCR's owner and an emergency wake-up timer, an iCR could spend hours in combinatorially explosive reasoning over variable frames, analyzing context bindings, and performing other off-line reasoning needed to prepare for the next wake cycle.

Mohyeldin et al. [315] point out that realizing reconfigurability of a system requires the extension of the service concept such that the logical architecture (i.e., function, task, or requirements) is changed. This is related to the frame problem [316, 317], the difficulty of relating a dynamically changing world using small, static models (frames). Over time, assumptions and abstractions become wrong even if they were valid at the time a model was constructed.

This can lead to errorful learning, such as misinterpreting a one-time exception as a new normal behavior. The user may let a child use an expensive wireless Internet game on the child's birthday, but the iCR could misinterpret that as permission for the child to download expensive games all the time. The process of adjusting models to reality is called calibration [315] and can be seen as an adaptation of the adaptation itself; even if the problem is not solvable, it may be avoided by changing the model, its context, or its grounding from time to time. With partial reconfiguration, there is always part of a model that cannot be changed such as runtime service proxies. Information in context can be used for ontology-based searches. Algorithms that analyze a day's behavior in order to adjust to such misinterpretations are best incorporated into a sleep cycle.

11.7 PITFALLS AND OPPORTUNITIES

While AM showed that automating discovery in mathematics was possible, its successors Eurisko and CYC demonstrated that it would take decades to apply automatic discovery to create commonsense knowledge. Having accrued two plus decades of technology, AACR takes on the commonsense domain of the <Radio/> <User/> again. While these technologies have promise, it is wise to remain skeptical and to pursue solid evolutionary paths. The relevance of the years of AI to AACR includes the following pitfalls and promises.

11.7.1 Pitfalls of Extensible Systems

Tieresias was developed by Professor Randall Davis as an extension to Mycin [318]. Although Tieresias worked, like many other AI systems, its technology was not fully deployed. Yet rule-based expert systems have been studied and popularized at length [13] and are readily applicable to the evolution of AACR, with concerns as follows:

Pitfalls of Rule-Based Systems

1. Rule chains suffer from the Gödel–Turing paradox that any Turing-capable self-referential system can express and attempt to compute expressions for which there is no computable answer, either crashing or consuming infinite computing resources in the search process.
2. Rule bases are typically constructed and maintained by people, and the cost of that highly skilled labor pool doomed many of the technical successes of the 1980s to economic failure.
3. Limited scope rule-based systems are embedded in mainstream software products like web browsers, email systems (e.g., rules for replying during an absence and for flushing spam), and eBusiness enterprise

systems (e.g., rules for access to data, for prioritizing orders, and for responding to problems in a supply chain).

4. Rules have been identified by the World Wide Web Consortium (W3C) as an enabling technology for the semantic web [319].

11.7.2 The Fragility of Machine Learning by Heuristics

Lenat's AM [320] learned mathematics by extending basic set theory, but it was confused as it moved too far from the foundation knowledge, so Eurisko tried to overcome those limitations. Eurisko led to CYC with over a million statements, but 100 million may be needed for "common sense." The general problem of the fragility of heuristics for extensibility remains.

CRs need to sustain relatively intelligent behavior from an initial body of knowledge by the interactive gathering of new "ground truth" data from the user or CWN. Michalski [321] describes other core ML algorithms relevant to AACR. CBR binds the current <Scene/> to prior known situations, and incremental schemes, not unlike Tieresias, to automatically formulate and manage internal knowledge. The CRA <Self/> can be used as schema-schema or as a specification for capabilities that can be programmed in conventional languages for adaptation by conventional means. While the <RF/> domain is relatively static, even with XG-class liberalization of spectrum-use rules, the <User/> domain is anything but, with combinatorially explosive possibilities and myriad uncertainties in what users say, do, and mean. So the implementation of <User/>-domain skills needs a degree of flexibility and robustness that isn't necessarily required for use-case realization in the <RF/> domain. The degree to which the CRA <Self/> (and similar alternatives) could be autonomously bootstrapped via machine learning to a robust AACR remains an open research question. Therefore, AACR evolution might best be served by initially regarding the CRA <Self/> as a description of facilities from which AACR systems engineers can pick and choose to achieve the use cases that the markets demand, leaving the more aggressive schema-schema interpretation and extension via AML to the research communities initially.

Yet, in order to increase the robustness of <User/>-domain perception and responsiveness even for relatively benign <RF/> applications like helping the user set up home wireless LAN networks, the AACR systems engineer can leverage advanced aspects of technologies introduced previously. ASR and video perception with reinforcement learning, dealing with uncertainty, and employing formal ontologies are the significant advanced topics of potential market significance. They are relevant whether one's objective is to autonomously bootstrap from <Self/> as schema-schema or just to hack in a few relevant UKOs and UDHs to meet the near-term AACR use case.

Even considering the pitfalls, the potential for integrating sensory perception and cognition capabilities into SDR to pursue AACR evolution is exciting. As the evolution unfolds, it will be increasingly important to connect to the semantic web community in a larger network-oriented undertaking that might be called semantic radio.

11.8 EXERCISES

11.1. It is easy to get an 80% solution to both NLP and ML with relatively small effort, but is this good enough? Go to the Web and find IBM's ViaVoice (or its various successors if that particular tool is not available now).

11.2. What do you think of the potential of such acoustic-channel language processing tools as your front-end processor to extract words for your CPDA? How will it do at home? How about in an acoustically noisy environment?

11.3. Suppose your CWPDA needs a UKO model of your communications patterns and you would like to be awakened by listening to your favorite news radio channel. Write the UKO that expresses this knowledge. Write a generic UKO for user behavior patterns, and factor the wake-up UKO into operations over the generic UKO. Program the updates to this blank UKO in a high order language (HOL), where the primitives operate over arbitrary data structures that you populate by hand. Refactor this UKO to operate on elements of CRA <Self/>. Refactor this UKO to abstract and embed the wake-up logic into a UDH. Test your UKO/UDH formulation by exposing it to your habits for waking, sleeping, use of commercial broadcast, TV, entering and leaving the house, starting the car, taking the kids to school, going to work, and returning at the end of the day to your favorite entertainment.

11.4. Regarding a trip to see your mother-in-law, your spouse says, "OK, I need you to take us there by 10 am to help her with her doctor's appointment." Generate the LCS-enhanced transcript and LCS primal sketch of this statement manually or using VISUALIZE. Does your UDH trigger on <Event> Appointment </Event> to update the UKO that captures your habits? Although the appointment has to do with your spouse, there may be implications for your own daily patterns. Enhance the appropriate UDH to accommodate the spouse's pattern. Test the UDH and update if necessary so that it determines that it needs to assist, and offers to create a wake-up call event that will turn your favorite radio station on at 8 am, your typical 2 hour lead time for getting to your mother-in-law's house. Compare a problem-specific set of rules to the generic UKO/UDH and experience through which it learns the rules. If your UDH learns parameters like time of day but not rules, then describe a further factoring in which your UDH contains no rules, but a more abstract UKO contains primitives from which a UDH can derive rules. (See [53] for inspiration.)

11.5. Compare NLP front end with UKO/UDH to other methods of acquiring data from users. Again, suppose you have a spouse. To know your spouse's name, the AACR must have some ability to acquire new data. Write a user interface that forces the user to enter the spouse's name in a field called "Spouse's name." Write an alternate user interface that passively parses speech dialogs from a speech recognizer to recognize dialog such as (woman's voice) "Is that your new PDA?" and the reply (male voice, probably <Owner/> 90%) "Yes,

Sweetie, it is." The appellation "Sweetie" certainly is a name. Write a UDH that will establish the correspondence among "Sweetie" and other kinds of name for the family member.

11.6. Too cute may not sell in important markets. So the CR should interact in a way that the user finds most satisfying. Establish interaction principles for Adam, such as: "Introductions should be brief and to the point." Write a UDH that implements this principle. Use that UDH to show that when introducing a new user to a CPDA, the experience may range from being introduced to a new colleague at work to being introduced to a new puppy. The new puppy isn't very talkative, but it is cute and somehow invites interaction.

11.7. Extend <Syntax/> of <Tag/> to (a) tags as delimiters and (b) tags with property lists.

11.8. What kinds of decisions undertaken by an AACR would be amenable to solution by linear dichotomizers? Each of the following represents a point in a one-dimensional space that divides that space into disjoint regions, each of which is associated with a different answer to a typical question. Consider the applicability of the dichotomizer, the source of the supervised training data, and the degree of variation needed in realistic applications with space–time variability. Consider the Mason–Dixon Line. "Are you a Southerner or a Northerner?" "Well, I live in Tampa, but grew up in New England. Why do you want to know?"

11.9. Search the Web for Java source code on supervised machine learning. Download Bayesian belief networks (BBNs) and implement a <User/>-domain capability. Find ANN source code in Java. What do the BBN and ANN tools you found have in common? How would you integrate them into your own AACR? What are the key implementation challenges in attempting to achieve the performance of AANs and BBNs via AML without those tools?

11.10. Define an iCR. How will you implement its internal perception data structures? What tools would be good for building, accessing, and updating such a data structure? Are building and updating the same thing? If not, why not? Is it a database? An object-oriented database? Is it enough to just add data, or is there other stuff that you are going to have to let the CR change by itself? In other words, will you have to write any code that writes new code? If so, how will you debug it? Rather, how will your CR test and debug the self-generated code?

CHAPTER 12

SEMANTIC RADIO

This chapter considers semantic radio, a near term step in AACR evolution supported by eBusiness networks and emerging semantic web technology.

12.1 CYC, eBUSINESS SOLUTIONS, AND THE SEMANTIC WEB

The problem of codifying human knowledge computationally was addressed by Douglas Lenat, the inventor of AM, an inspiration for the CRA <Self/> to include knowledge objects and domain heuristics. AM led to CYC, a public, commonsense on-line ontology. Open-source ontologies like CYC contribute to AACR in several ways. In one page of code at AAAI in 2005, Jim Hendler defined feline leukemia from NIH cancer and CYC "cat." Although CYC represents general world knowledge, the treatment of domain-specific topics like <RF/>, which is central to AACR, falls short of what is needed. The ways in which it falls short are instructive and shed light on the prognosis and pitfalls for AACR, RXML, and the semantic web.

One is regularly faced with the question of whether to work within one of the several existing frameworks toward domain-specific knowledge or to create one's own domain-specific ontology, which typically is much less expensive, can be accomplished quicker, and has the potential of placing one in a unique position in a marketplace. The question of ontologies for AACR has been an active research topic since 1997. Today's Web knowledge collections may contribute much to the evolution of industrial strength AACR. Business

Cognitive Radio Architecture: The Engineering Foundations of Radio XML

process systems from eBusiness companies like Autonomy® Acxiom®, BEA Systems®, Engenium®, Informatica®, Mapamundi®, NetMiner®, RetrievalWare®, Insightful Corp®, Oracle®, VivoMind®, Webtas® [322], and numerous others have adopted and exploited computational ontologies, search engines, rule-based systems, PROLOG, fourth generation databases, and XML to create waves of eBusiness solutions as these software tools are often called. The good news is that the tools for identifying, extracting, and exploiting knowledge from unstructured data sets now have strong commercial markets. There are several kinds of bad news. Nearly all of these solutions are both expensive to license and require large, expensive computing platforms. Almost none are completely portable to PDAs, although most have thin clients to access the corporate eBusiness support systems. Most eBusiness solutions, offer the following relevant to AACR:

1. XML semantic tagging of unstructured and semistructured text.
2. Automated creation of the XML tag itself by processing related information.
3. Aggregate data from enterprise sources and convert to unified XML format.
4. Enterprise applications use standard XML rather than source formats.

Many solutions embed proprietary versions of CYC and no company makes its understandably proprietary version public. Although the semantic web has been a potential eBusiness enabler since before 2002, it has been slow to develop by Web timetables. Professor Jim Hendler has been a consistent advocate of knowledge interchange languages including earlier thrusts like KQML and now the semantic web [323], typically using RDF, RDF-Schema, the DARPA Agent Markup Language (DAML), the Ontology Interchange Language (OIL), and the Ontology Web Language (OWL). These languages can express nearly anything, and 5000 domain-specific ontologies existed in August 2005, but semantic radio had not yet been proposed to W3C. The CRA <Self/> version of RXML is a candidate for evolution to semantic radio, but much needs to be done. Therefore, this chapter develops lessons learned from the disconnects between general purpose ontologies and <RF/> specific ontologies to lay groundwork for future convergence.

12.2 CYC CASE STUDY

The open-source version of CYC [324] offers radio knowledge in an Alphabetized List of CYC Constants, an Upper Ontology Flat File, an Artifact and Device Vocabulary, a Fundamental Vocabulary, and a Top-Level Vocabulary.

In its Alphabetized List of CYC Constants [325], CYC lists #$RadioWave. Searching these concepts, the substring "radio" occurs exactly once as such.

The Constants data structure consists of an alphabetic list of 2560 constants with widely disparate concepts as general as #$Thing and as nuanced as #$thermalConductivityOfObject.

12.2.1 Upper Ontology

In its Upper Ontology Flat File, CYC refers to radio 11 times, mostly in comment fields, some of which appear to have the weight of definitions but little relationship to radio as a radio engineer would know it. The treatment reveals the perspective of the interested nonexpert.

1. CYC first refers to radio as an electrical device, which is a type of physical device, but only in a footnote, as follows:
   (a) (#$genls #$ElectricalDevice #$PhysicalDevice).
   (b) (#$comment #$ElectricalDevice "A collection of physical devices; the most general collection of electrical devices. Such devices require an input of electrical current (as #$energySource) in order to perform their intended functions. Instances of #$Electrical Device include both complex devices (e.g., elements of the collections #$StereoSystem or #$Computer) and simpler ones (e.g., elements of #$ElectricalComponents and #$Electronic Components).
   (c) Note: in some contexts, crystal **radios** might be classified as inert (unpowered) electrical devices; the same for some sorts of passive radar detectors. These are exceptional cases, but still elements of this collection. In other contexts, the power for these devices can be viewed as being supplied from the outside, hence they are clearly 'powered' in such contexts.") #$Electromagnetic Radiation.
2. Further along the Upper Ontology Flat File notes that a radio wave is a type of electromagnetic radiation
   (a) (#$isa #$ElectromagneticRadiation #$WavePropagationType).
      (i) (#$isa #$ElectromagneticRadiation #$DefaultDisjointScript Type).
      (ii) (#$genls #$ElectromagneticRadiation #$WavePropagation).
      (iii) (#$comment #$ElectromagneticRadiation "A collection of events; a subset of #$WavePropagation. Each element of #$ElectromagneticRadiation is an event that arises from the interaction of an electrical field and a magnetic field. Examples include the elements of the collections #$VisibleLight, #$**RadioWaves**, and #$XRays.")
3. It notes that a radio wave is an #$InformationBearingThing #$IBT, which is an interesting construct borrowed from CYC for RXML as <Information> <Thing/> </Information>.

(a) #$IBTGeneration.
(b) (#$isa #$IBTGeneration #$TemporalObjectType).
(c) (#$genls #$IBTGeneration #$Action).
(d) (#$genls #$IBTGeneration #$TransferOut).
(e) (#$genls #$IBTGeneration #$InformationTransferEvent).
(f) (#$comment #$IBTGeneration "A collection of information transfer events. Each element of #$IBTGeneration is an event which creates some information-bearing thing—thus, an event in which some idea or information is expressed. In elements of #$IBT-Generation, the particular IBT (i.e., element of #$Information-BearingThing) which is created may be either a transient wave phenomenon (e.g., made of sound, light, or **radio** waves), or it may be a relatively long-lasting instance of #$InformationBearing-Object (cf. #$IBOCreation). Humans frequently generate such IBTs as spoken language, gestures, and handwritten notes. It is irrelevant for elements of #$IBTGeneration whether there is another agent who immediately (or, indeed, ever) accesses the resulting IBTs. Note the difference: reading is NOT an IBT generation event, but writing (usually) is. IBTs may be generated intentionally or unintentionally. Also, every communication act starts with an instance of #$IBTGeneration. See also #$Communicating and its subsets, esp. #$CommunicationAct-Single.")

4. CYC repeats this assertion with a twist in a related collection of ideas, again in a comment. CYC "radio signals" are different from "radio waves" but this ontology doesn't convey technical differences (e.g., signals versus noise, such as lightning; and that signals can occur inside a radio device and the signal-bearing waves can propagate in waveguide):
   (a) ;;; #$InformationBearingWavePropagation.
   (b) (#$isa #$InformationBearingWavePropagation #$TemporalStuff Type).
   (c) (#$genls #$InformationBearingWavePropagation #$Wave Propagation).
   (d) (#$genls #$InformationBearingWavePropagation #$Information BearingThing).
   (e) (#$comment #$InformationBearingWavePropagation "A collection of information bearing things (IBTs). Each element of #$InformationBearingWavePropagation is a #$WavePropagation (q.v.) event that carries information, for an interpreter which understands its conventions. Examples of #$InformationBearing-WavePropagation include sounds, **radio signals**, and images of visible light. These event-like IBTs should be contrasted with the relatively static, persistent, object-like IBTs in the collection #$InformationBearingObject.")

5. Later, the upper ontology links radio to news but in a comment, as follows:
   (a) ;;; #$News.
   (b) (#$isa #$News #$StuffType).
   (c) (#$genls #$News #$PropositionalInformationThing).
   (d) (#$comment #$News "A collection of abstract (intangible) informational items. Each element of #$News consists of some factual information about recent events in the world (or #$geographicalSubRegions thereof). News is commonly embodied in newspapers and communicated through **radio** and television news broadcasts.")
6. Radio Waves finally get their own class in the RadioWaves collection. If one takes the comment about radio waves as a definition, then we have a definition of radio that starts at 100 Hz, which isn't too bad, but that stops at 3 GHz or 100 MHz, hardly the extent that an engineer would attribute to the radio spectrum. This is as good as it gets in open source.
   (a) ;;; #$**RadioWave**
   (b) (#$isa #$**RadioWave** #$TemporalStuffType)
   (c) (#$isa #$**RadioWave** #$ScriptType)
   (d) (#$genls #$**RadioWave** #$ElectromagneticRadiation)
   (e) (#$comment #$**RadioWave** "A collection of events; a subset of #$ElectromagneticRadiation. Each element of #$RadioWave is an instance of electromagnetic radiation having a wavelength in the range from approximately 1 centimeter (1 × 10^8 #$Angstrom) to 3,000,000 meters (3 × 10^18 #$Angstrom), and a frequency of approximately 10^8 #$Hertz to 10^2 #$Hertz. This includes the spectrum for **RadioWave**-UHF, **RadioWave**-VHF, **RadioWave**-FM, **RadioWave**-AM, and several other types of common use #$ElectromagneticRadiation.")
7. The upper ontology goes on to include the notion of receiving a wave, such as viewing a scene using light as the mediating wave. In this case, the radio telescope example of the comments seems hardly definitional.
   (a) ;;; #$ReceivingAWave.
   (b) (#$isa #$ReceivingAWave #$TemporalStuffType).
   (c) (#$genls #$ReceivingAWave #$Receiving).
   (d) (#$comment #$ReceivingAWave "A collection of events; a subset of #$Receiving. Each element of #$ReceivingAWave is an event in which an instance of #$WavePropagation is received at a #$toLocation. For example, my CD player receiving an infrared signal from the remote control; hearing a sound of distant thunder;

a **radio** telescope receiving signals from a celestial body. See also #$WavePropagation.").

8. Next, the upper ontology describes how to rotate a radio dial, which is a nonperiodic motion, as follows:
   (a) ;;; #$Rotation-NonPeriodic.
   (b) (#$isa #$Rotation-NonPeriodic #$TemporalObjectType).
   (c) (#$genls #$Rotation-NonPeriodic #$Movement-NonPeriodic).
   (d) (#$genls #$Rotation-NonPeriodic #$Movement-Rotation).
   (e) (#$comment #$Rotation-NonPeriodic "The set of all rotational movements in which rotation occurs in a nonperiodic fashion; e.g., the turning of a knob on a kitchen appliance or a **radio** dial, or movements of a trackball. See also #$Rotation-Periodic for the context-sensitive nature of this dichotomy.")
9. The next occurrence of radio in the CYC upper ontology is that of a clock radio as an example of a class that they call a sibling disjoint collection. To express this idea to their satisfaction, CYC offers some predicate calculus.
   (a) ;;; #$SiblingDisjointCollection.
   (b) (#$isa #$SiblingDisjointCollection #$Collection).
   (c) (#$genls #$SiblingDisjointCollection #$Collection).
   (d) (#$comment #$SiblingDisjointCollection "#$SiblingDisjoint-Collection captures a very important concept, but one that is rarely given a name. There are many sets of sets for which any two member sets either will be disjoint (i.e., have no intersection) or else one will be a subset of the other. For instance, consider the various types (i.e., sets) of animals in the usual Linnaean taxonomy: Vertebrate, Bird, Dog, Mammal, Invertebrate, Person, etc.; Vertebrates and Invertebrates are mutually disjoint, while Bird, Mammal, Dog, and Person are all subsets of Vertebrate. Dog and Person are disjoint with each other, but each of them is a subset of Vertebrate. All of the Linnaean sets, or collections, of animals can be grouped together into one set, or collection, of sets, which in turn is an instance of #$SiblingDisjointCollection. It turns out that the real situation—and the real definition of #$SiblingDisjoint-Collection—is slightly more complicated than that. Consider types of appliances: toasters, cars, shavers, clocks, etc. Is the set of such appliance-types a #$SiblingDisjointCollection, the way we defined it above, for types of animals? Almost, but not quite. One could have an appliance-type '**ClockRadio**', which would be the set of all clock **radios**, and clearly each clock **radio** is both a clock and a **radio**, yet neither #$Clock nor #$**RadioReceiver** is a subset of the other. So if we have some item that purports to be both a clock and radio, that is okay if one of the following three conditions is met:

(i) (1) the collection #$Clock is known to be a subset of #$**RadioReceiver**;
(ii) (2) the collection #$**RadioReceiver** is known to be a subset of #$Clock;
(iii) (3) there is already defined a collection X which is a subset of both #$Clock and #$**RadioReceiver**.

(e) More formally, the axiom that defines #$SiblingDisjointCollection is as follows: SIB is an element of #$SiblingDisjointCollection if and only if:
(i) (#$implies
(ii) (#$and
(iii) (#$isa C1 SIB)
(iv) (#$isa C2 SIB) (#$isa C1-EL C1)
(v) (#$different C1 C2))
(vi) (#$or
(vii) (#$not (#$isa C1-EL C2))
(viii) (#$thereExists C3 (#$and (#$genls C3 C2) (#$genls C3 C1) (#$isa C1-EL C3))))).

(f) That axiom, together with the minimization of #$genls, gives us the following characterization of our concept: If we have a collection SIB that is an element of #$SiblingDisjointCollection, and if we take two elements C1 and C2 of that collection SIB, then each element of C1 which is not an element of a common specialization (C3) of C1 and C2, MUST NOT BE an element of C2.

(g) In cases where there are a few exceptions—that is, a couple of elements of SIB might have some overlap—but it is undesirable to explicitly create a new reified constant (like '**ClockRadio**', above) for that intersection, CYC allows you to use an explicit mechanism to override the #$SiblingDisjointCollection constraints for a particular C1 and C2; namely, you would assert to CYC (#$siblingDisjointExceptions C1 C2). See also #$siblingDisjoint Exceptions.")

These are good examples of the complexity that arises when competing factions like military radio experts and commercial radio experts meet in settings like the SDR Forum or OMG for AACR and semantic radio. Logic is put to the test, twisted up, and sometimes compromised nearly out of existence until consensus is reached as necessary in a broadly based standards body. Software tools that automate such processes are generally lacking. While UML generates models, few tools automate compilation of semantics of such models (e.g., from WordNet), or visual references (e.g., for primal sketches), or first principles from a core ontology like the CRA <Self/>.

10. The next example shows another limitation of the CYC approach. Radios share many properties with physical objects, such as orientation. Radio towers have a vertical orientation that distinguishes the object. If an algorithm should use this ontology to represent everyday knowledge, then the algorithm can find instances "radio." Reasoning from the comment, an algorithm may infer that a radio tower has VerticalOrientation. Associated with the string "VerticalOrientation" should be chunks of algorithm that manipulate data associated with things that have a vertical orientation. The assertion is shallow and chunks of algorithm needed for skills are not present.
    (a) ;;; #$VerticalOrientation (SIC)
    (b) (#$isa #$VerticalOrientation #$OrientationAttribute)
    (c) (#$comment #$VerticalOrientation "(#$orientation OBJECT #$VerticalOrientation) means that OBJECT is vertical with respect to the current instance of #$FrameOfReference. A linear (#$LongAndThin) object is vertical if and only if its longest dimension is perpendicular to horizontal (#$HorizontalDirection). A planar (#$SheetShaped) object has #$VerticalOrientation if and only if its planar surface is perpendicular to the current horizontal plane. Typically, vertical objects include window panes, skyscrapers, trees, **radio** towers, and walls.")
11. Then radio is viewed as the source of waves, as follows:
    (a) ;;; #$waveSource
    (b) (#$isa #$waveSource #$TernaryPredicate)
    (c) (#$arg1Isa #$waveSource #$SomethingExisting)
    (d) (#$arg2Isa #$waveSource #$SomethingExisting)
    (e) (#$arg3Isa #$waveSource #$WavePropagationType)
    (f) (#$arg3Genl #$waveSource #$WavePropagation)
    (g) (#$comment #$waveSource "The predicate #$waveSource is used to indicate that a type of wave is travelling between a source and a reception point. (#$waveSource SOURCE ENDPOINT WAVETYPE) means that there is a #$WavePropagation of type WAVETYPE propagating between the #$fromLocation SOURCE and the #$toLocation ENDPOINT. For example, (#$waveSource VoiceOfAmerica-Seoul #$CityOfBeijingChina #$**RadioWave**).")

That is the upper ontology directly related to radio.

12.2.2 Artifact and Device Vocabulary

In the Artifact and Device Vocabulary "radio" occurs only once, as follows:

1. #$ElectricalDevice A collection of physical devices; the most general collection of electrical devices. Such devices require an input of electrical current (as #$energySource) in order to perform their intended functions. Instances of #$ElectricalDevice include both complex devices (e.g., elements of the collections #$StereoSystem or #$Computer) and simpler ones (e.g., elements of #$ElectricalComponents and #$ElectronicComponents).
2. Note: in some contexts, crystal radios might be classified as inert (unpowered) electrical devices; the same for some sorts of passive radar detectors. These are exceptional cases, but still elements of this collection. In other contexts, the power for these devices can be viewed as being supplied from the outside, hence they are clearly "powered" in such contexts.
3. isa: #$ExistingObjectType
4. genls: #$PhysicalDevice #$PoweredDevice #$PartiallyTangibleProduct #$SolidTangibleThing
5. some subsets: #$ElectronicDevice #$ElectricalComponent (plus 460 unpublished subsets)

Presumably, radio occurs in one of the unpublished subsets.

12.2.3 Fundamental Vocabulary

In CYC's Fundamental Vocabulary, radio is mentioned as an example of a sibling disjoint collection, a derivative of Thing:

1. #$Thing #$Thing is the universal set: the collection of everything! Every CYC constant in the Knowledge Base is a member of this collection; in the prefix notation of the language CycL, we express that fact as (#$isa CONST #$Thing). Thus, too, every collection in the Knowledge Base is a subset of the collection #$Thing; in CycL, we express that fact as (#$genls COL #$Thing). See #$isa and #$genls for further explanation of those relationships.
2. Note: There are even a few collections, such as #$CharacterString and #$Integer, which have a #$defnSufficient that recognizes non-constants (such as strings and numbers) as instances of #$Thing.
3. isa: #$Collection
4. some subsets: #$Path-Generic #$Intangible #$Individual #$Simple-SegmentOfPath #$Path-Simple #$MathematicalOrComputational-Thing #$IntangibleIndividual #$Product #$TemporalThing #$Spatial Thing #$Situation #$EdgeOnObject #$FlowPath #$ComputationalObject #$Microtheory (plus 1488 more public subsets, 13568 unpublished subsets).

#$Thing and #$Path are fundamentals in the CRA and LCS. The object-oriented semantics of Individual, Collections, Predicate, isa, and genls (generalizes) encroach on similar computer science constructs.

12.2.4 Top-Level Vocabulary

The CYC Top-Level Vocabulary expands the Fundamental Vocabulary into a metalevel dictionary of human experience, defining situations, events, roles, intangibles, Objects, Stuff, Time (a kind of Stuff), Temporal Things, Tangible and Intangible Things, Creation, Destruction, Transformation, Transfers, Movement, Information Transfer, and Exchanges. In these data structures, radio is a type of receiver. These constructs are similar to RXML <Abstraction/>s.

1. #$Receiving A collection of events; a subset of #$GeneralizedTransfer. Each element of #$Receiving is an event in which something "comes in" to an object. Typically, a receiving has associated with it an element of #$Translocation; a particular receiving and its associated translocation(s) are related by the predicate #$transferInSubEvent. If the thing which "comes in" is an instance of #$PartiallyTangible (such as a baseball, or a SCUD missile), then its reception belongs to the specialized subset, #$ReceivingAnObject (q.v.). If the translocation associated with the receiving is an instance of #$WavePropagation (such as a radio broadcast, or heat radiation from the Sun), then the receiving belongs to the subset #$ReceivingAWave (q.v.).
2. isa: #$TemporalObjectType
3. genls: #$TransferIn
4. some subsets: #$ReceivingAWave #$ReceivingAnObject

12.3 CYC IMPLICATIONS

The fact that CYC doesn't provide a radio engineer's view of radio suggests the degree to which people are domain oriented. If I am a Carpenter and you are a Lady, then I know a lot more about wood, tools to shape wood, and nails that make a pile of sticks into a house, than you. If you are a Lady, you probably care a lot more about end items like rooms and furniture than about how I made them.

The same is true for computational ontology. Although there have been numerous attempts to organize all knowledge computationally, as knowledge becomes even a little specific, knowledge differences grow exponentially between generalists and specialists or even hobbyists. Radio has many specializations from GSM and 4G to the nuances of SATCOM and HF Ham radio.

12.3.1 Ontology Normalization

Radio ontology may be suited either to generalists (commercial users) or to a group of specialists. Normalization is the social process of agreeing on a single ontology even though many arise naturally. This unnatural act often has limited success. In the 1960s and 1970s, for example, the U.S. military decided that there were too many different abbreviations in use in military communications. To fix this "once and for all," the precursor to the Defense Information Systems Agency (DISA), the Army, and the Navy created a Joint Army Navy standard, JANAP 128 and a Defense Operating Instruction, DOI 103, with standardized abbreviations and symbols that almost nobody uses. The Army has adhered to some graphics standards. A box with an X across and two XX's on top represents an infantry division now as it did then. Man-portable surface-to-air missiles didn't even exist at the time the symbols were standardized, and thousands of other standard abbreviations languish unused. Thus, within very constrained domains like military map symbols, it is possible to normalize ontologies, and if one organizational entity has the authority to enforce conformance, the knowledge standards endure. Most human undertakings are not so well structured.

12.3.2 Ontology Mapping

When domain ontologies can't be normalized, they must be mapped. If you call it "less filling" and I say it "tastes great" we both know we are talking about "Miller beer." That TV commercial was about ontology mapping. What is important to me will be strongest in my vocabulary, and it may not match your perception of the identical object. Therefore, AACR evolution needs technology for the graceful extensibility of concepts, knowledge, and skills readily mapping among ontologies as the AACR developers, network operators, web infrastructure suppliers, and many users communities interact.

Expression 12-1 Ontology Mapping

```
<Path> <From> <My-Ontology> <Beer> Miller-Lite
  <Quality> <First> Less-filling </First> </Quality> </Beer>
    </My-Ontology> </From>
  <To> <Your-Ontology> <Beer> Miller-Lite
  <Quality> <First> Tastes-great </First> </Quality> </Beer>
    </Your-Ontology> </To>
  </Path>
```

The path of Expression 12-1 says we agree to disagree. If there were some way in which my concept and yours were identical or mutually compatible, it would be a resolvent of the path, such as <Drink-beer/>. While it is relatively easy to resolve #$Thing to <Thing/>, the semantics do not match as well.

<Superman/> is an <Abstraction/> but not a <Thing/>, while in CYC Superman is a #$Thing. This ontological stance differs substantially from most web languages.

12.4 WEB LANGUAGES

The definition and use of ontologies is facilitated by markup languages. Yolanda Gil and Varun Ratnakar characterized the trade-offs among markup languages for knowledge representation, comparing XML (eXtensible Markup Language), RDF (Resource Description Framework), and DAML (DARPA Agent Markup Language) as summarized in Table 12-1 [326].

In Gil and Ratnakar's terminology, range specifies the kinds of values (elements/classes/datatypes) that a property or element can have. Domain specifies which elements or classes can have that property. XML enables one to define <Transitive/>, <Not/>, and so on with one's own semantics, but most such tagging for ontology-related functions that are explicit in DAML are not explicit in XML, by design. Negation implies the absence of some element (e.g., no Car is a person). Necessary and sufficient conditions for class membership specify a class definition that can be used (1) to determine (or recognize) whether an instance belongs to that class or (2) to determine (or specify) whether the class is a subclass of another class. Their treatment is of the level of detail needed for industrial strength radio ontology development for the Web.

12.5 RADIO XML

In Radio XML, <Radio/> defines "the domain of natural and artificial knowledge and skill having to do with the creation, propagation, and reception of radio signals from sources natural and artificial." That's pretty much how RXML, defined in terms of the use of XML <Tag/>s as schema-schema, was envisioned, within an open framework for general world knowledge needed for AACR. RXML recognizes critical features of microworlds not openly addressed in any of the eBusiness or semantic web languages yet:

1. Knowledge often is procedural.
2. Knowledge has a source that often establishes whether it is authoritative or not, or its degree of attributed voracity.
3. Knowledge takes computational resources to store, retrieve, and process.
4. A chunk of knowledge fits somewhere in the set of all knowledge and knowing more or less where that knowledge fits can help an algorithm reason about how to use it.

TABLE 12-1 Ontology-Related Markup Languages

Aspect	XML (Schema)	RDFS (RDF Schema)	DAML
Namespace	xmlns:<label>	xmlns:rdfs =	xmlns:daml =
Namespace anchors	xmlns:xsd = "www.w3.org/2001/ XMLSchema"	Syntax:"http://www.w3.org/1999/ 02/22-rdf-syntax-ns#" Schema:"http://www.w3.org/2000/ 01/rdf-schema#"	"http://www.daml.org/ 2001/03/daml+oil#"
Ontology reference	Ontology may be an XML schema	Ontology may be a resource	Import ontology to use ontology classes
Classes	No classes only Elements	Resource top-level class	"Things" subclasses
Properties	No properties only Types	RDFS.html#property	Object and Datatype
Inheritance	Subtypes	subClassOf, subPropertyOf	Subclass, subproperty
Range	Global, local	Global <rdfs:range . . . .>	Global and local <daml:Restriction>
Domain	Element where defined	Conjunctive <rdfs:domain..>	Same as RDFS
Cardinality	minOccurs, maxOccurs	See OIL Ontology	min max cardinality
Datatypes	Numeric, temporal, string	"Literals," which are strings	Allows XML schema
Enumeration of property values	Use the <enumeration> tag	Not possible	<one of rdf:ParseType = "daml:collection".>
Ordering	<sequence> tag	<rdf:Seq> tag	<daml:List> tag
Bounding	Not inherent	Not possible	<daml:collection>
Transitive	Not inherent	Can't specify	<TransitiveProperty>
Negation	Not inherent	Not possible	<complementOf> tag
Disjunction	<union> tag	Bag for unions of properties	<disjointUnionOf . . .>
Class membership	<unique> tag	No	sameClassAs UnambiguousProperty

Thus, RKRL offers the following constructs:

RXML	Web Languages
1. <Tag/>	Identifier
2. Body	Statements
3. Models	Relationships such as “isa”
4. Context	Similar to namespace, but space–time annotated
5. Source	Web languages use namespace to attribute knowledge to sources
6. Resources	No equivalent; not concerned with concrete complexity

12.5.1 Elements of RXML

RXML includes the following:

1. A <Tag/>ed XML syntax using <Tag/>s as scoping operators for frames.
2. Lexical semantic models of time, space, entities, and communications among entities.
3. Entities (people, places, and things) occupy subsets of physical (Space × Time).
4. An initial set of knowledge, the CRA <Self/> with representation sets, definitions, conceptual models, and radio-domain models including the AACR functional architecture, and the cognition cycle
5. Mechanisms for extending the lexical semantics, modifying and extending RXML.

12.5.2 RXML Syntax

In RXML each <Tag/> is a schema-schema, and its uses specify the classes of data structure that may be derived from the tag. In other words, each tag is both a syntactic device for tagging content and an ontological primitive that makes statements about the role of that which is tagged in radio ontology of the CRA <Self/> to which each such AACR subscribes.

Each <Tag> that delimits a <Body/> of content </Tag> specifies a <Frame/>, a computational structure within which names, objects, procedures, assertions, logic, states, references, and other computational necessities may be constrained without distorting the semantics of the <Self/>, <RF/>, or <User/> domains of the CRA.

Models expressed in RXML specify the relationships among tagged elements in the body of a frame. The body may function as predicates, neural networks, if–then rules, CBR cases, and so on depending on how the frame is interpreted. These dynamic semantics also may be defined with <Tag/>s.

A <Context/> is a path to the root of RXML's universal ontology. All context roots originate with the string "RXML." This specifies that this frame belongs to RXML. The context root is a path from RXML to the frame through set membership (via contains) in the microworlds hierarchy. Context also includes the source of the material in the frame (e.g., the author) and the time and place at which the content of the frame was generated. This additional metalevel frame context supports machine reasoning about the validity of the frame. If the frame is out of date or was provided by a deprecated source, the cognitive agent can take appropriate action to update the frame.

A <Resource/> specifies software-radio-domain computational resources (e.g., processing time, computational capacity, memory, and/or interconnect bandwidth). This metalevel knowledge supports machine reasoning about the computational resources needed to accomplish a task. For example, the resources associated with an equalizer could specify the time within which the equalizer should converge [327]. The CR can apply computational constraints by setting a watchdog timer before invoking a procedure. Resources need not be computational. Frames concerned with antennas, for example, might specify the maximum RF power as a <Resource/> that can be supplied to the antenna. Without such <Resource/> constraints, CR could not reliably reason introspectively about changes to the <Self/> like downloads.

12.5.3 Heterogeneous Skills in RXML

In most software systems, knowledge and reasoning are represented homogeneously. All rule bases use rules, predicate calculus systems use Horn clauses, typically interpreted by a PROLOG engine, and CBR systems tend to use databases. Object-oriented technology makes it easy to employ multiple representations for the same information by attaching alternate representations to the object's slots. But C+, Java objects, and C++ objects are all declared differently. RXML's <Model/> primitive tags such expressions for interpretation by the appropriate compiler. The language-independent representation of the set-theoretic and model structure of such objects is mappable to any of them. RXML thus supports heterogeneous knowledge representation as illustrated in Figure 12-1. Frames may incorporate one of the following model classes:

Model Class	Interpretation Mechanism
Ontological (e.g., Scope, Contains, Set)	Set-theoretic tools
Natural language (e.g., Definition)	Human ASR, or VoiceXML interpretation
Axiomatic (e.g., Time, Now, Place, Location Predicates, Horn Clauses)	PROLOG, Rule-interpreters
Stimulus–response (the default interpretation)	serModel, neural networks, sums
Reserved (e.g., Excel, Outlook, OPRs)	Associated proprietary tool

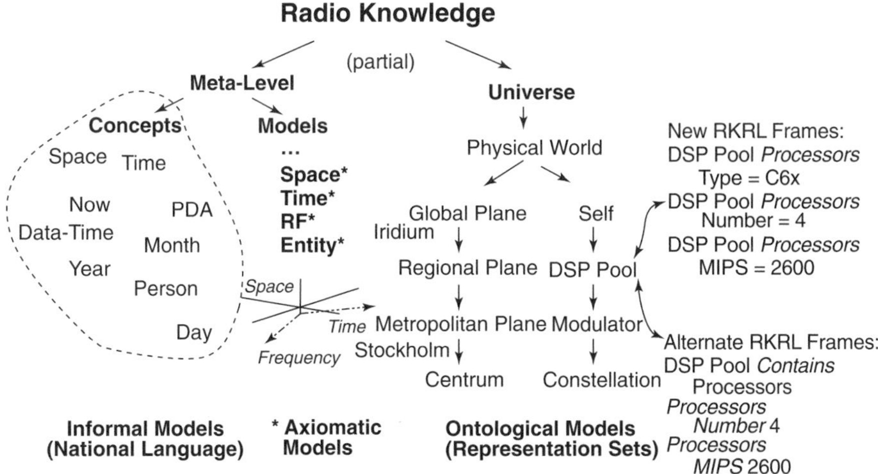

FIGURE 12-1 Heterogeneous skill representation in RXML.

RXML does not prefer one form over another. However, it does specify that if there is a model, then that model is identified in the Models metaworld. Space, time, RF (e.g., radio signals and propagation), and entities are axiomatized in LCS, but that is just one of many alternative axiomatizations.

In addition, the semantics of that model has to be specified somewhere in the inference hierarchy, for example, by specifying the API. Each model slot in a RXML frame implies the related interpretation mechanism. For example, the contains model entails set-theoretic expansion of the parent frame to contain child frames that inherit the parent's context. Natural language models rely on human interpretation of language. These may be as formal as the KQML microworld or as informal as any User microworld, as the use context indicates. RXML is developed further in the companion CD-ROM/ web site.

12.6 CONCLUSIONS

The convergence of speech recognition, vision, and robotics with SDR and radio spectrum liberalization portends a wireless Web where VoIP from the PDA displaces conventional cellular service providers in urban centers. On the other hand, conventional cellular service providers may leverage AACR technologies to regain profitability. Either way, the use cases, enabling technologies, radio knowledge, KOs, DHs, and web technologies introduced in this book will make exciting contributions to academic, government, and commercial interests for decades to come.

GLOSSARY

AAAI American Association for Artificial Intelligence
AACR aware, adaptive, and cognitive radio
AAR aware–adaptive radio
ABC always best connected
ABI always better informed
ADC analog-to-digital converter
AGC automatic gain control
AIKA autonomous incremental knowledge acquisition
ALE automatic link establishment
AML autonomous machine learning
ANN artificial neural network
Antecedent the precondition of an If–Then logic structure
APIs applications–programmer interfaces
ARQ automatic repeat request
ARs aware radios
ASR Automatic speech recognition
ASIC application-specific integrated circuit
AV-ASR audio-visual automatic speech recognition
AWGN additive white Gaussian noise
BER bit error rate

BIT built-in test
BLoS beyond line-of-sight
BPSK binary phase shift keying
BSC base station controller
BTS base transmission station
CAD computer-aided design
CASE computer-aided software engineering
CBR case-based reasoning
CCI co-channel interference
CDAAs circularly disposed array antennas
CDMA code division multiple access
CFAR constant false alarm rate
CFGs context-free grammars
CIR carrier-to-interference ratio, a measure of radio signal strength
CNs cognitive networks
COMSEC communications security
CORBA common object request broker architecture
CR cognitive radio
CRA cognitive radio architecture
CRO cognitive radio ontology
CVSDM continuously variable slope delta modulation
CWN cognitive wireless network
CWPDA cognitive wireless personal digital assistant
DAB digital audio broadcast
DAC digital-to-analog converter
DBSs direct broadcast satellites
DECT Digital European Cordless Telephone
DH domain heuristics
DLL dynamic link library
DoA direction of arrival
DP dynamic programming
DSP digital signal processor
DSSS direct sequence spread spectrum
DVB direct video broadcast
ECCMs electronic counter-countermeasures
EIRP effective isotropic radiated power
EM electromagnetic
EMP electromagnetic pulse

ETSI European Telecommunications Standards Institute
FCC Federal Communications Commission (U.S.A.)
FDM frequency division multiplexing
FEC forward error control
FFT fast Fourier transform, a means of estimating the power spectrum
FH frequency hop
FOPC first-order predicate calculus
FPGA field programmable gate array
FSK frequency shift keyed
FSLs finite state languages
FSM finite state machine
FSO free-space optics
3G third generation mobile cellular systems (e.g., based on CDMA)
GA genetic algorithm
GoS grade of service
GPP general purpose processor
GUI graphical user interface
HDR hardware-defined radio
Hidden nodes receivers not detected by potential interferers
HLR home location register
HLT human language technology
HMMs hidden Markov models
HTML Hypertext Markup Language
IBR instance-based reasoning
iCR ideal cognitive radio
IDL CORBA Interface Definition Language
IFF if and only if; identification friend or foe
IMU inertial measurement unit
INFOSEC information security
IRDA infrared data access
ISA instruction set architecture
ISAPI information services applications–programmer interface
ISM Instrumentation, Scientific and Medical—a set of low power RF bands
ISO International Standards Organization
ISP Internet service provider
ITU International Telecommunications Union
IVHSs intelligent vehicle highway systems

JADE Java Agent Development Environment
JESS Java Expert System Shell
JTIDS Joint Tactical Information Distribution System
JTRS Joint Tactical Radio System
KB knowledge base
KDD knowledge discovery and data mining
KOs knowledge objects
KQML Knowledge Query and Manipulation Language
KS knowledge source
KTH Kungle Tekniska Hogskolan
LCS lexical conceptual semantics
LEO low Earth orbit
LLC logical link control
LNA low noise amplifier
LPC Linear Predictive Coding
LPI low probability of intercept
LSB lower sideband
LSI latent semantic indexing
LUF lowest usable frequency
LVHF lower very high frequency
MAC media access control
MBA multibeam antenna array
MBC meteor burst communications
MBMMR multiband multimode radio
Method a procedure attached to a software object
MIMO multiple input multiple output
ML machine learning
MSC message sequence chart
MTSO mobile telephone switching office
MUF maximum usable frequency
NL natural language
NLP natural language processing
NoI Notice of Inquiry
NPRM Notice of Proposed Rule Making
NVI near vertically incident
OBR ontology-based radio
ODP open distributed processing
OEM original equipment manufacturer

OFDM orthogonal frequency division multiplexing

OMG Object Management Group

Ontology branch of metaphysics that studies the nature of existence or being

OOK on–off keying

OOT object-oriented (OO) technology

OSI open systems interconnect

ORB object request broker, software that dispatches remote procedure calls, and so on

OTA over the air

OWL Ontology Web Language

PAN personal area network

PDA personal digital assistant

PDA DL PDA design language

PDANode object that contains a srModel and related slots and methods

PDL Program Design Language

PDR programmable digital radio

PHY physical layer of the ISO protocol stack

PIM platform-independent model

PPM pulse position modulation

PSD power spectral density—a spectrogram, power of signals and noise versus RF

PSK phase shift keying

PSTN Public Switched Telephone Network

PVMs parallel virtual machines

PWM pulse width modulation

QoI quality of information

QoS quality of service

QPSK quaternary phase shift keying

R&O Rule and Order

RA regulatory authority

RAP radio access protocol

RBL relevance-based learning

RDHs radio-domain heuristics

RDO radio-domain ontology

Reinforcement assigning an increased degree of belief or relevance to a stimulus, plan, and so on

RF radio frequency

RHSs reinforced hierarchical sequences

RISC reduced instruction set computer architectures

RKRL Radio Knowledge Representation Language

RL reinforcement learning

RSSI received signal strength indication

RXML Radio XML

SCA software communications architecture

SCPC single channel per carrier

SDL Specification and Description Language (ITU Recommendation Z.100)

SDR software-defined radio

<Self/> ontological self in RXML

serModel stimulus–experience–response model

SKOs spatial knowledge objects

SLI situated language interpreter

Slot data structure that is part of a software object

SNR signal-to-noise ratio

SOI signal operating instruction

SOP standard operating procedure

SRA software radio architecture

srModel stimulus–response model (deprecated)

SSs skill sets

SVMs support vector machines (an IBR technology)

SWR software radio

Taxonomy science dealing with description, identification, naming, and classification

TD time difference

TDL temporal difference learning

TDM time division multiplexed

TIA Telecommunications Industries Association (U.S.A.)

TRANSEC transmission security

TV television

UDHs user-domain heuristics

UML Unified Modeling Language (an object-oriented design language)

USB upper sideband

UWB ultra-wideband

VAD voice activity detector

VCM virtual channel management

VoiceXML Voice version of the eXtensible Markup Language (XML)

VoIP Voice over IP

VOX voice operation transition (same as VAD)

VSAT very small aperture terminal

VSB vestigial sideband

VTC video teleconference

W3C World Wide Web Consortium

W-CDMA wideband CDMA, the emerging international 3G standard

WiFi IEEE 802.11 wireless network

WPDA wireless PDA

WRC World Radio Conference

XG DARPA neXt Generation spectrum-use policy and language

XML eXtensible Markup Language

REFERENCES

1. Semantic Web Home Page, www.w3.org/2001/sw.
2. Ontology Web Language (www.wc3.org) June 2004.
3. Dr. Mark McHenry, personal communication.
4. M. Mouly and M. B. Pautet, *The GSM System for Mobile Communications* (Plaiseau, France: published by the authors) 1983.
5. http://www.europa.eu.int/comm/research/nfp/pdf/instruments.pdf.
6. Petri Mähönen, "Overview and Goals," *Dagstuhl Workshop* (Aachen, Germany: RWTH-Aachen) Oct. 2004.
7. Ontology, *Merriam Webster's Collegiate Dictionary* (Britannica 2003 Deluxe Edition, CD-ROM).
8. N. Guarino and P. Giaretta, "Ontologies and Knowledge Bases: Towards a Terminological Clarification." In: N. J. I. Mars (Ed.), *Towards Very Large Knowledge* (Padova, Italy: Ios Pr.) 1995.
9. www.landcglobal.com.
10. Theo Kanter, "Transparent Cognitive Radio," *Dagstuhl Workshop on Cognitive Radio* (Aachen, Germany: RWTH-Aachen) Oct. 2004.
11. George A. Miller, *Language and Communication* (New York: McGraw-Hill) 1951.
12. *Handbook of Artificial Intelligence* (Palo Alto, CA: Tioga Publishing) 1982.
13. Stuart Russell and Peter Norvig, *Artificial Intelligence: A Modern Approach,* 2nd ed. (Upper Saddle River, NJ: Prentice Hall) 2003.
14. Ian Watson, *Applying Case-Based Reasoning* (San Francisco, CA: Morgan Kaufmann) 1997.

Cognitive Radio Architecture: The Engineering Foundations of Radio XML
By Joseph Mitola III

15. Andrew Moore, *Instance-Based Learning Course Notes* (Pittsburgh, PA: Carnegie Mellon University) 2001.
16. A. Aamodt, *Case-Based Reasoning* (Norway: Norwegian University of Science and Technology) 2001.
17. N. Cristianini and J. Shawe-Taylor, *Support Vector Machines* (Cambridge, UK: Cambridge University Press) 2000.
18. *Transduction (Machine Learning) Guide, Meaning, Facts, Information and Description*, www.e-paranoids.com/t/tr/transduction__machine_learning_.html.
19. J. Mitola III, email to Professor Gerald Q. Maguire, recommending the title of the Licentiate as Cognitive Radio, Jan. 1998.
20. *Inquiry Regarding Software Defined Radios*, ET Docket No. 00-47 (Washington, DC: U.S. Federal Communications Commission) 21 Mar. 2000.
21. FCC 03-322 (Washington, DC: U.S. Federal Communications Commission) 19 Dec. 2003.
22. *Facilitating Opportunities for Flexible, Efficient, and Reliable Spectrum Use Employing Cognitive Radio Technologies*, ET Docket No. 03-108 (Washington, DC: U.S. Federal Communications Commission) 30 Dec. 2003.
23. Walter Tuttlebee, Regulatory Authority's (RA) Regulatory Round Table (RRT) on Software Defined Radio (SDR) (UK: Virtual Center of Excellence) 11 Sept. 2004.
24. *Proceedings of the E2R Workshop* (Brussels: EC) Sept. 2004.
25. *Proceedings of the Dagtsuhl Workshop on Cognitive Radios and Networks* (Aachen, Germany: RWTH-Aachen) Oct. 2004.
26. Jorge Periera, "Cognitive Radio and Networks EC Perspective," *Proceedings Dagstuhl Workshop on Cognitive Radios and Networks* (Aachen, Germany: RWTH-Aachen) Oct. 2004.
27. www.sdrforum.org.
28. Michael Chartier, "Intel Cognitive Radio," SDR Forum Working Group, Sept. 2003.
29. Preston Marshall, *Remarks to the SDR Forum* (Rome, NY: SDR Forum) Sept. 2003.
30. Dr. Anthony Tether, Director of the Defense Advanced Research Projects Agency (DARPA), personal communication, Sept. 2003.
31. Ignas Niemegeers, "Cognitive Networks and Radios," *Dagstuhl Workshop on Cognitive Radio* (Aachen, Germany: RWTH Aachen) Oct. 2004.
32. ITT Corporation, "Mesh Networks," private communication, 2003.
33. Ignas Niemegeers, *Dagstuhl Workshop on Cognitive Radio* (Aachen, Germany: RWTH Aachen) Oct. 2004.
34. Alistair Munro, *Dagstuhl Workshop on Cognitive Radio* (Aachen, Germany: RWTH Aachen) Oct. 2004.
35. Keith Devlin, *Logic and Information* (Cambridge, UK: Cambridge University Press) 1991, p. 60.
36. A. G. Hamilton, *Logic for Mathematicians* (Cambridge, UK: Cambridge University Press) 1978, pp. 54–69.

37. Judea Pearl, *Causality* (Cambridge, UK: Cambridge University Press) 2000, p. 346.
38. eTrepid web site, www.etrepid.com.
39. Marc Zissman, personal communication (Lexington, MA: MIT Lincoln Laboratory) May 1999.
40. T. Mowbray and R. Zahavi, *The Essential CORBA* (Hoboken, NJ: John Wiley & Sons) 1995.
41. James Allen, *Natural Language Understanding* (San Francisco, CA: Benjamin Cummings) 1987; 2nd edition, 1995.
42. Mark Burstein and Drew McDermott, "Ontology Translation for Interoperability Among Semantic Web Services," *AI Magazine* (Menlo Park, CA: AAAI) Spring 2005.
43. *IEEE Spectrum* (New York: IEEE Press) Aug. 2004.
44. ITU Z.100, *Object GEODE* (Paris, France: Verilog) 1998.
45. B. P. Lathi, *Signals, Systems, and Communication* (Hoboken, NJ: John Wiley & Sons) 1965.
46. David Luenberger, *Optimization by Vector Space Methods* (Hoboken, NJ: John Wiley & Sons) 1969.
47. E. Yourdon, *Structured Design* (Englewood Cliffs, NJ: Yourdon Press) 1978.
48. H. D. Mills, "Stepwise Refinement and Verification in Box-Structured Systems," *IEEE 0018-9162/88/0600-0023* (New York: IEEE Press) June 1988.
49. G. Booch, *Object-Oriented Design with Applications* (San Francisco, CA: Benjamin Cummings) 1991.
50. P. Coad and E. Yourdon, *Object-Oriented Design* (Englewood Cliffs, NJ: Prentice Hall/Yourdon Press) 1991.
51. B. Douglass, *Real-Time UML* (Boston, MA: Addison Wesley) 1998.
52. J. Ellis, *Objectifying Real-Time Systems* (New York: SIGS Books) 1994.
53. R. Sun and T. Peterson, "Some Experiments With a Hybrid Model for Learning Sequential Decision Making," *Information Sciences 111 83-07* (Amsterdam, The Netherlands: Elsevier) 1998.
54. James Culbertson, *Consciousness: Natural and Artificial* (Roslyn Heights, NY: Libra Publishers) 1982.
55. Mark Ridley, *Evolution* (Cambridge, MA: Blackwell Science) 1996.
56. S. Kaufman, *At Home in the Universe* (Santa Fe, NM: The Santa Fe Institute) 1992.
57. "Beyond 3G," *IEEE Communications Magazine*, June 2004.
58. www.sdrforum.org.
59. M. Minsky and S. Papert, *Artificial Intelligence Project* (Palo Alto, CA: Stanford University) 1957.
60. Open Procedural Reasoning System (OPRS) (www.oprs.org).
61. www.omg.org.
62. *The Handbook of Artificial Intelligence* (Palo Alto, CA: Tioga Publishing) 1982.

63. Brendan Frey, *Graphical Models for Machine Learning and Digital Communications* (Cambridge, MA: MIT Press) 1998.
64. Order FCC 03-322, Notice of Proposed Rule Making (Washington, DC: U.S. Federal Communications Commission) Dec. 2003.
65. ACM Special Interest Group on Knowledge Discovery and Data Mining (www.acm.org/sigkdd) 2003.
66. Embedded ViaVoice Enterprise Edition (www.ibm.com) Sept. 2004.
67. R. Eberhart et al., *Computational Intelligence PC Tools* (Boston, MA: AP Professional) 1996.
68. Sean Luke, *Evolutionary Computation in Java* (cs.gmu.edu/~eclab/projects/ecj) Mar. 2004.
69. John Koza et al., *Genetic Programming IV DVD* (Los Altos, CA: Genetic Programming, Inc.) 2004.
70. SDR Forum, *Technical Report 1.0* (www.sdrforum.org) 1998.
71. Matlab Digest (www.mathworks.com) 1999.
72. RFCAD 2.3 (www.comm-data.com) 1999.
73. WRAP Software Tool (Vaxjo, Sweden: Enator Communications AB) 1999.
74. R. Aler et al., "Genetic Algorithms and Deductive–Inductive Learning: A Multistrategy Approach," *Proceedings of the International Conference on Machine Learning* (San Francisco, CA: Morgan Kaufmann) 1998.
75. Christian Rieser, "Biologically Inspired Cognitive Wireless L12 Functionality," FCC (VA Tech), 19 May 2003.
76. Thomas W. Rondeau, "Cognitive Radios With Genetic Algorithms: Intelligent Control of Software Defined Radios," *Proceedings of SDR Technical Symposium 2004* (Rome, NY: SDR Forum) Nov. 2004.
77. Friedrich K. Jondral, "Parameter Controlled Software Defined Radio," *IEEE Communications Magazine* (NY: IEEE Press) Jun 2004.
78. Friedrich K. Jondral, "Parameterization—A Technique for SDR Implementation." In: W. Tuttlebee (Ed.), *Software Defined Radio: Enabling Technologies* (London: Wiley) 2002, pp. 233–256.
79. Joint Tactical Radio System Web Site (www.jtrs.saalt.army.mil) Oct. 2002.
80. Yolanda Gil and Varun Ratnakar, *Markup Languages: Comparison and Examples* (www.usc.edu: TRELLIS project) 2004.
81. www.darpa.mil/XG.
82. T. Finin, "KQML—A Language and Protocol for Knowledge and Information Exchange," *International Conference on Building and Sharing of Very Large-Scale Knowledge Bases*, Tokyo, Dec. 1993.
83. Community of Agent Based Systems (CoABS).
84. www.mindswap.org/2004/owl-s/api/.
85. www.w3.org/1999/xhtml.
86. Emmanuel Lavinal, Thierry Desprats, and Yves Raynaud, "A Conceptual Framework For Building Cim-Based Ontologies," *Proceedings IEEE IM, 2003* (New York: IEEE Press) 2003.

87. Jorge E. López de Vergara, Víctor A. Villagrá, and Julio Berrocal, "Applying the Web Ontology Language to Management Information Definitions," *IEEE Communications Magazine*, July 2004.
88. J. Wang, M. Kokar, K. Baclawski, and D. Brady, *Achieving Self-Awareness of SDR Nodes Through Ontology-Based Reasoning and Reflection* (Boston, MA: Northeastern University) www.ece.neu.edu/groups/scs/publications.htm, 2005.
89. Pat Winston, *Artificial Intelligence* (Cambridge, MA: MIT Press) 1983.
90. P. Hayes-Roth, *Pattern-Directed Inference Systems* (Upper Saddle River, NJ: Prentice Hall) 1978.
91. Nils Nillson, *Artificial Intelligence Programming* (Palo Alto, CA: Tioga Publishing) 1980.
92. Richard S. Sutton and Andrew G. Barto, *Reinforcement Learning: An Introduction* (Cambridge, MA: MIT Press) 1998.
93. Pat Langley, *Elements of Machine Learning* (San Francisco, CA: Morgan Kaufmann) 1996.
94. S. Albayrak (Ed.), *Intelligent Agents for Telecommunications Applications* (Berlin: IOS Press) 1998.
95. *Proceedings of the 2004 Software Defined Radio Technical Conference* (Rome, NY: SDR Forum) Nov. 2004.
96. Julius Bendat and Allan Piersol, *Measurement and Analysis of Random Data* (Hoboken, NJ: John Wiley & Sons) 1966.
97. R. Haralick et al., "Computer Vision Update," *The Handbook of Artificial Intelligence, Volume IV* (Boston, MA: Addison Wesley) 1989.
98. C. J. C. Burges, *A Tutorial on Support Vector Machines for Pattern Recognition* (Whippany, NJ: Bell Laboratories, Lucent Technologies) Dec. 1999.
99. Laveen Kanal, "Patterns in Pattern Recognition," 1968–79 *IEEE Trans. Information Theory*, 17–20 (6), Nov. 7, pp. 697–722.
100. *Mathematical Statistics* (Boston, MA: Addison Wesley) 1969.
101. D. Tebbs and G. Collins, *Real Time Systems* (Berkshire, UK: McGraw-Hill) 1977.
102. W. H. Press et al., *Numerical Recipes in C* (Cambridge, UK: Cambridge University Press) 1988.
103. David E. Goldberg, *Genetic Algorithms* (Boston, MA: Addison Wesley) 1989.
104. Tandra Pal, "SOGARG: A Self-Organized Genetic Algorithm-Based Rule Generation Scheme for Fuzzy Controllers," *IEEE Transactions on Evolutionary Computing* Vol. 7, No. 4, pp. 397–415, Aug. 2003.
105. D. Traum and P. Dillenbourg, "Miscommunication in Multi-modal Collaboration," *Proceedings of the AAAI Workshop on Detecting, Preventing, and Repairing Human–Machine Miscommunication* (Palo Alto, CA: AAAI) 1996.
106. James Reggia et al., "Diagnostic Expert Systems Based on a Set Covering Model," *International Journal of Man-Machine Studies*, Vol. 19, p. 437, 1983.
107. "Semantic Integration," *AI Magazine*, Vol. 26, No. 1, Spring 2005.
108. "Brownian Motion," *Encyclopedia Britannica 2003 CD-ROM*.

109. Xiaoli Ma et al., "Iterative Decoding for Differential MIMO Systems With Near Coherent Performance," *Proceedings SDR Technology Symposium* (Rome, NY: SDR Forum) Nov. 2004.

110. B. Hochwald and W. Sweldens, "Differential Unitary Spacetime [sic] Modulation," *IEEE Transactions on Communications*, Vol. 48, No. 12, pp. 2041–2052, Dec. 2000.

111. B. L. Hughes, "Differential Space-Time Modulation," *IEEE Transactions on Information Theory*, Vol. 46, No. 7, pp. 2567–2578, Nov. 2000.

112. Carlo Di Nallo et al., "Enabling Antenna Technologies for the SDR," *Proceedings SDR Technology Symposium* (Rome, NY: SDR Forum) Nov. 2004.

113. J. Nossek, "Adaptive Antennen" (in German), *German Workshop on Future Mobile Communications* (Bonn, Germany: Regulatory Authority of Germany) 30 May 2001.

114. Le Guen and Ali Mansour, "Automatic Recognition Algorithm for Digitally Modulated Signals Based on Statistical Approach in Time-Frequency Domain," *Proceedings of the Sixth Baiona Workshop on Signal Processing in Communications* (Baiona, Spain: University of Vigo) Sept. 2003.

115. Thomas W. Rondeau et al., "Cognitive Radios With Genetic Algorithms: Intelligent Control of Software Defined Radios," *Proceedings SDR Forum Technical Symposium* (Rome, NY: SDR Forum) 2004.

116. "Islands of Understanding" was first used in computer science in this way by Dr. Reddy on the Hearsay II Speech Understanding Project in the 1970s.

117. Ian Watson, *Applying Case-Based Reasoning* (San Francisco, CA: Morgan Kaufmann) 1997.

118. D. Bertsekas and J. Tsitsiklis, *Neural-Dynamic Programming* (Belmont, MA: Athena Scientific) 1996.

119. C. Watkins and P. Dayan, "Q-Learning," *Machine Learning*, Vol. 8, p. 279, 1992.

120. D. Bertsekas and J. Tsitsiklis, *Neural-Dynamic Programming* (Belmont, MA: Athena Scientific) 1996.

121. H. Tong and T. Brown, "Adaptive Call Admission Control Under Quality of Service Constraints: A Reinforcement Learning Solution," *IEEE JSAC*, Vol. 18, No. 2, pp. 209–221. (New York: IEEE Press) Feb. 2000.

122. L. M. G. Goncalves, "Multimode Stereognosis," *Autonomous Agents 99* (Seattle, WA: ACM) 1999.

123. Gene F. Franklin et al., *Digital Control of Dynamic Systems* (Boston, MA: Addison Wesley) 1990.

124. G. G. L. Meyer and R. C. Raup, "On the Structure of Cluster Point Sets of Iteratively Generated Sequences," *International Journal of Optimization Theory and Applications*, Vol. 28, No. 3, pp. 353–362. July 1997.

125. J. R. Quinlan, "Learning and Efficient Classification Procedures and Their Application to Chess End Games," *Machine Learning* (Palo Alto, CA: Tioga Publishing) 1983.

126. T. Finin, "KQML—A Language and Protocol for Knowledge and Information Exchange," *International Conference on Building and Sharing of Very Large-Scale Knowledge Bases*, Tokyo, Dec. 1993.

127. www.daml.org.
128. jade-develop@sharon.cselt.it.
129. http://www.networkinference.com 2004.
130. Matti Keijola, *The BRIEFS Knowledge Base and the WAP Pilot* (Helsinki, Finland: HUT/TAI Research Centre) July 2000.
131. *WordNet 2.0* (Princeton, NJ: Princeton University) 2003.
132. Ted Pedersen, *WordNet: Similarity (v0.09) and Supporting Software* (tpederse@d.umn.edu, http://search.cpan.org/dist/WordNet-Similarity) 20 July 2004.
133. *Proceedings of the Seventh European Conference on Case-Based Reasoning (ECCBR-04)* (Madrid, Spain) 2004.
134. S. Louis et al., *A Case Study in Object Oriented Modeling, Architecting, and Designing an Enterprise Monitoring Application N00014-03-1-0104* (Washington, DC: Office of Naval Research).
135. Klaus-Dieter Althoff et al., *A Review of Industrial Case-Based Reasoning Tools* (http://wwwagr.informatik.uni-kl.de/~lsa/CBR/toolreview.html) 2004.
136. Juyang Weng, *Technology Review* (Boston, MA: MIT) (08/03); Rebecca Zacks, "Teachable Robots," Michigan State University.
137. James S. Albus and Alexander M. Meystel, *Engineering of Mind* (Hoboken, NJ: John Wiley & Sons) 2001.
138. H. Eriksson and M. Penker, *UML Toolkit* (Hoboken, NJ: John Wiley & Sons) 1998.
139. T. Mowbray and R. Malveau, *CORBA Design Patterns* (Hoboken, NJ: John Wiley & Sons) 1997.
140. Wireless World Research Forum (www.wwrf.com) 2004.
141. Jim Wolf, "Pentagon Seeks to Sort, Store Lifetime Experience," *Reuters* (www.reuters.com) 28 May 2003.
142. J. Mitola III, "Software Radio Architecture: A Mathematical Perspective," *IEEE JSAC,* Vol. 17, No. 4, pp. 514–538 (New York: IEEE Press) Apr. 1999.
143. R. Hennie, *Introduction to Computability* (Boston, MA: Addison Wesley) 1997.
144. Joseph Mitola III, *Software Radio Architecture* (Hoboken, NJ: Wiley-Interscience) 2000.
145. Joseph Mitola III, *Cognitive Radio: An Integrated Agent Architecture for Software Defined Radio* (Stockholm: KTH, The Royal Institute of Technology) June 2000.
146. *Cognitive Vision* (Palo Alto, CA: AAAI) June 2004.
147. SNePS (Internet: ftp.cs.buffalo.edu:/pub/sneps/) 1998.
148. C. Koser et al., "read.me" (www.cs.kun.nl: The Netherlands: University of Nijmegen) Mar. 1999.
149. The XTAG Research Group, *A Lexicalized Tree Adjoining Grammar for English,* Institute for Research in Cognitive Science (Philadelphia, PA: University of Pennsylvania) 1999.
150. PC-KIMMO Version 1.0.8 for IBM PC, 18 Feb. 1992.
151. Walter Tuttlebee, *Software Defined Radio* (Chichester, UK: John Wiley & Sons) 2002.

152. Preston Marshall, *Remarks to the SDR Forum* (Rome, NY: SDR Forum) Sept. 2003.
153. C. Phillips, "Optimal Time-Critical Scheduling," *STOC 97* (www.acm.org: ACM) 1997.
154. Xiaoli Ma et al., "Iterative Decoding for Differential MIMO Systems With Near Coherent Performance," *Proceedings SDR Technology Symposium* (Rome, NY: SDR Forum) Nov. 2004.
155. Bruce Fette, "SDR Technology Implementation for the Cognitive Radio," FCC 19 May 03 (Falls Church, VA: General Dynamics) 2003.
156. Federal Communications Commission, *Public Forum on Secondary Markets* (Washington, DC: FCC) 21 May 2000.
157. Joseph Mitola III, "Cognitive Radio for Flexible Mobile Multimedia Communications," *Mobile Multimedia Communications* (MoMUC 99) (New York: IEEE Press) Nov. 1999.
158. www.it.kth.se/~jmitola.
159. Friedrich K. Jondral and Timo A. Weiss, Universität Karlsruhe, "Spectrum Pooling: An Innovative Strategy for the Enhancement of Spectrum Efficiency," *IEEE Communications Magazine* Mar. 2004.
160. Stephan Lang, Raghu Mysore Rao, and Babak Daneshrad, "Design and Development of a 5.25 GHz Software Defined Wireless OFDM Communication Platform," *IEEE Communications Magazine*, June 2004.
161. John Sydor of CRC, john.sydor@crc.ca.
162. John Sydor, "5 GHz Cognitive Radio: An Approach to Rural Community Broadband Access" (www.crc.ca/fr/html/milton/home/iee_comm_engarticle_april2004.pdf) July 2004.
163. U.K. Regulatory Authority Workshop (London, UK) 15 Sept. 2004.
164. harada@crl.go.jp.
165. S. Kaufman, *At Home in the Universe* (Hoboken, NJ: John Wiley & Sons) 1995.
166. "UCSD, VA and Cal-$(IT)^2$ Wireless Technology to Enhance Mass Casualty Treatment in Disasters," *UCSD Health Sciences News* (San Diego, CA: University of California at San Diego) 23 Oct. 2003.
167. Hironori Hiraishi and Fumio Mizoguchi, "A Cellular Telephone-Based Application for Skin-Grading to Support Cosmetic Sales," *Proceedings AAAI 2002* (Menlo Park, CA: American Association for Artificial Intelligence) 2004.
168. Arc Explorer (http://www.esri.com/mapobjects: Environmental Research Institute Inc.) 2003.
169. P. Withington, "Impulse Radio Overview" (www.timedomain.com: Time Domain, Inc.) 1999.
170. *Reference Data for Engineers* (Carmel, IN: Sams Publishing) 1993.
171. L. Wagner, J. Goldstein, W. Meyers, and P. Bello, "The HF Skywave Channel: Measured Scattering Functions for Mid-latitude and Auroral Channels and Estimates for Short-Term Wideband HF Rake Modem Performance," *MILCOM 89* (New York: IEEE Press) Oct. 15–18, 1989.
172. *Jane's Military Communications 1992–93* (Surrey, UK: Jane's Information Group) 1992.

173. "Digital Cellular Technologies (as of 1 April 99)" (http://home.intekom.com) Sept. 1999.

174. R. Ziemer and R. Petersen, *Digital Communications and Spread Spectrum Systems* (New York: Macmillan) 1985.

175. R. E. Crochier, "Sub-Band Coding," *Bell Systems Technical Journal,* Vol. 60, No. 7, pp. 1633–1653, Sept. 1981.

176. G. Kaiser, *A Friendly Guide to Wavelets* (Berlin: Birkhauser) 1994.

177. X. Tanaka et al., "Urban Multipath Propagation Delay Characteristics in Mobile Communications," *Transactions IEICE* (*B-II*), pp. 772–778, Nov. 1990.

178. Enator Communications AB, Radio Systems, Ljungadalsgatan 2 SE-351 80 Växjö, Sweden, 1998.

179. Communications Data Services (CDS), 1998.

180. S. Affes and P. Mermelstein, "Spatio-Temporal Array-Receiver for Multipath Tracking in Cellular CDMA," *ICC 97* (New York: IEEE Press) 1997.

181. AN/GRC 103 (Toronto: Marconi Canada) 1995.

182. FHM9104 (Paris: SAT Communications) 1996.

183. R. Ziemer and R. Peterson, *Digital Communications and Spread Spectrum Systems* (New York: Macmillan) 1985.

184. Office of the Under Secretary of Defense for Acquisition, Technology, and Logistics, Washington, DC, *Report of the Defense Science Board Task Force on Wideband Radio Frequency Modulation Dynamic Access to Mobile Information Networks* (Washington, DC: U.S. DoD) Nov. 2003.

185. TFH950S Product Description (Paris: Alcatel) 1997.

186. August W. Rihaczek, *High Resolution Radar* (Upper Saddle River, NJ: Prentice Hall) 1970.

187. G. Comparetto, "Satellite Communications—Current Features and Future Trends," *WESCON 96*, p. 233.

188. R. Leopold, "Low-Earth Orbit Global Cellular Communications Network," *ICC 91* (New York: IEEE Press) 1991.

189. Dong-Hee Lee et al., "A Network Architecture for the Integration of IRIDIUM and CDMA Systems," *ICC97* (New York: IEEE Press) 1997.

190. *Satellite Tool Kit* (*STK*) (Valley Forge, PA: Analytical Graphics) 2002.

191. *Advanced Communications Technology Satellite* (*ACTS*) (New York: IEEE Press) 1997.

192. Ye Tian, Kai Xu, and Nirwan Ansari, "TCP in Wireless Environments," *IEEE Communications Magazine* Mar. 2005.

193. S. O'Keefe, "MPLS: High Octane IP" (www.hottech.com) 1998.

194. www.ieff.org/html.charters/mpls-charter.html.

195. Yongkang Xiao, Xiuming Shan, and Yong Ren, "Game Theory Models for IEEE 802.11 DCF in Wireless Ad Hoc Networks," *IEEE Communications Magazine*, Mar. 2005.

196. Nils Nilsson, *Principles of Artificial Intelligence* (Palo Alto, CA: Tioga Publishing) 1980.

197. P. E. Ross, "Silicon Shows Its Mettle," *IEEE Spectrum*, Mar. 2003.

198. Naomi Sager, *Natural Language Information Processing* (Boston, MA: Addison Wesley) 1981.
199. T. Winograd, *Language as a Cognitive Process* (Boston, MA: Addison Wesley) 1983.
200. Ernest J. Friedman-Hill, *JESS, The Java Expert System Shell* (herzberg.ca.sandia.gov/jess) 30 June 1999.
201. Bruno Haible and Michael Stoll, CLISP (haible@ilog.fr) 1999.
202. *Common Lisp* (New York: American National Standards Institute) 1999.
203. Fabio Bellifemine (www.jade.org\src\examples\jess\JessAgent.java CSELT) 2004.
204. K. Clark and S.-A. Tarnlund (Eds.), *Logic Programming* (London: Academic Press) 1982.
205. W. F. Clocksin and C. S. Mellish, *Programming in Prolog* (Berlin: Springer-Verlag) 1981.
206. Paul Tarau, *BinProlog* (Denton, TX: BinNet Corp.) 1999.
207. *Amateur Radio Relay League (ARRL) Handbook* (Reston, VA: ARRL) 1999.
208. David Goodman, Cooperative Communications: A New Paradigm for Wireless Networking? (goodman web site Polytechnic University) 30 Nov. 2003.
209. N. Sloane and A. Wyner (Eds.), *Claude Elwood Shannon Collected Papers* (New York: IEEE Press) 1993.
210. Timothy Kennedy and Eric Kandel, "Memory and Learning, Molecular Basis of." In: Robert A Meyers (Ed.), *Molecular Biology and Biotechnology* (New York: VCH Publishers) 1995.
211. R. Abeles, P. Frey, and W. Jencks, *Biochemistry* (Boston, MA: Jones and Bartlett Publishers) 1992.
212. J. Fodor, *Modularity of Mind* (Cambridge, MA: MIT Press) 1983.
213. G. K. Zipf, "The Meaning–Frequency Relationship of Words," *Journal of General Psychology*, Vol. 33, pp. 251–256, 1945.
214. Paul Garvin, *Natural Language and the Computer* (New York: McGraw-Hill) 1963.
215. E. V. Condon, "Statistics of Vocabulary", *Science*, 1928.
216. Richard Crandall, *Projects in Scientific Computation* (New York: Springer-Verlag) 1994.
217. Sixth Text Retrieval Conference (TREC-6), *NIST Special Publication 500-240* (Gaithersburg, MD: National Institute of Standards and Technology) Aug. 1998.
218. James Storer, *Data Compression* (Rockville, MD: Computer Science Press) 1988.
219. E. Davis, "The Naïve Physics Perplex," *AAAI Magazine*, Vol. 19, No. 4, Winter 1998.
220. John Riedl, Anthony Jameson, and Joseph Konstanl, "*AI Techniques for Personalized Recommendations*," (http://dfki.de/~jameson/aaai04-tutorial).
221. *IEEE Communications Magazine*, May 2004.
222. Bob Edgar, *The VoiceXML Handbook* (New York: CMP Books) 2001.
223. From the 10, 11 June 2004 Cognitive Radios course in Alexandria, VA.
224. Participants in the 10, 11 June 2004 Cognitive Radios course in Alexandria, VA.

225. N. O. Bernsen, H. Dybkjaer, and L. Dbykjaer, *Designing Interactive Speech Systems* (London: Springer) 1998.
226. K. Goodman and S. Nirenburg, *The KBMT Project: A Case Study in Knowledge-Based Machine Translation* (San Francisco, CA: Morgan Kaufmann) 1991.
227. R. Berwick, *The Acquisition of Syntactic Knowledge* (Cambridge, MA: MIT Press) 1985.
228. J. Allen, *Natural Language Processing* (Upper Saddle River, NJ: Prentice Hall) 1997.
229. "IBM ViaVoice Online Companion Empowers Internet Applications With Speech Technology" (http://www.speechtechnology.com/news/04029901.shtml) 2 Apr. 2000.
230. R. Rodman, *Computer Speech Technology* (Boston, MA: Artech House) 1999.
231. E. De Mori, *Computer Models of Speech Using Fuzzy Algorithms* (New York: Plenum Press) 1983.
232. J. Allen, "Natural Language Understanding," *The Handbook of Artificial Intelligence, Volume IV* (Boston, MA: Addison Wesley) 1989.
233. *WordNet 2.0* © 2003 Princeton University.
234. Ted Pedersen, *WordNet CD* (Duluth: University of Minnesota) 2004.
235. George A. Miller, Richard Beckwith, Christiane Fellbaum, Derek Gross, and Katherine Miller, *Introduction to WordNet: An On-line Lexical Database* (www.wordnet.org) 2004.
236. SNePS (Internet: ftp.cs.buffalo.edu:/pub/sneps/) 1998.
237. C. Koser et al., "read.me," (www.cs.kun.nl: The Netherlands: University of Nijmegen) Mar. 1999.
238. The XTAG Research Group, *A Lexicalized Tree Adjoining Grammar for English*, Institute for Research in Cognitive Science (Philadelphia, PA: University of Pennsylvania) 1999.
239. PC-KIMMO Version 1.0.8 for IBM PC, 18 Feb. 1992.
240. Speech Recognition Grammar Specification Version 1.0 W3C Recommendation 16 Mar. 2004 (www.w3.org/TR/2004/REC-speech-grammar-20040316) 2004.
241. Speech Synthesis Markup Language (SSML) W3C Recommendation (www.w3.org/TR/speech-synthesis/) 2004.
242. ftp://ftp.cs.rochester.edu/pub/papers/ai/96.tn5.Design_and_implementation_of_TRAINS-96_system.ps.gz.
243. ThingFinder (www.inxight.com) 2004.
244. James F. Allen, *The TRAINS Parsing System Version 4.0 A User's Manual* (cs.rochester.edu) 2002.
245. SAPI Universal Phone Set (www.microsoft.com) 2003.
246. *AGC/VOX Software for TMS320C6201* (Hillsboro, OR: RadiSys Corporation) 2001.
247. Mastor (www-5.ibm.com/de/) 2005.
248. www.microsoft.com/speech/evaluation/casestudies/.
249. H. Gish, *GTE Natural Language Processing Systems* (Cambridge, MA: GTE BBN) 1999.

250. J. Naik, "Speaker Verification: A Tutorial," *IEEE Communication Magazine*, Jan. 1990.

251. E. Monte, J. Hernando, X. Miró, and A. Adolf, *Text Independent Speaker Identification on Noisy Environments by Means of Self Organizing Maps* (Barcelona, Spain: Universitat Politécnica de Catalunya) 2003.

252. Emre Sevinc, "Morphix-NLP" (ileriseviye.org) 2003-12-08.

253. Speech Recognition Grammar Specification Version 1.0, W3C Recommendation 16 Mar. 2004. (www.w3.org/TR/2004/REC-speech-grammar-20040316/) 2004.

254. Daniel C. Burnett, Mark R. Walker, and Andrew Hunt, *Speech Synthesis Markup Language (SSML) Version 1.0*, W3C Recommendation 7 Sept. 2004 (www.w3.org/TR/speech-synthesis/) 2005.

255. www.divio.com.

256. Goeffrey Barrows, *Summary of Capabilities for Cognitive Arthropods* (Washington, DC: Centeye) Oct. 2003.

257. Goefffrey Barrows, Centeye (www.centeye.com) 2005.

258. R. Dutta and M. A. Snyder, "Robustness of Correspondence-Based Structure from Motion," *IEEE CH 2934-8* (New York: IEEE Press) 1990.

259. Uwe R. Zimmer, *Adaptive Approaches to Basic Mobile Robot Tasks*, PhD Thesis (Karlsruhe, Germany: Universitat Karlsruhe) July 1995.

260. Nelson Johnson, *Advanced Graphics in C* (Berkeley, CA: Osborne McGraw-Hill) 1987.

261. Craig Lindley, *Practical Image Processing in C* (Hoboken, NJ: John Wiley & Sons) 1991.

262. Hanan Samet, *The Design and Analysis of Spatial Data Structures* (Boston, MA: Addison Wesley) 1990.

263. Roger Stevens, *Practical Programming and Ray Tracing with C++* (San Mateo, CA: M&T Books) 1990.

264. James Albus, *Robot Control System* (Gaithersburg, MD: National Institute of Standards and Technology) 2001.

265. G. Granlund and A. Moe, "Unrestricted Recognition of 3D Objects for Robotics Using Multilevel Triplet Invariants," *AAAI Magazine*, Summer 2004.

266. Gerasimos Potamianos et al., "Audio-Visual Automatic Speech Recognition: An Overview." In: G. Bailly, E. Vatikiotis-Bateson, and P. Perrier (Ed.), *Issues in Visual and Audio-Visual Speech Processing* (Cambridge, MA: MIT Press) 2004.

267. www.roomba.com.

268. Kazuo Murano, President Fujitsu Laboratories, *IEEE Spectrum*, Nov. 2004.

269. Anne Watzman, "Robotic Achievements: GRACE Successfully Completes Mobile Robot Challenge at Artificial Intelligence Conference" (Pittsburg, PA: Carnegie Mellon Views) 09/06/02.

270. U.S. Naval Research Laboratory (Washington, DC: NRL) 2004.

271. www.palantir.swarthmore.edu/GRACE.

272. *AAAI Magazine*, May 2004.

273. Author's experience interacting with GRACE at AAAI 04, San Jose, CA.
274. Ray Jackendoff, *Semantic Structures* (Cambridge, MA: MIT Press) 1990.
275. John R. Bender, *Connecting Language and Vision Using a Conceptual Semantics*, Masters Thesis (Cambridge, MA: Massachusetts Institute of Technology) 2001.
276. Hans-Hellmut Nagel, "Steps Toward a Cognitive Vision System," *AAAI Magazine*, Summer 2004.
277. H. Kamp and U. Reyle, *From Discourse to Logic* (Dordrecht, The Netherlands: Kluwer Academic) 1993.
278. John Kelleher, *A Perceptually Based Computational Framework for the Interpretation of Spatial Language*, Doctoral Dissertation (Dublin, Ireland: Dublin City University) 2003.
279. Roger C. Schank, "Conceptual Dependency: A Theory of Natural Language Understanding," *Cognitive Psychology*, Vol. 3, No. 4, 1972.
280. Ronald Brachman, *Knowledge Representation and Reasoning* (San Francisco, CA: Morgan Kaufmann) 2002.
281. www.xbow.com.
282. Petri Mahonen, comments at the Dagstuhl Workshop on Cognitive Wireless Networks (Dagstuhl, Germany: University of Karlsruhe) 2004.
283. Kenneth De Jong, *Evolutionary Computation: Theory and Applications* (Cambridge, MA: MIT Press).
284. Paul Harmon and David King, *Expert Systems* (Hoboken, NJ: John Wiley & Sons) 1985.
285. Alex Hayzelden and Rachel Bourne, *Agent Technology for Communication Infrastructures* (Hoboken, NJ: John Wiley & Sons) 2001.
286. John Hopcroft and Jeffrey Ullman, *Introduction to Automata Theory, Languages, and Computation* (Boston, MA: Addison Wesley) 1979.
287. Richard Korf, *Learning to Solve Problems by Searching for Macro-Operators* (Boston, MA: Pittman) 1985.
288. J. Larson, *Inductive Inference in the Variable-Valued Predicate Logic System VL21: Methodology and Computer Implementation* (Urbana, IL: University of Illinois) 1977.
289. Ryszard Michalski, Jamie Carbonell, and Tom Mitchell, *Machine Learning, Volume II* (San Francisco, CA: Morgan Kaufmann) 1986.
290. M. Gallab, D. Nau, and P. Traverso, *Automated Planning Theory and Practice* (San Francisco, CA: Morgan Kaufmann) 2004.
291. Nils Nillson, *Learning Machines* (Upper Saddle River, NJ: Prentice Hall).
292. B. Kulwich, *AI Magazine,* Vol. 18, No. 2, pp. 37–45, 1997.
293. D. McClelland and J. Rumelhardt, *Explorations in Parallel Distributed Processing* (Cambridge, MA: MIT Press) 1988.
294. David Van derLinde, *Stochastic Optimal Control* (Course Notes) (Baltimore, MD: The Johns Hopkins University) 1973.
295. Edward J. McShane and Truman A. Botts, *Real Analysis* (Princeton, NJ: Van Nostrand) 1959.

296. Judea Pearl, *Causality* (Cambridge, UK: Cambridge University Press) 2000.

297. *Analytica Users Guide*, © Lumina Decision Systems (Denver, CO: Decisioneering) 1998.

298. B. Williams and J. Cagan, "Activity Analysis: The Qualitative Analysis of Stationary Points for Optimal Reasoning," *AAAI 94* (Menlo Park, CA, AAAI) 1994.

299. Brian Williams, "Qualitative Analysis of CMOS Circuits." In: Daniel G. Bobrow (Ed.), *Qualitative Reasoning About Physical Systems* (Cambridge, MA: MIT Press) 1985.

300. P. Cohen, "Neo: Learning Conceptual Knowledge by Sensorimotor Interaction with the Environment," *Proceedings, Autonomous Agents 97* (Seatlle, WA: ACM) 1997.

301. Subject: Re: [UAI] Causal_vs_Functional Models?; Date: Thu, 20 Nov 2003 10:06:24 -0800; From: "Lotfi A. Zadeh" zadeh@cs.berkeley.edu; To: uai@cs.orst.edu.

302. Irvin Goodman and Hung Nguyen, *Uncertain Models for Knowledge-Based Systems* (Amsterdam: North Holland) 1985.

303. Subject: [UAI] Fuzzy versus probability—An example in practice; Date: Sat, 19 Jul 2003 10:59:54 -0700; From: young@ncsu.edu; To: uai@cs.orst.edu.

304. D. Gilbert and M. Schroeder, *FURY: Fuzzy Unification and Resolution Based on Edit Distance* (London, UK: City University Northampton Square) 2000.

305. Yau-Hwang Kuo and Shiuh-Chu Lee, *Software Development Tool for Fuzzy Object Knowledge Systems* (Taiwan, ROC: National Cheng Kung University) 1997.

306. J. Baltzersen, *Qunilan's Algorithms and the Rough Sets Approach Project Report 1995* (Trondheim, Norway: The Norwegian Institute of Technology) 1995.

307. uai@cs.orst.edu.

308. www.pmr.poli.usp.br/ltd/Software/EBayes/index.html.

309. D. Dowe, The Mixture Modeling Home Page (Internet: http://www.csse.monash.edu.au/~dld/mixture.modelling.page.html) 1998.

310. Sixth Text Retrieval Conference (TREC-6), *NIST Special Publication 500-240* (Gaithersburg, MD: National Institute of Standards and Technology) Aug. 1998.

311. C. P. Mah and R. J. D'Amore, *DISCIPLE Final Report*, PAR Report #83–121 (New Hartford, NY: PAR Technology Corporation) 28 Oct. 1983.

312. ThingFinder (RetrievalWare) 2000.

313. D. Traum and P. Dillenbourg, "Miscommunication in Multi-modal Collaboration," *Proceedings of the AAAI Workshop on Detecting, Preventing, and Repairing Human–Machine Miscommunication* (Palo Alto, CA: AAAI) 1996.

314. G. G. L. Mason, *Theory of Algorithms* (Course Notes) (Baltimore, MD: The Johns Hopkins University) 1974.

315. E. Mohyeldin et al., "A Generic Framework for Negotiations and Trading in Context Aware Radio," *Proceedings of the 2004 Software Defined Radio Technical Conference* (Rome, NY: SDR Forum) Nov. 2004.

316. Christopher Lueg, "Operationalizing Context in Context-Aware Artifacts: Benefits and Pitfalls," *Informing Science*, Vol. 5, No. 2, pp. 43–47 2002. (http://informingscience.com/Articles/Vol5/v5n2p043-047.pdf.)

317. Daniel C. Dennett, "Cognitive Wheels: The Frame Problem of AI." In: C. Hookway (Ed.), *Minds, Machines, and Evolution* (Cambridge, UK: Cambridge University Press) 1984, pp. 129–151.

318. R. Davis, *The Use of Meta-Level Knowledge in the Construction and Maintenance of Large Knowledge Bases*, PhD Dissertation (Palo Alto, CA: Stanford University) 1982.

319. Rule ML web site (http://www.dfki.uni-kl.de/ruleml/) Nov. 2003.

320. Douglas Lenat, *A Mathematician* (*AM*) (Palo Alto, CA: Stanford University).

321. R. Michalski, *Machine Learning* (Palo Alto, CA: Tioga Publishing) 1983.

322. See Google.com and search for the product by name typically www.<name-provided>.com.

323. www.cs.umd.edu/users/hendler/sciam/walkthru.html.

324. www.cycorp.com Nov. 1997.

325. Alphabetized List of CYC® Constants www.cyc.com (Austin, TX 78731: Cycorp, 3721 Executive Center Dr.) 10/14/97.

326. Yolanda Gil and Varun Ratnakar, *Markup Languages: Comparison and Examples* (www.usc.edu: TRELLIS project) 2004.

327. M. Kokar, *Ontology Based Radio* (Boston, MA: Northeastern University) 2004.

328. www.wwrf.org.

INDEX

INDEX